Lecture Notes in Networks and Systems

Volume 433

Series Editor

Janusz Kacprzyk, Systems Research Institute, Polish Academy of Sciences, Warsaw, Poland

Advisory Editors

Fernando Gomide, Department of Computer Engineering and Automation—DCA, School of Electrical and Computer Engineering—FEEC, University of Campinas—UNICAMP, São Paulo, Brazil

Okyay Kaynak, Department of Electrical and Electronic Engineering, Bogazici University, Istanbul, Turkey

Derong Liu, Department of Electrical and Computer Engineering, University of Illinois at Chicago, Chicago, USA
Institute of Automation, Chinese Academy of Sciences, Beijing, China

Witold Pedrycz, Department of Electrical and Computer Engineering, University of Alberta, Alberta, Canada
Systems Research Institute, Polish Academy of Sciences, Warsaw, Poland

Marios M. Polycarpou, Department of Electrical and Computer Engineering, KIOS Research Center for Intelligent Systems and Networks, University of Cyprus, Nicosia, Cyprus

Imre J. Rudas, Óbuda University, Budapest, Hungary

Jun Wang, Department of Computer Science, City University of Hong Kong, Kowloon, Hong Kong

AF895613

The series "Lecture Notes in Networks and Systems" publishes the latest developments in Networks and Systems—quickly, informally and with high quality. Original research reported in proceedings and post-proceedings represents the core of LNNS.

Volumes published in LNNS embrace all aspects and subfields of, as well as new challenges in, Networks and Systems.

The series contains proceedings and edited volumes in systems and networks, spanning the areas of Cyber-Physical Systems, Autonomous Systems, Sensor Networks, Control Systems, Energy Systems, Automotive Systems, Biological Systems, Vehicular Networking and Connected Vehicles, Aerospace Systems, Automation, Manufacturing, Smart Grids, Nonlinear Systems, Power Systems, Robotics, Social Systems, Economic Systems and other. Of particular value to both the contributors and the readership are the short publication timeframe and the world-wide distribution and exposure which enable both a wide and rapid dissemination of research output.

The series covers the theory, applications, and perspectives on the state of the art and future developments relevant to systems and networks, decision making, control, complex processes and related areas, as embedded in the fields of interdisciplinary and applied sciences, engineering, computer science, physics, economics, social, and life sciences, as well as the paradigms and methodologies behind them.

Indexed by SCOPUS, INSPEC, WTI Frankfurt eG, zbMATH, SCImago.

All books published in the series are submitted for consideration in Web of Science.

For proposals from Asia please contact Aninda Bose (aninda.bose@springer.com).

More information about this series at https://link.springer.com/bookseries/15179

Marcelo V. Garcia · Félix Fernández-Peña ·
Carlos Gordón-Gallegos
Editors

Advances and Applications in Computer Science, Electronics, and Industrial Engineering

Proceedings of the Conference on Computer Science, Electronics and Industrial Engineering (CSEI 2021)

Springer

Editors
Marcelo V. Garcia
Universidad del País Vasco
Bilbao, Spain

Carlos Gordón-Gallegos
FISEI
Universidad Técnica de Ambato
Ambato, Ecuador

Félix Fernández-Peña
FISEI
Universidad Técnica de Ambato
Ambato, Ecuador

ISSN 2367-3370 ISSN 2367-3389 (electronic)
Lecture Notes in Networks and Systems
ISBN 978-3-030-97718-4 ISBN 978-3-030-97719-1 (eBook)
https://doi.org/10.1007/978-3-030-97719-1

© The Editor(s) (if applicable) and The Author(s), under exclusive licence
to Springer Nature Switzerland AG 2022
This work is subject to copyright. All rights are solely and exclusively licensed by the Publisher, whether the whole or part of the material is concerned, specifically the rights of translation, reprinting, reuse of illustrations, recitation, broadcasting, reproduction on microfilms or in any other physical way, and transmission or information storage and retrieval, electronic adaptation, computer software, or by similar or dissimilar methodology now known or hereafter developed.
The use of general descriptive names, registered names, trademarks, service marks, etc. in this publication does not imply, even in the absence of a specific statement, that such names are exempt from the relevant protective laws and regulations and therefore free for general use.
The publisher, the authors and the editors are safe to assume that the advice and information in this book are believed to be true and accurate at the date of publication. Neither the publisher nor the authors or the editors give a warranty, expressed or implied, with respect to the material contained herein or for any errors or omissions that may have been made. The publisher remains neutral with regard to jurisdictional claims in published maps and institutional affiliations.

This Springer imprint is published by the registered company Springer Nature Switzerland AG
The registered company address is: Gewerbestrasse 11, 6330 Cham, Switzerland

Preface

The present book comprises the best-selected papers accepted for presentation and discussion at the 2021 International Conference on Computer Science, Electronics, and Industrial Engineering (CSEI 2021). This conference had the support of Springer, Lecture Notes in Networks and Systems, and the Technical University of Ambato. It took place at Ambato, Tungurahua, Ecuador, October 25–29, 2021.

The International Conference on Computer Science, Electronics, and Industrial Engineering (CSEI 2021) is an international scientific conference especially dedicated to computer science, electronics, and industrial engineering and its solutions. The aim of the CSEI conference is to provide a forum for researchers, developers, and practitioners to review and discuss the most recent trends in the area and share innovative research directions.

CSEI 2021 is built on the successes of First CSEI 2019 and Second CSEI 2020 held in Ambato, Ecuador. The Program Committee of CSEI 2021 was composed of a multidisciplinary group of more than 150 experts in different areas in information systems, electronics, and industrial engineering. They have had the responsibility for evaluating, in a double "blind-review" process, the papers received for each of the main themes proposed for the conference: (A) computer science, (B) electronics, and (C) industrial engineering.

We were honored to invite 6 experts to give the impressive and excellent keynote speeches: Dr. Luis Enrique Garcia Muñoz from Carlos III of Madrid University, Spain; Dr. Lucas Teixeira from Universidade Federal de Santa Maria, Brazil; Dr. Olatz Arbelaitz from Pais Vasco University, Spain; Dr. César Guevara, from Indoamerica University, Ecuador; Dr. Manuel Diaz-Madroñero from Polytechnic University of Valencia, Spain; and Dr. Daniel Jerez from Technical University of Ambato, Ecuador.

CSEI 2021 was a successful, fruitful online academic event. It received about 150 contributions from 10 countries around the world. The program included keynote speeches, oral presentations, and discussions covering a wide range of subjects from computer system, electronics engineering to industrial engineering. From the exhaustive and exigent review process, 20 best papers were accepted for presentation and discussion at the conference. The accepted papers are published by

Springer in the series "Lecture Notes in Networks and Systems (LNNS)" in one volume and will be submitted for indexing by ISI, Ei Compendex, Scopus, DBLP, and/or Google Scholar, among others.

It is important and necessary to thank the organizers who made this conference another unforgettable experience and also the authors for their valuable contributions and to the attendees for their active participation. Moreover, we would like to express our gratitude to the reviewers, who provided constructive criticism and stimulating comments and suggestions to improve the authors' contribution. Then, we are grateful to the conference chair, co-chairs and host, and all members of the technical program committee for their active contribution to the execution of CSEI 2021. Finally, we are also grateful to the internationally renowned scientists who acted as keynote speakers and provide an invaluable speech at the CSEI 2021 Conference.

October 2021

Marcelo V. Garcia
Félix Fernández-Peña
Carlos Gordón-Gallegos

Organization

General Chair

Marcelo Vladimir García Sánchez	Technical University of Ambato, Ecuador

Co-chairs

Carlos Nuñez	Technical University of Ambato, Ecuador
Félix Fernández	Technical University of Ambato, Ecuador
Carlos Gordón	Technical University of Ambato, Ecuador

Organizing Committee

Pilar Urrutia	Technical University of Ambato, Ecuador
Carlos Sánchez	Technical University of Ambato, Ecuador
Clay Aldás	Technical University of Ambato, Ecuador
Christian Mariño	Technical University of Ambato, Ecuador
Geovanni Brito	Technical University of Ambato, Ecuador

Reviewer Committee

António Abreu	ISEL - Instituto Superior de Engenharia de Lisboa
Victor Guachimbosa	Technical University of Ambato, Ecuador
Patricia Acosta-Vargas	Universidad de Las Américas, Ecuador
Norah Alanazi	Aljouf University, Saudi Arabia
Jorge Barbosa	University of Vale do Rio dos Sinos, Brazil
Pablo Bengoa	TECNALIA, Basque Research and Technology Alliance (BRTA), Spain
Piotr Borkowski	Maritime University of Szczecin, Poland

Matteo Bottin	University of Padova, Italy
Pablo Burneo	Universidad San Francisco de Quito, Ecuador
Flaviana Calignano	Politecnico di Torino, Italy
Jorge Cardoso	Universidade de Coimbra, Portugal
Enrique Carrera	Universidad de las Fuerzas Armadas ESPE, Ecuador
Gonçalo Carvalho	University of Coimbra, Portugal
Francesco Cauteruccio	University of Calabria, Italy
Heitor Costa	Federal University of Lavras, Brazil
Claudia Durán	Universidad Tecnológica Metropolitana, México
Krzysztof Ejsmont	Warsaw University of Technology, Poland
Abdelkrem Eltaggaz	University of Guelph, Canada
Fábio Fernandes	Universidade de Aveiro, Portugal
Evangelos Gavalas	ELKEME S.A., Greece
Rosa Galleguillos Pozo	Universidad Politecnica de Cataluña, Spain
Joan Garcia	Universitat Autònoma de Barcelona, Spain
Carlos A. Garcia	Universidad Politecnica de Cataluña, Spain
Alain Gil-Del-Val	Tecnalia Corporación Tecnológica, Spain
Vasileios Gkioulos	Norwegian University of Science and Technology (NTNU), Norway
Gabriel Gomes de Oliveira	Laboratory of Visual Communications, Spain
Miguel Gutierrez Gaitan	CISTER/ISEP, Politécnico do Porto, Portugal
Anand Hareendran S.	Muthoot Institute of Technology and Science, India
Zahir Hussain	University of Kufa, Iraq
Alexander Ibarra	Universidad de las Fuerzas Armadas - ESPE, Ecuador
Oscar Andrés Jaramillo	CONACYT-Instituto de Energías Renovables, Mexico
K. Selvakumar	Anna University, India
Sokratis Katsikas	Norwegian University of Science and Technology, Norway
Panagiotis Kyratsis	University of Western Macedonia, Greece
Maohan Liang	Wuhan University of Technology, China
José Lousado	Polytechnic Institute of Viseu, Portugal
Pablo Marchetti	Universidad Tecnológica Nacional, Argentina
Nuno Martins	Universidade de Lisboa, Portugal
Nestor Maya	Instituto Superior Universitario Central Técnico, Ecuador
Maad Mijwil	Baghdad College of Economic Sciences University, Iraq
Tahir Nawaz Minhas	Blekinge Institute of Information Technology, Sweden
Amgad Muneer	Universiti Teknologi PETRONAS, Malaysia

Resmi N. G.	Muthoot Institute of Technology and Science, India
Jose E. Naranjo	Technical University of Ambato, Ecuador
Anand Nayyar	Duy Tan University, Vietnam
Haitham Nobanee	Abu Dhabi University, United Arab Emirates
Mahalingam P. R.	Muthoot Institute of Technology and Science, India
Szczepan Paszkiel	Opole University of Technology, Poland
Ratko Pilipovic	University in Ljubljana, Slovenia
Ivan Miguel Pires	Universidade da Beira Interior, Portugal
Filipe Portela	University of Minho, Portugal
Walter Priesnitz Filho	Universidade Federal de Santa Maria, Brazil
Maikel Lázaro Pérez Gort	Ca' Foscari University of Venice, Italy
Guanqiu Qi	Arizona State University, USA
Dewar Rico	Universidad Francisco de Paula Santander, Colombia
David Romero	Tecnológico de Monterrey, México
Gandhiya Vendhan S.	Bharathiar University, India
Asier Salazar-Ramirez	University of the Basque Country, Spain
Sandra Sanchez-Gordon	Escuela Politécnica Nacional, Ecuador
Sebastian Saniuk	University of Zielona Góra, Poland
Bharath Singh	R.M.K. Engineering College, India
Özgür Tonkal	Samsun Üniversitesi, Turkey
Franklin Tchakounté	University of Ngaoundéré, Cameroon
Elías Todorovich	Universidad Nacional del Centro de la Provincia de Buenos Aires, Argentina
Davide Torre	Istituto per le Applicazioni del Calcolo "Mauro Picone", Italy
Josef Tomas	Aalen University of Applied Science, Germany
Tadele Belay Tuli	University of Siegen, Germany
M. Mujiya Ulkhaq	Diponegoro University, Indonesia
Miguel Vargas-Lombardo	Technological University of Panama, Panama
Leanna Yovita	Telkom University, Indonesia
Qinghe Zheng	Shandong University, China
Dimitris Ziouzios	University of Western Macedonia, Greece
Giambattista Gruosso	Politecnico di Milano, Italy
Mateus Mendes	Polytechnic Institute of Coimbra, Portugal
Phuc Vu	University of Economics Ho Chi Minh City, Vietnam

Contents

Computer Science

Electronics Engineering

Industrial Engineering

An Approach to the Morphological Quality of Fruits with Applying Deep Learning, A Lustrum of Analysis

Murillo-C Felix M., Gonzalez Renan A., Rodriguez Nancy P., and Hidalgo Giovanni J.

Abstract The agricultural sector plays a key role in current development. The task of fruit classification is vital in the agricultural industry since the current market demands high quality standards in its products. However, the fruit grading adopted by the operating personnel is inefficient, labor intensive and error prone. An automatic classification system with optical image processing helps to analyze images and videos to obtain the valuable information. Thus, this document proposes a review of recent technologies and the theoretical concept that explains the development of artificial vision applying deep learning.

Keywords Deep learning · Neural networks · Image processing · Computer vision · Agriculture sector

1 Introduction

Around the world, agriculture constitutes one of the most important parts of the global economy [1]. In most Agro-industrial industries, generally the quality inspection of fruits is carried out by qualified and trained personnel under product acceptance criteria pre-established by standards and applying techniques of good manufacturing practices [2].

This manual classification process generates time in its operations and suffers from the problem of incoherence and inaccuracy in judgment by the person in the inspection line [3]. With the advent of fast and high-precision machine vision technologies,

M.-C. Felix M. (✉) · G. Renan A. · R. Nancy P. · H. Giovanni J.
Technical University of Ambato, 180103 Ambato, Ecuador
e-mail: fmurillo8409@uta.edu.ec

G. Renan A.
e-mail: rgonzalez6505@uta.edu.ec

R. Nancy P.
e-mail: np.rodriguez@uta.edu.ec

H. Giovanni J.
e-mail: ghidalgo1491@uta.edu.ec

© The Author(s), under exclusive license to Springer Nature Switzerland AG 2022
M. V. Garcia et al. (eds.), *Advances and Applications in Computer Science, Electronics, and Industrial Engineering*, Lecture Notes in Networks and Systems 433,
https://doi.org/10.1007/978-3-030-97719-1_1

it is expected that the automation of the classification process reduces the cost of labor, improves the efficiency and accuracy of the classification process [4].

The Machine vision and image processing techniques have been found increasingly useful in the fruit, especially for applications in quality inspection and defect sorting applications [5]. Fruits produced in harvest fields are classified according to size, color, shape, and the presence and type of skin defects. Maturity and defect are the most important factors that determine the quality of the harvested fruits [6].

An automatic sorting system with optical fruit sorting image processing helps both consumers and growers by providing good quality fruits in the market [7]. Therefore, this paper proposes to develop a review of artificial vision systems using image processing techniques.

An automatic classification system with optical processing allows analyzing images and video to obtain necessary information from the process and to understand information about digital image processing [8].

Although numerous research works have been carried out in the area of artificial vision and it has led to significant improvements in the field of morphological classification of fruits, it is still a challenge due to the classification conditions such as the lighting of the site etc. [9].

The constant advance of the advancement of technology in recent decades, in automation and neural networks, have encouraged the agribusiness sector to develop new knowledge and methodologies to implement technologies that facilitate the task of quality classification and reduce fatigue of the quality operator, [10].

For this reason, the factories have begun to take the first step of great importance, which is the obtaining and treatment of the source image, and trained through neural networks to perform the fruit quality classification in real time [4].

It is important to mention that, the human eye can only see between 60 and 70% of the defects in a quality inspection line, in addition to that, visual fatigue and external distractions reduce concentrations and influence the decision criteria [11].

Therefore, the human presence of the mind is crucial as a distraction can result in considerable loss. Furthermore, the challenges have increased from an industrial point of view.

One of the main challenges for agribusiness is to maintain the quality of its products, as well as to maintain its reliability in the market [3]. Therefore, operating beings cannot perform this quality inspection task for a long time with the required precision [12–16].

The development of the field of optical image processing provides a useful tool for decision making and environment detection during the quality review process, [17–20]. This is done by extracting useful information from an image or video from the fruits. However, the extraction of useful or desired information from an image or video is not an easy task [21], this is because, sometimes when the image is captured, there is poor or extreme lighting on the object. , diffuse surroundings, relative movement between the object and the camera during the image or video capture process, etc. [22].

Therefore, the captured image must be previously processed before being sent to the recognition and deep learning system [23].

This document proposes a review of recent technologies and the theoretical concept that explains the development of artificial vision applying deep learning and in this way provides an overview of the application characteristics for the morphological recognition of fruits in the agricultural sector, such as described in previous paragraphs, recent technologies and the theoretical concept can be analyzed where the development of artificial vision is explained applying deep learning and in this way provide an overview of the application characteristics for the morphological recognition of fruits in the agricultural sector.

This article consists of 5 sections, including the introduction, which is constituted in the way, Sect. 2 presents the following methodology followed to select the articles on Deep Learning, later in Sect. 3 the review of the literature is detailed and the results obtained on the subject, Sect. 4 shows the benefits, solutions and future work of DL, and finally Sect. 5 details the conclusions.

2 Methodology

This research was conducted by selecting the most recent research works from 2015 to 2020, selecting relevant research within databases of high bibliographic quality such as, The Institute of Electrical and Electronic Engineers (IEEE), Science Direct, Springer Link , Elsevier, Google Scholar, Computer Science, The Royal Society and the Scopus database, and consisted of 4 steps: i) Research Questions, ii) Document Search, iii) Paper Selection and iv) Data Extraction.

2.1 Research Questions

4 research questions were established. The purpose of these questions is to cover the subject of the development of computer vision applying Deep Learning in the Agro-industrial sector. It is worth mentioning that, to carry out the analysis, three points

Table 1 Research questions (RQ)

Item	Research questions (RQ)	Motivation
RQ1	What role does Deep Learning play in Agro-industrial development?	Identify the main contributions of deep learning in the industrial sector
RQ2	What types of technologies are used for the morphological recognition of fruits in the agricultural sector?	Identify other technologies, used for the morphological recognition of patterns
RQ3	Which devices are used for the fruit sorting process in the industrial sector?	Identify how neural network products are classified
RQ4	What are the benefits of using quality inspection applications with Deep Learning?	Identify the advantages of virtual reality in quality classification and defect detection within the production line

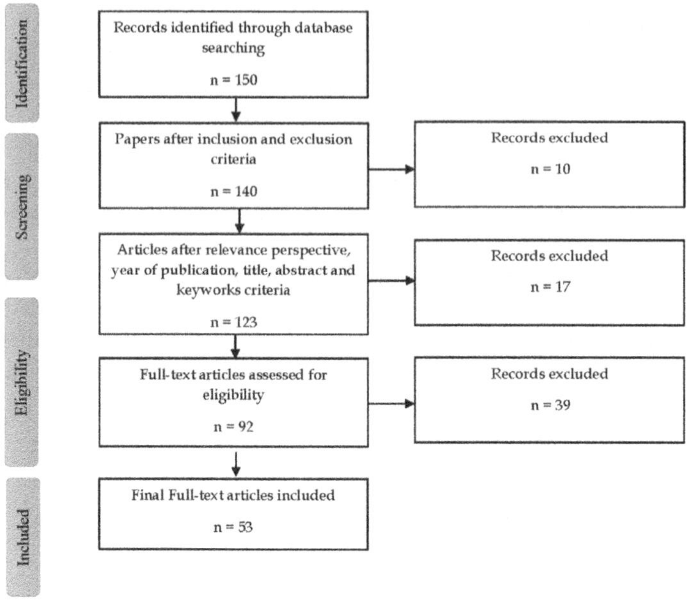

Fig. 1 PRISMA flow diagram

of view have been taken into account: VP1) Development of deep learning within the agroindustrial environment, VP2) Evolution of technologies for morphological recognition of figures and VP3) Quality inspection and classification of defects with deep learning, In Table 1, shows the questions asked.

2.2 *Document Search*

A systematic bibliographic search was carried out in which studies were included from 2015 to 2020. This time interval was considered since the advancement of artificial vision and Deep learning technologies advance in an accelerated manner, so that all researchers considered that This time interval is adequate to study the most current advances in this technology and to understand future approaches and applications in the field of techniques used for optical image treatment and their own characteristics based on the need to solve real problems in the industry, in addition to understanding how neural networks perform in image acquisition, fulfilling safety and operational criteria.

Specific terms were used according to the three perspectives described in the previous section. By VP1 ((“Development” OR “Evolution” OR “Application” OR “Utilization”) AND (“Deep Learning” OR “Artificial Vision”)) were established, for VP2 ((“Technologies” OR “Production Process” OR “continuous improvement” OR “Recognition”) AND (“Morphological” OR “pattern recognition” OR “image

Table 2 Inclusion and exclusion criteria

Number	Inclusion	Exclusion
C1	Related articles pattern recognition in the agribusiness sector	Articles not related to Deep learning applied in the industrial sector
C2	Articles published within the years 2015 and 2020	Similar articles from different databases from years after 2015
C3	Articles written in English and high impact magazines	Articles that are not published in scientific journals
C4	Articles related to the purposes of classifying products with neural networks	Studies published in areas of knowledge other than image acquisition and neural networks
C5	Articles related to Deep learning	Study articles and applied with applied study cases

acquisition”)) were used. Finally, for VP3, ((“benefits” OR “use” OR “applications” OR “inspection”) AND 75 were selected (“classification with computer vision” OR “product quality” OR “neural networks”)). Based on titles and abstracts, the papers were reviewed in detail by three reviewers dedicated to the sampling and filtering of articles according to the topic described.

2.3 *Paper Selection*

We worked on this section dividing this section into four parts. Thus, in the first part, inclusion and exclusion criteria were applied, where the following aspects were taken into account, such as the year of publication, the journal where the application was published, and the relevance to the research topic., Among others (See Table: 2). In the second phase, the investigations were ordered by relevance, keywords and abstract, this included a more efficient review of each of the articles studied. Then, in the third part, the analysis was focused on the introduction and conclusions, which allow us to have support for the research questions posed. And finally, the references were reviewed through detailed information that has its own context for the investigation with a correct perspective on the subject. Next, it presents a summary of the documents studied in each of the parts according to the PRISMA guidelines. See Fig. 1.

2.4 *Data Extraction*

In this section, the 53 final articles were reviewed again by 3 members of the research team, ensuring that the selection of the articles was correct. Table 3 shows the articles ordered by year of publication, and the information extracted from these articles allows answering the research questions. Several aspects have been considered in which Deep learning is developed within the agro-industrial field, allowing fruit classification efficiently using neural networks.

Table 3 Selected papers

Code	Title	Database	Year	Viewpoint	Authors	Objective
G1	Mobile system with real time route learning using Hardware Artificial Neural Network	IEEE	2015	VP1	Alin Mazare; Laurentiu-Mihai Ionescu; Adrian-Ioan Lita; Gheorghe Serban; Marin Ionut	This article presents a solution for tracking a mobile system using an artificial neural network. The mobile system collects data from the environment using an ultrasonic transmitter and receiver then data is processing using a binary artificial neural network. Some templates have been pre-loaded into the system to avoid blockages or additional routes. The solution is implemented on a SOC manufactured by Xilinx: Zync7000 Artix which consists of an FPGA and an ARM processor
G2	Genesis of the Large Hadron Collider	The Royal Society	2015	VP2	Chris Llewellyn Smith	This paper describes the scientific, technical and political genesis of the Large Hadron Collider (LHC). It begins with an outline of the early history of the LHC, from first thoughts and accelerator and detector developments that underwrote the project, through the first studies of the LHC and its scientific potential and the genesis of the experimental programme, to the presentation of the proposal to build the LHC to the CERN Council in December 1993
G3	Blurred Image Splicing Localization by Exposing Blur Type Inconsistency	IEEE	2015	VP1	K. Bahrami and A. C. Kot and L. Li and H. Li	In a tampered blurred image generated by splicing, the spliced region and the original image may have different blur types. Splicing localization in this image is a challenging problem when a forger uses some postprocessing operations as antiforensics to remove the splicing traces anomalies by resizing the tampered image or blurring the spliced region boundary. Such operations remove the artifacts that make detection of splicing difficult. In this paper, we overcome this problem by proposing a novel framework for blurred image splicing localization based on the partial blur type inconsistency

(continued)

Table 3 (continued)

Code	Title	Database	Year	Viewpoint	Authors	Objective
G4	Quality assessment for out-of-focus blurred images	IEEE	2015	VP3	Yutao Liu; Guangtao Zhai; Xianming Liu; Debin Zhao	During the process of image acquisition, images are often subject to out-of-focus or defocus blur because of the improper adjustment of the camera's focal length, this image blur will degrade the image quality. However, in the literature, image quality assessment (IQA) methods dedicated to evaluating the quality of images with out-of-focus blur remain few. Therefore, in this paper, we focus our attention on the quality assessment of images that suffer from out-of-focus blur and propose an objective quality assessment method accordingly
G5	Visual Saliency Detection With Free Energy Theory	IEEE	2015	VP2	K. Gu and G. Zhai and W. Lin and X. Yang and W. Zhang	Visual saliency can be thought of as the product of human brain activity. Most existing models were built upon local features or global features or both. Lately, a so-called free energy principle unifies several brain theories within one framework, and tells where easily surprise human viewers in a visual stimulus through a psychological measure. We believe that this "surprise" should be highly related to visual saliency, and thereby introduce a novel computational Free Energy inspired Saliency detection technique (FES)
G6	Advancing human pose and gesture recognition	Computer Science	2015	VP3	Tomas Pfister	This thesis presents new methods in two closely related areas of computer vision: human pose estimation, and gesture recognition in videos. In human pose estimation, we show that random forests can be used to estimate human pose in monocular videos. To this end, we propose a co-segmentation algorithm for segmenting humans out of videos, and an evaluator that predicts whether the estimated poses are correct or not

(continued)

Table 3 (continued)

Code	Title	Database	Year	Viewpoint	Authors	Objective
G7	A new combination model for short-term wind power prediction	IEEE	2015	VP3	Zizhong Yin; Dongyang Yin; Zhong Chen; Qi Li	Short-term wind power prediction is important to the dispatch and operation of power system. A prediction model based on the rough set, principal component analysis (PCA) and Elman neural network (ElmanNN) is constructed for short-term wind speed forecasting to improve the prediction accuracy of short-term wind power. The wind speed prediction model is established by using ElmanNN, and PCA is used to extract the feature of wind speed data, which optimizes the inputs of ElmanNN
G8	Where to Buy It: Matching Street Clothing Photos in Online Shops	IEEE	2015	VP3	M. H. Kiapour and X. Han and S. Lazebnik and A. C. Berg and T. L. Berg	In this paper, we define a new task, Exact Street to Shop, where our goal is to match a real-world example of a garment item to the same item in an online shop. This is an extremely challenging task due to visual differences between street photos (pictures of people wearing clothing in everyday uncontrolled settings) and online shop photos (pictures of clothing items on people, mannequins, or in isolation, captured by professionals in more controlled settings)
G9	Deep Retinal Image Understandin	SpringerLink	2016	VP2	Maninis, K.; Pont, J.; VanGool, L.	It uses deep convolutional neural networks (CNN), which have proven to be revolutionary in other fields of computer vision, such as object detection and image classification, and we bring their power to the study of fundus imaging

(continued)

Table 3 (continued)

Code	Title	Database	Year	Viewpoint	Authors	Objective
G10	Image Processing, Computer Vision, and Deep Learning: new approaches to the analysis and physics interpretation of LHC event.	ScienceDirect	2016	VP1	Schwartzman, A.; Kagan, M.; Mackey, L.; Nachman, B.	Presents recent developments in the application of image processing, computer vision and neural networks
G11	Approaches of Artificial Intelligence In Biomedical Image Processing	IEEEXplore	2016	VP3	Shukla, S.; Agarwal, A.; Lakhmani, A.	It focuses on understanding artificial intelligence, its concepts and the various models available for image classification, their applications, advantages and disadvantages, and also results
G12	Computer Vision Based Fruit Grading System for Quality Evaluation of Tomato in Agriculture industry	ScienceDirect	2016	VP2	Arakeri, M.; Lakshmana	Develop hardware and software to capture the tomato image and classify the fruit to the appropriate containers without manual intervention. The software is developed using image processing techniques to analyze the fruit for defects and ripeness
G13	A Patch Memory System For Image Processing and Computer Visio	IEEEXplore	2016	VP1	Clemons, J.; Cheng, Ch.; Frosio, I.; Johnson, D.; Keckler, S.	It improves the performance of the artificial vision system and processing of images captured by multidimensional targeting for industrial applications
G14	Machine Learning for Plant Phenotyping Needs Image Processing	ScienceDirect	2016	VP3	Tsaftaris, S.; Minervini, M.; Scharr, H.	It employs a process of segmentation and delimitation of the figures with background treatment for morphological recognition of plants and adapts according to the complexity according to the imaging conditions and the specific task in agro-industrial applications

(continued)

Table 3 (continued)

Code	Title	Database	Year	Viewpoint	Authors	Objective
G15	Mapping hard real-time applications on many-core processors	Google Scholar	2016	VP3	Quentin Perret, Pascal Maurere, Eric Noulard, Claire Pagetti, Pascal Sainrat, Benoît Triquet	Many-core processors are interesting candidates for the design of modern avionics computers. Indeed, the computational power offered by such platforms opens new horizons to design more demanding systems and to integrate more applications on a single target. However, they also bring challenging research topics because of their lack of predictability and their programming complexity. In this paper, we focus on the problem of mapping large applications on a complex platform such as the Kalray mppa R-256 while maintaining a strong temporal isolation from co-running applications We propose a constraint programming formulation of the mapping problem that enables an efficient parallelization and we demonstrate the ability of our approach to deal with large problems using a real world case study
G16	Computer Vision Based Fruit Grading System for Quality Evaluation of Tomato in Agriculture industry	Google Shcolar	2016	VP1	Megha. P. Arakeri and Lakshmana	Agriculture sector plays a key role in the economic development of India. The task of fruit grading is vital in the agricultural industry because there is a great demand for high quality fruits in the market. However, fruit grading by human is inefficient, labor intensive and prone to error. The automated grading system not only speeds up the time of processing, but also minimizes error. There is a great demand for tomatoes in both local and foreign markets. Thus, this paper proposes an automatic and effective tomato fruit grading system based on computer vision techniques
G17	Image mixed blur classification and parameter identification based on cepstrum peak detection	IEEE	2016	VP1	Ying-jie Li; Xiao-guang Di	Motion blur and out-of-focus blur are two main components of image blur in most cases. Blur classification and parameter identification are very important for image processing such as image restoration. In this paper, a novel method based on cepstrum peak detection is presented to classify and identify three types of image blur including motion blur, out-of-focus blur and mixed blur. First, frequency information of detected image is utilized to determine the image blur

(continued)

Table 3 (continued)

Code	Title	Database	Year	Viewpoint	Authors	Objective
G18	Intelligent Traffic Management System for Cross Section of Roads Using Computer Vision	IEEEXplore	2017	VP2	Osman, T.; Shahjahan, S.; Shafi, F.; Zaman, U.	Design and implementation of an intelligent and automated traffic control system that takes advantage of computer vision and image processing techniques, along with conventional computer vision techniques, this research focuses on factors, image processing low cost and analysis of results
G19	Light Field Image Processing: An Overview	IEEEXplore	2017	VP1	Wu, G.; Masia, B.; Jarabo, A.; Zhang, Y.; Wang, L.; Dai, Q.	Presents an overview of aspects of light field image processing, including light field representation and basic theory, acquisition, resolution, depth estimation, compression, editing, processing algorithms for light field visualization, and applications light field data computer vision
G20	Computer vision approaches based on deep learning and neural networks: Deep neural networks for video analysis of human pose estimation	IEEE	2017	VP2	Eralda Nishani; Betim Çiço	Deep architectures with convolution structure have been found highly effective and commonly used in computer vision. With the introduction of Graphics Processing Unit (GPU) for general purpose issues, there has been an increasing attention towards exploiting GPU processing power for deep learning algorithms. Also, large amount of data online has made possible to train deep neural networks efficiently. The aim of this paper is to perform a systematic mapping study, in order to investigate existing research about implementations of computer vision approaches based on deep learning algorithms and Convolutional Neural Networks (CNN)
G21	Object recognition using cellular simultaneous recurrent networks and convolutional neural network	IEEE	2017	VP2	Md Zahangir Alom Department of Electrical and Computer Engineering, University of Dayton, OH, USA; M. Alam; Tarek M. Taha; K. M. Iftekharuddin	In recent years, Convolutional Neural Networks (CNNs) have become very popular and have achieved great success in many computer vision tasks - particularly in object recognition. Partially inspired by neuroscience, CNNs share many properties with the visual system of the brain. However, the filters of convolutional layers play a vital role on overall accuracy of CNNs. In this paper, the Cellular Simultaneous Recurrent Networks (CSRNs) are applied to generate initial filters of Convolutional Networks (CNs) for features extraction and Regularized Extreme Learning Machines (RELM) are used for classification

(continued)

Table 3 (continued)

Code	Title	Database	Year	Viewpoint	Authors	Objective
						Furthermore, Deep Belief Networks (DBN), CNNs with random and Gabor filters are implemented to evaluate the overall performance against the proposed CSRN's filters based CNs with RELM. Experiments were conducted on three popular datasets for object recognition (such as face, pedestrian, and car) to evaluate the performance of the proposed system
G22	Prediction of Laboratory Equipment Demand Based on Grey Relation Analysis and Elman Neural Network	IEEE	2017	VP3	S. Ding and Q. Wu and X. Chang and F. Zhang	To deal with low accuracy problem of prediction of laboratory equipment demand, a new demand prediction model for laboratory equipment was established combining grey relation analysis and Elman (GRA-Elman) neural network. First, the correlation between various influence factors was calculated using grey relation analysis method. Then three factors with higher correlation were chosen as training data of Elman neural network. The original 6-12-1 Elman network model is reduced to a 3-6-1model. The algorithm was tested based on the actual demand of a certain equipment, and the prediction accuracies of GRA-Elman network model, single Elman network model and single BP network model were compared
G23	A Hierarchical, Bulk-Synchronous Stochastic Gradient Descent Algorithm for Deep-Learning Applications on GPU Clusters	IEEE	2017	VP2	G. Cong and O. Bhardwaj	The training data and models are becoming increasingly large in many deep-learning applications. Large-scale distributed processing is employed to accelerate training. Increasing the number of learners in synchronous and asynchronous stochastic gradient descent presents challenges to convergence and communication performance. We present our hierarchical, bulk-synchronous stochastic gradient algorithm that effectively balances execution time and accuracy for training in deep-learning applications on GPU clusters

(continued)

Table 3 (continued)

Code	Title	Database	Year	Viewpoint	Authors	Objective
G24	Implementation of deep-learning based image classification on single board computer	IEEE	2017	VP3	Hasbi Ash Shiddieqy; Farkhad Ihsan Hariadi; Trio Adiono	In this paper, a deep-learning algorithm based on convolutional neural-network is implemented using python and tflearn for image classification. A large number of different images which contains two types of animals, namely cat and dog are used for classification. Two different structures of CNN are used, namely with two and five layers. It is shown that the CNN with higher layer performs classification process with much higher accuracy. The best CNN model with high accuracy and small loss function deployed in single board computer
G25	AF classification from a short single lead ECG recording: The PhysioNet/computing in cardiology challenge 2017	IEEE	2017	VP3	G. Clifford and Chengyu Liu and Benjamin Moody and L. Lehman and Ikaro Silva and Q. Li and Alistair E. W. Johnson and R. Mark	The PhysioNet/Computing in Cardiology (CinC) Challenge 2017 focused on differentiating AF from noise, normal or other rhythms in short term (from 9–61 s) ECG recordings performed by patients. A total of 12,186 ECGs were used: 8,528 in the public training set and 3,658 in the private hidden test set. Due to the high degree of inter-expert disagreement between a significant fraction of the expert labels we implemented a mid-competition bootstrap approach to expert relabeling of the data, levering the best performing Challenge entrants' algorithms to identify contentious labels
G26	Short term load forecast method using artificial neural network with artificial immune systems	IEEE	2017	VP2	R. Alonso and A. Chávez	Although there are widely used methods as Genetic Algorithms, Fuzzy Logic and Artificial Neural Network, the Optimization Based Tools are considered the future of the systems of information. This issue is about Artificial Neural Network (ANN) used in Short Term Load Forecast (STLF). It proposes that the method is valid to predict STLF and how important it is on demand scheduling, contingency analysis, power flow analysis, planning and maintenance of power systems or distribution networks

(continued)

Table 3 (continued)

Code	Title	Database	Year	Viewpoint	Authors	Objective
G27	An image processing and computer vision framework for efficient robotic sketching	ScienceDirect	2018	VP2	Tiwari, M.; Lamba, S.; Gupta, Bh.	It proposes a method that initially eliminates the defects of the unfavorable environmental conditions present in the image/video and later presents a suitable computer vision method for the extraction of information
G28	In Defense of Classical Image Processing: Fast Depth Completion on the CPU	IEEEXplore	2018	VP1	Ku, J.; Harakeh, A.; Waslander, S.	Present an algorithm based on neural networks for data analysis through complex methods. Based on basic image processing operations to realize in-depth completion of deep data and its contribution to today's industry
G29	Impact of Deep Learning in Image Processing and Computer Vision	Springer	2018	VP1	Goswami, T.	Systematic review of the current state of deep learning in the industrial sector, focusing on recent developments in the domain of remote sensing, understanding of the retinal image and understanding of the scene based on recently proposed deep architectures
G30	Suitability of recent hardware accelerators (DSPs, FPGAs, and GPUs) for computer vision and image processing algorithms	Elsevier	2018	VP2	HajiRassouliha, A.; Taberner, A.; Nash, M.; Nielsen, P.	It presents and discusses important considerations when selecting suitable hardware accelerators for computer vision and imaging tasks, as well as proposing a comprehensive review of hardware accelerators. In addition, it discusses the nuts and bolts of chip architectures, available tools and utilities, development time, and the relative advantages and disadvantages of using DSP, FPGA, and GPU

(continued)

Table 3 (continued)

Code	Title	Database	Year	Viewpoint	Authors	Objective
G31	DAG Scheduling Algorithm for a Cluster-Based Many-Core Architecture	IEEE	2018	VP3	Yuto Kitagawa; Tasuku Ishigooka; Takuya Azumi	This paper proposes a directed acyclic graph (DAG) scheduling algorithm for cluster-based many-core architecture. Most of DAG scheduling methods that consider multiple processors and communication delays use a heuristic approach because it is difficult to shorten a schedule length (i.e., makespan). Unfortunately, existing heuristic algorithms do not consider tasks that require a large number of computational resources. Such tasks typically need to be offloaded to a large-scale computational resource such as a many-core system or graphics processing unit (GPU)
G32	A study on blind image restoration of blurred images using R-map	IEEE	2018	VP1	Satoshi Motohashi; Takahiro Nagata; Tomio Goto; Reo Aoki; Haifeng Chen	Image restoration which restores a clear image from a single blur image is a difficult problem of estimating two unknowns: a point spread function (PSF) and its ideal image. In this paper, we propose a novel blind deconvolution method to alternately estimate PSF and the latent image. And we incorporate the gradient reliability map (R-map) that enables edge selection appropriate for PSF estimation processing. This method improves restoration performance by excluding noise that adversely affects the estimation, and the experimental results show that robustness is improved in our proposed method
G33	Comparative Analysis of Stairways Detection Based on RGB and RGB-D Image	IEEE	2018	VP3	Md. Khaliluz-zaman; Mohammad Yakub; Niloy Chakraborty	Stairways detection from RGB and RGB-D stair image is the challenging work in the computer vision research area. The detection system provides topnotch solution with greater portability in assisting visually impaired people and guiding the autonomous navigation system at the smart environments in the real world. In this paper, a framework is introduced to compare the stairways detection performance based on the RGB and RGB-D image. Here, the stairways candidate region is detected through the geometrical feature of a stair, i.e., stair steps are appeared in the concurrent sorted order

(continued)

Table 3 (continued)

Code	Title	Database	Year	Viewpoint	Authors	Objective
G34	Classfication of Categorized KMUTT-BKT's Landscape Images Using RGB Color Feature	IEEE	2018	VP3	Siriphan Phetnuam; Thaweesak Yingthawornsuk	This study proposes how to implement the computer based program able to process the image, estimate and classify images the landscape images based on adaption of RGB Color Space *technique into the procedure of image processing* The database consisted of total 705 satellite images categorized into main four groups of landscapes: Building, Agricultural, Forest and Water. The procedure of study included image segmentation, histogram approximation for RGB color intensity, feature estimation and classification
G35	Landscape Classification with Deep Neural Networks	Google Scholar	2018	VP3	Buscombe, Daniel and Ritchie, Andrew	The application of deep learning, specifically deep convolutional neural networks (DCNNs), to the classification of remotely-sensed imagery of natural landscapes has the potential to greatly assist in the analysis and interpretation of geomorphic processes. However, the general usefulness of deep learning applied to conventional photographic imagery at a landscape scale is, at yet, largely unproven. If DCNN-based image classification is to gain wider application and acceptance within the geoscience community, demonstrable successes need to be coupled with accessible tools to retrain deep neural networks to discriminate landforms and land uses in landscape imagery
G36	Artificial Intelligence in Agriculture	ScienceDirect	2019	VP3	Ireri, D.; Belal, E.; Okinda, C.; Makange; N.; Ji, Ch.	It proposes an online tomato grading tool by means of an artificial vision system for grading tomatoes based on RGB images, this to ensure that quality standards are met and maintained
G37	Applications of Computer Vision in Plant Pathology: A Survey	Springer	2019	VP2	ChouhaN, S.; Singh, U.; Jain, S.	Review of articles that adopt the concept of computer vision and computational methods for the identification and classification of plant leaf diseases in the agro-industrial sector, in addition to presenting the state of the art of the concepts, applications and theories associated with the digital image processing methodologies and soft computing

(continued)

Table 3 (continued)

Code	Title	Database	Year	Viewpoint	Authors	Objective
G38	A Visual Similarity Recommendation System using Generative Adversarial Networks	IEEE	2019	VP3	Betül Ay; Galip Aydın; Zeynep Koyun; Mehmet Demir	The goal of content-based recommendation system is to retrieve and rank the list of items that are closest to the query item. Today, almost every e-commerce platform has a recommendation system strategy for products that customers can decide to buy. In this paper we describe our work on creating a Generative Adversarial Network based image retrieval system for e-commerce platforms to retrieve best similar images for a given product image specifically for shoes. We compare state-of-the-art solutions and provide results for the proposed deep learning network on a standard data set
G39	A Deep Learning Based Distributed Smart Surveillance Architecture using Edge and Cloud Computing	IEEE	2019	VP3	Halil Can Kaskavalci; Sezer Gören	Smart surveillance is getting increasingly popular as technologies become easier to use and cheaper. Traditional surveillance records video footage to a storage device continuously. However, this generates enormous amount of data and reduces the life of the hard drive. Newer devices with Internet connection save footage to the Cloud. This feature comes with bandwidth requirements and extra Cloud costs. In this paper, we propose a deep learning based, distributed, and scalable surveillance architecture using Edge and Cloud computing. Our design reduces both the bandwidth and as well as the Cloud costs significantly by processing footage prior sending to the Cloud
G40	Operation Scheme of Multi-Layer Neural Networks Using NAND Flash Memory as High-Density Synaptic Devices	Database	2019	VP3	Halil Can Kaskavalci; Sezer Gören	Smart surveillance is getting increasingly popular as technologies become easier to use and cheaper. Traditional surveillance records video footage to a storage device continuously. However, this generates enormous amount of data and reduces the life of the hard drive. Newer devices with Internet connection save footage to the Cloud

(continued)

Table 3 (continued)

Code	Title	Database	Year	Viewpoint	Authors	Objective
						This feature comes with bandwidth requirements and extra Cloud costs. In this paper, we propose a deep learning based, distributed, and scalable surveillance architecture using Edge and Cloud computing. Our design reduces both the bandwidth and as well as the Cloud costs significantly by processing footage prior sending to the Cloud
G41	Model of Information Flows of the Information System at the Stage of Modernization	IEEE	2019	VP2	Michael Strelbitsky; Ivan Katerynchuk	Initial data have been presented and analytical dependences of information flows in the integrated information and telecommunication system of the State Border Guard Service of Ukraine at the stage of modernization have been formulated. The mathematical model of information flows of the information system at the stage of modernization and the method of determining the rational sequence of modernization of the information systems elements have been developed
G42	A ConvNet-Based Approach Applied to the Gesticulation Control of a Social Robot	Springer	2020	VP1	Arias, E.; Encalada, P.; Tigre, F.; Granizo, C.; Gordon, C.; Garcia, M.V.	Application of a Convolutional Neural Network algorithm for the gesticulation of a robot incorporating an auditory communication system for Human-Robot interaction, coordinating the movement of the robot's mouth with the processed audio of the text converted to the robot's voice
G43	An Algorithm Oriented to the Classification of Quinoa Grains by Color from Digital Images	Springer	2020	VP3	Quispe, M.; Arroyo, J.; Kemper, G.; Soto, J.	The objective is to carry out a quality control of the production based on the statistics of the coloration of the grain, using an image processing algorithm for its color through digital images and provide the corresponding statistics for the quality control of the production. The algorithm uses RGB, HSV and YCbCr color models, thresholding, segmentation by binary masks, erosion, connectivity, labeling and sequential classification based on 8 colors established by agronomists

(continued)

Table 3 (continued)

Code	Title	Database	Year	Viewpoint	Authors	Objective
G44	Computer Vision for Image Understanding: A Comprehensive Review	Springer	2020	VP1	Jácome, L.; Realpe, R.; Chamba, L.; Viñán, M.	Collect information on the latest techniques to understand and describe images. These new applications have the potential to increase human-machine interaction to new levels of usability and user satisfaction
G45	Modeling Groundwater Spring Potential of Selected Geographical Area Using Machine Learning Algorithms	Springer	2020	VP2	Husen, S.; Bhalchandra, Kh.; Lohkhande, S.; Tamsekar, P.	Develop groundwater spring potential maps for the selected geographic area using machine learning models
G46	Ship Image Classification Using Deep Learning Method	Springer	2020	VP1	Pati, P.; Ranganathan, M.; Patil, H.	Propose a deep learning framework that you use to ensure accuracy in classifications according to various categories using neural networks and artificial vision
G47	Customer Feedback Through Facial Expression Recognition Using Deep Neural Network	Springer	2020	VP3	Dandgawhal, T.; Momin, B.	Proposes using facial reference points to train our create a new model to train a neural network, he proposes a deeper architecture of the neural network to address the FER issue in various facial data sets
G48	Customer Feedback Through Facial Expression Recognition Using Deep Neural Network	Springer	2020	VP3	Dandgawhal, T.; Momin, B.	Proposes using facial reference points to train our create a new model to train a neural network, he proposes a deeper architecture of the neural network to address the FER issue in various facial data sets
G49	Accurate Contention Estimate Scheduling Method Using Multiple Clusters of Many-core Platform	IEEE	2020	VP3	Shingo Igarashi; Yuto Kitagawa; Takuro Fukunaga; Takuya Azumi	Embedded systems such as self-driving systems require a computing platform with high computing power and low power consumption. Multi-/many-core platforms satisfy exactly these requirements. However, for hard real-time applications, multiple demands on shared resources can hinder real-time performance. Memory is among the resources that can most dramatically impair the desired performance. Therefore, we addressed contentions induced by the shared memory

(continued)

Table 3 (continued)

Code	Title	Database	Year	Viewpoint	Authors	Objective
						We improve the predictability of contentions by dividing tasks into the memory access phase and the execution phase using a Directed Acyclic Graph (DAG)
G50	LocAnalyzer: A computer vision method to count locules in tomato fruits	ScienceDirect	2020	VP3	Flavio E. Spetale and Javier Murillo and Dana V. Vazquez and Paolo Cacchiarelli and Gustavo R. Rodríguez and Elizabeth Tapia	Fruit production represents an important economic resource for many countries. The size and shape of fruits are crucial in production systems because they affect the market value. Particularly, in tomatoes (Solanum lycopersicum), the number of seed-containing cavities within each fruit, called locules, affects these attributes. The number of locules is also relevant for genotype-phenotype association studies designed to accelerate tomato breeding programs
G51	A microcontroller based machine vision approach for tomato grading and sorting using SVM classifier	ScienceDirect	2020	VP3	S. Dhakshina Kumar and S. Esakkirajan and S. Bama and B. Keerthiveena	Advances in computer vision have led to the development of promising solutions for challenging problems in agriculture. Fruit grading and sorting are complex problems which require a great deal of human expertise. In this paper, we propose a non-destructive system for sorting and grading tomatoes, which is confounding even for expert human sorters. The proposed system performs classifications of tomatoes in three stages with digital images of samples captured in an experimental setup deployed using microcontroller
G52	Evaluation of low-cost depth cameras for agricultural applications	ScienceDirect	2020	VP2	Isabella C.F.S. Condotta and Tami M. Brown-Brandl and Santosh K. Pitla and John P. Stinn and Késia O. Silva-Miranda	Low-cost depth-cameras have been used in many agricultural applications with reported advantages of low cost, reliability and speed of measurement. However, some problems were also reported and seem to be technology-related, so understanding the limitations of each type of depth camera technology could provide a basis for technology selection and the development of research involving its use

(continued)

Table 3 (continued)

Code	Title	Database	Year	Viewpoint	Authors	Objective
						The cameras use one or a combination of two of the three available technologies: structured light, time-of-flight (ToF), and stereoscopy. The objectives were to evaluate these different technologies for depth sensing, including measuring accuracy and repeatability of distance data and measurements at different positions within the image, and cameras usefulness in indoor and outdoor settings
G53	RGBD-Dog: Predicting Canine Pose from RGBD Sensors	IEEE	2020	VP2	S. Kearney and W. Li and M. Parsons and K. I. Kim and D. Cosker	The automatic extraction of animal 3D pose from images without markers is of interest in a range of scientific fields. Most work to date predicts animal pose from RGB images, based on 2D labelling of joint positions. However, due to the difficult nature of obtaining training data, no ground truth dataset of 3D animal motion is available to quantitatively evaluate these approaches. In addition, a lack of 3D animal pose data also makes it difficult to train 3D pose-prediction methods in a similar manner to the popular field of body-pose prediction. In our work, we focus on the problem of 3D canine pose estimation from RGBD images, recording a diverse range of dog breeds with several Microsoft Kinect v2s, simultaneously obtaining the 3D ground truth skeleton via a motion capture system. We generate a dataset of synthetic RGBD images from this data

3 Results

In this section, the article focuses on the analysis of several works carried out by authors in order to obtain the main characteristics, techniques that they use in solving the problems that arise in the application of neural networks, with acquisition techniques of images, tests and thus obtain quality in the selection of fruits.

3.1 Guide Conducted Studies - Literature Review

Artificial Neural System. Artificial Neural System (ANS), is an massively interconnected networks in parallel [24] of simple elements with adaptability and hierarchical organization, which try to interact like the objects of the real world in the same way that the biological nervous system.

Although, in [25] there are widely used methods such as Genetic Algorithms, Fuzzy Logic and Artificial Neural Network, Tools Based on Optimization are considered the future of information systems. When upgrading the components of an integrated system, information flows have a mixed structure of both new and old versions of software and hardware, like see in [26]. The principal characteristic of ANS is the learning (knowledge), changing the behavior in reason to environment, a set of inputs and adjust themselves to produce consistent outputs.

The most commonly used propagation (information) rule is to combine the inputs (x) and synaptic weights (w) linearly, obtaining:

$$h_i(x_1, ..., x_n, w_{i_1}, ..., w_{i_n}) = \sum_{i=n}^{n} w_{ij}x_j \tag{1}$$

Architecture of ANS, in, [27] the author propose a system for quality evaluation of tomato in the agriculture industry, consist in design the drawback with a manual technique first of all (1. fruit with hand, 2. module of process images). Similar work, like [28] the system is used to move the fruit on the convey for the image process/acquisition. This image process are analyzed with the two position, a. defective, and b. not faulty, the number of holes containing seed within union of each fruit/tomato [28, 28], in helped in computer vision have led to the development of a lot of solutions and changed problems in detection object like vegetable.

ANS Architecture, for development of the image process is needed different software with architecture GPU's, DSP's, or more, this each can used as a hardware accelerator [29] using the maximum of speed in computation. To update and increase this capabilities using the last technique, technology advances, see in [30].

The solution proposed for Artificial Neural Network in [31], using a binary artificial neural network, for tracking any mobile system based-on an artificial neural network, in which the data is collected in the environment work, through of ultrasonic transmitters and receivers. All of decisions of the neural network are convey to the unit control with dealing motors management in that case the movements of the

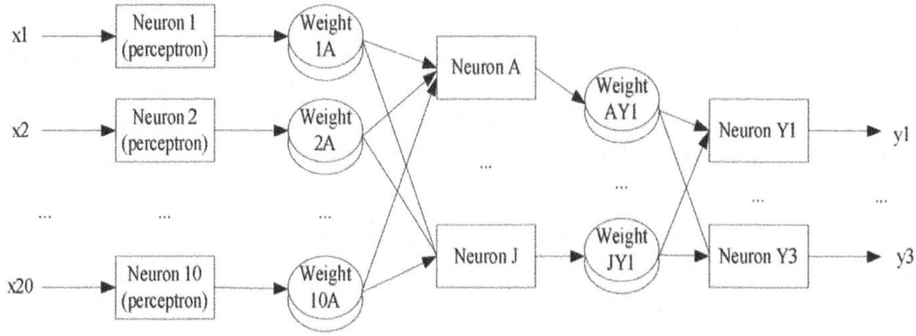

Fig. 2 Neural network (x), (y)

motor is in the different direction, identify many obstacles in the environment based proximity sensing and distance measurement [32].

The technique used in [31] is designed an a neural network binary (Boolean) with 0 and 1, like a feed-forward and back-propagation learning algorithm in the control of the data acquisition, see Fig. 2.

In equation 1, identify the most important rule in ANS and combine the inputs (x) and synaptic weights (y) linearly. The solution weights in [33] are stored in finite state machines with pseudo-random transitions activation [32] for desired mapping from (x) input to (y) output.

The algorithm with the solution in [34] using the points in the weight space for ANS and ANN, is a simple neural network like a computation manageable, that has (2) input neurons, (3) hidden neurons and (2) output neurons in order to has (2) bias terms; the number of weight in NN was (17) for training process is similar or equivalent to find out a solution for (17) dimensional space. The back propagation algorithm is used for training and the decision rule is to select the class corresponding to the output neuron with the largest output, see [35].

Neural Network/Memory: When a problem is presented in which the solution points must be obtained in the weight space, similar or new, the information on the weight distribution to provide good starting weight points and to guide the training precedent includes understanding the distribution of weights, applying unsupervised or approximate groupings to one million weight points [34, 36] of solution with pattern classification problems and selecting 100 groupings whose means can be used as initial weight points.

The embedded systems like as self driving, require a full computing platform with highest computing power tools and low cost power consumption [36], however the analysis is work scheduling on multi, many core of platforms for deciding to mapping for execute the task, see in [37, 38].

When Load using an artificial neural network, in (ANN) method is applied to forecast the short-term load for a large power system [39], the load of data in a model with structure in witch inputs and output have a two covert layers and tested a lot of combinations of neurons, similarity of Fig. 2.

Image Processing, Computer Vision, and Deep Learning. The recent development for image process, is the Large Hadron Collider (LHC) it is built based on a particle detector [37], exploring at the energy frontier similar functionality of nature environment, for analysis the principle of LHC data is used a jet particles in stream. However, identify an individual jets with a high energy interactions (explosion) [40], see Fig. 3.

Image Process. In, [39] the design of a system is presented in which the image is processed on the server or cpu, it is more profitable which will allow the system to process a multitude of images simultaneously, in a short time of action.

Application of HSV Conversion, is uses the RGB values that were separated based [10] on the intensity of the gray scale of each of which, in the lathes, take into account the variations in the intensity: light due to changes in the environment, such as shadows and rain.

The functionality of HSV with image RGB, are divided by 255 in order the range from 0 to 1, see the code in Python and MatLab using reshape (ones(64,1), and addition detailed inspection of the HSV color space, create a synthetic RGB image.

The maximum and minimum values in RGB in Fig. 4, combination are calculated with delta value and subtracting this values.

RGB to HSV converted image is shown in Fig. 7 (Fig. 5)

Blur Detection - Fusión SVD (M-SVD), the blur detection using SVD - multi scale (M-SVD), in [41], the details of image blur region detection with an introduce a spectrum residual, based in method get-a-coarse blur on map of each image, next step is an iterative upgrade mapping mechanic to define the blur mapping from-coarse-to-fine.

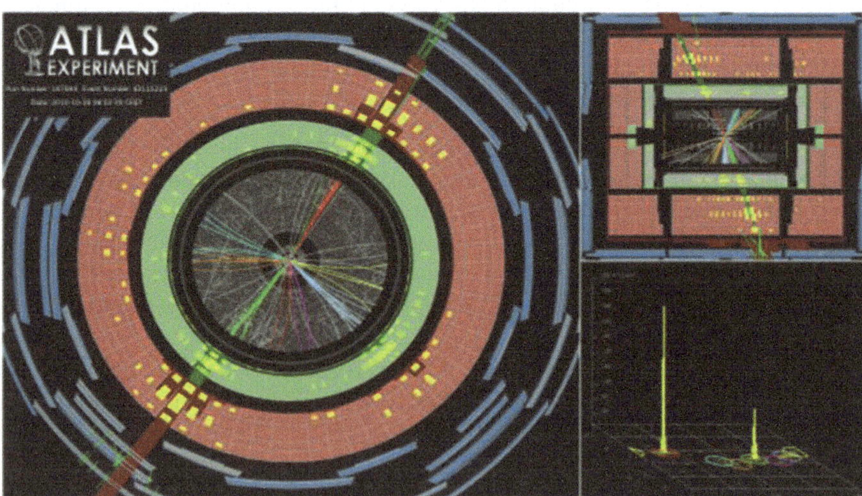

Fig. 3 ATLAS event display containing two high pT jets [37]

Fig. 4 Image converted BGR - RGB

Fig. 5 Image converted HSV

The blur process [42], meaning a lot of causes for loss of particularly details of an image, or an image damaged, therefore the difference is small between blur image and refocused blurred image.

The quality assessment in [43] divided in 2 group: Image Materials, Subjective Experiment; firstly, the data based contained 150 images divided into categories and each category depict one scenery and contains out of focus blur image in five different degrees. The order of image was randomized for each subject, like in [44].

Similar work in [45] present the restored images are achieved from the blur degradation model, and the restoration was classified as non-blind deconvolution and blind deconvolution, see the Eq. 2

$$c = x \bigotimes k + noise \tag{2}$$

The data meaning as follow: blurred image (c), convolution of latent image (x), point spread function (k), adding the noise.

Images in RGB FULL, and images in RGB-D format have greater distinction, greater entropy in the derivatives in space (scale) [45], RGB are measures of amplitude that depend on lighting. The depth of the image, the values of each pixel means more distance from the camera to the objects in focus [46].

In [47] the performance of object detection (stairs) based on RGB and RGB-D image is compared RGB Color Space technique [48], the region of each object is detected through the geometric characteristic, that is, they appear in the concurrent classification order, in addition the characteristic of the concurrent parallel horizontal edges is extracted and from these the image of the single edge is extracted.

Computer Vision. Background, in used to solve problems in the correct and organized approach to the movements of the human body and classify them [49], it is the recognition of patterns that are based on the contrasting geometric properties of each part of the body, see Fig. 6.

Celular simultaneous with images in NN, invariant distortion image recognition in [51], lead to an algorithm similar to the biological algorithm, and it is capable of capturing the information in a temporary and recurrent way, reducing the amount of weight in the neural network combining the NNs and focusing on the Elman Neural Network [52].

The ENN, that has feedback that interconnects the inputs with their outputs, with a capacity much higher than most static networks, it is flexible and with a speed of convergence due to its shorter training time and less computational complexity.

Algorithm of Elman Neural Network, the ENN it is generally divided into four layers: input, hidden, receiving and output layers, most of them are nonlinear state space expression, see the Eq. 3, 4, 5, 6.

$$y(k) = g(w^3 i(k)) \tag{3}$$

$$i(k) = f(w_1 i_c(k) + w_2(u(k-1))) \tag{4}$$

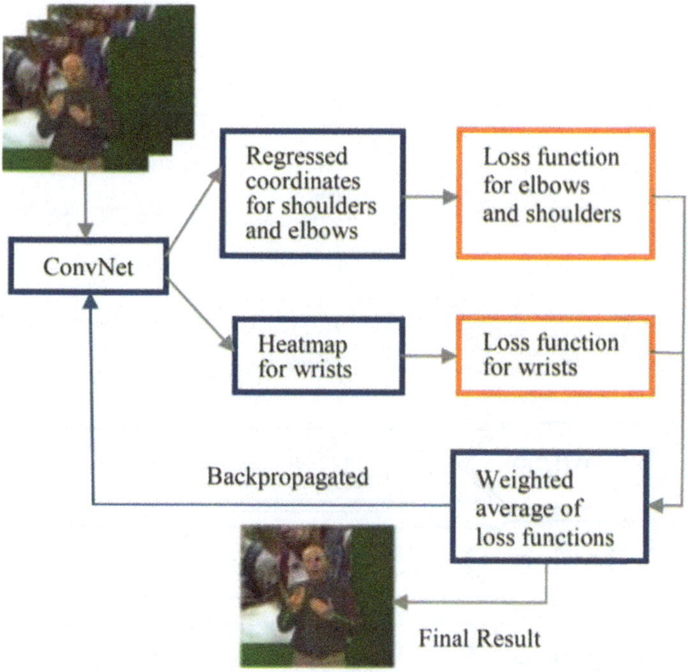

Fig. 6 Background Convet [50]

$$(i_c(k) = (i(k-1)) \tag{5}$$

$$ENN(w) = \sum_{k=1}^{n}[y_n(w) - y'_k w]^2 \tag{6}$$

The connection from intermediate layer to out layer is interconnected with middle and the receiving layer, see in Eq. (5) (6), the weight (w_1), (w_2), (w_3).

The transfer function (f()) in Eq. 4 is the inter layer neurons, and y' is the error in the system NN, like [53].

Deep Learning. In, [54] presents a case of study in almost every e-commerce platform maintained a recommendation system strategy for many products with a deep learning strategy, in helpful to decide when the people can chose in order to buy; however, in [55] the study contributed in (3) steps, first of all: 1. Introduction of the Exact Street to Shop task, 2. Development and evaluation of deep learning feature based characteristic and similarity of learning focus, and 3. Human evaluations.

Approaches studies, it is based on the algorithm that retrieves characteristics calculated in all the images of each region of the objects, the work [56] explain the way to train most of the models using stochastic gradient [57] descent with a batch size of 128 examples, momentum of 0.9, and weight decay of 0.0005.

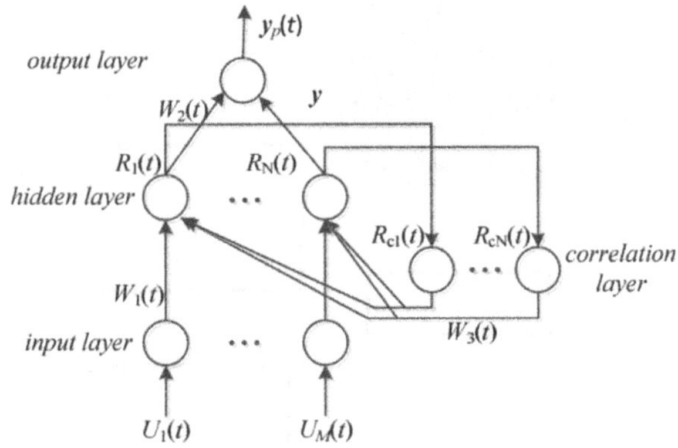

Fig. 7 ENN 4 layers

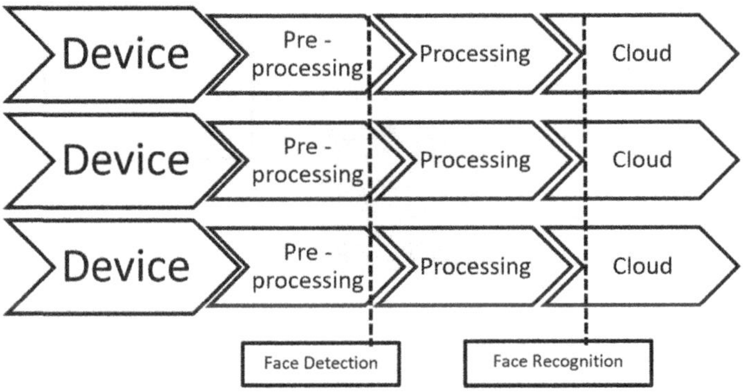

Fig. 8 Deep learning architecture

Deep Learning based, distributed, and scalable, however, in [58] propose an architecture based on some modifications and enhancements, see Fig. 8.

1. Device Layer (DL), in [58] contain array of device like (IP-camera, web cam), multiples client using the same channel for this communication and OpenCV has native support to stream.
2. Pre - processing layer (PPL), capture with the device (faces) from non-pre-processing images, this item includes in a cluster of nodes in one or more server, PL reduce the band-with and sending to process layer using a multiples sensor [59].
3. Processing Layer (PL), accept forwarding from pre - processing layer, detected faces and non-process-images and other data, such as the exposure of the capture [60].
4. Cloud layer (CL), using the recognition results and transform to another (face) recognition, this layer all of the images and object are storage [61].

3.2 Research Questions

The 53 previously selected articles contain the information necessary to understand how the application of deep learning for image recognition and product classification has progressed, its advances, due to its versatility and technologies, and how these can be applied in different areas of the Agri-food industry. Next, the answer to the RQs presented in Subsect. 2.1 are shown below.

RQ1: What role does Deep Learning play within Agro-industrial Development? The use of DL is relative new in Agro-industrial field, nowadays deep learning techniques are applied to various Agro-industrial problems such as detection or identification, fruit or plants classification, fruit counting and others.

In [7] is propose a deep learning framework that use to ensure accuracy in classifications according to various categories using neural networks and artificial vision.

The paper [54], describe a work on creating a Generative Adversarial Network based image retrieval system for e-commerce platforms to retrieve best similar images for a given product image specifically it could help to improve the Agro-Indutrial Sector.

In paper [59] is propose a deep learning based, distributed, and scalable surveillance architecture using Edge and Cloud computing. Our design reduces both the bandwidth and as well as the Cloud costs significantly by processing footage prior sending to the Cloud. With this technology can reduce the cost that implies to buy specialized hardware to implement deep learning.

With DL we can employ models which have a source of the data and with the use of according hardware we the possibility of real-time application to establish an integration with autonomous robotic platforms to develop new ways of doing things in Agro-industrial domain.

The articles show that deep learning provides high accuracy results, and its an alternative to the traditional image processing techniques in terms of accuracy.

RQ2: What types of technologies are used for the morphological recognition of fruits in the Agricultural Sector? It is still difficult to recognize the morphology of fruit because of different colors, shapes, and textures. Nowadays the researchers are using images from many fruit data set and in this research they use four steps: pre-processing, boundary extraction, feature extractions, and classification.

In [62] employs a process of segmentation and delimitation of the figures with background treatment for morphological recognition of plants and adapts according to the complexity according to the imaging conditions and the specific task in agro-industrial applications.

The most useful sorter are: Naive Bayes (NB), Logistic Regression (LR), Support Vector Machine (SVM), Decision Tree (DT), K-Nearest Neighbor (KNN), and Random Forest (RF), which is the better for this application.

Automated or robot-assisted which combines the aspects of computer vision and machine intelligence technology is use for morphological recognition of fruits.

RQ3: Which devices are used for the fruit sorting process in the industrial sector? Some sorting systems are use in industrial sector including a weight system and a vision-based sorting and grading unit which also measures size, it comes with a user interface that enables definition of classification parameters, reconfiguration of the outputs and maintenance of production statistics.

In this work [28], an automatic computer vision method for the identification and counting of the number of fruit images, called LocAnalyzer, is presented. The aim of LocAnalyzer is to speed up the processing time and reduce the subjectivity relative to the traditional manual approach for counting. The method was tested on two real tomato datasets. Promising results in terms of accuracy, precision, and recall measures were obtained, suggesting the potential usefulness of the approach for the development of a tool for the automatic measurement of other internal tomato attributes

In this paper [29] is propose a non-destructive system for sorting and grading tomatoes, which is confounding even for expert human sorters. The proposed system performs classifications of tomatoes in three stages with digital images of samples captured in an experimental setup deployed using microcontroller.

A machines that use a special architectures can find on market , a powerful processors boards, hardware implementation of special purpose algorithm, and other important characteristics.

Also, we can indicate that the hardware improves very fast, providing powerful electronics and low cost architectures to use for many purposes and trying to meet the requirements of the system, mainly in speed and accuracy of measurements.

RQ4: What are the benefits of using quality inspection applications with deep learning and defect classification? The learning machine process has reached a remarkable progress and offer reliable results in many cases. DL is used to learn high-level generic features and applied to a wide range of textures or defects that are difficult to detect. This characteristic make that DL is well fit for automated defect identification in surface integration inspection. The experimental results show that CNN works properly with different types of defects on textured or non-textured surfaces [63].

In [5] is proposed an automatic and effective tomato fruit grading system based on computer vision techniques. The proposed quality evaluation method consists of two phases: development of hardware and software. The hardware is developed to capture the image of the tomato and move the fruit to the appropriate bins without manual intervention. The software is developed using image processing techniques to analyze the fruit for defects and ripeness. Experiments were carried out on several images of the tomato fruit. It was observed that the proposed method was successful with 96.47% accuracy in evaluating the quality of the tomato.

In this paper [64] is perform a quality control of production based on the statistics of grain coloration. Through this work, an algorithm is proposed oriented to classify the grains of the acquired samples by their color via digital images and provide corresponding statistics for the quality control of the production. The algorithm uses the color models RGB, HSV and YCbCr, thresholding, segmentation by binary masks, erosion, connectivity, labeling and sequential classification based on 8 colors established by agronomists. The obtained results showed a performance of the proposed algorithm of 91.25% in relation to the average success rate.

3.3 Paper Selection Analysis

In Fig. 9 it can be seen that of the 13 selected articles, 24.53% corresponded to viewpoint 1 (Development of deep learning within the Agro-Industrial environment), 30.19% focused on viewpoint 2 (Evolution of technologies for morphological recognition of figures) and finally, 45.28% concentrated on viewpoint 3 (Quality inspection and classification of defects with deep learning).

As can be observed, most articles point to show the use of deep learning in quality inspection to get a useful tool for classification of defects. The use of DL is improving the classification of defects in many industrial fields but specially in agriculture, it help to decrease the faults in classification, although the training takes a considerable time, the precision that is obtained in the end is very high.

In Fig. 10 , it can be seen a numerical analysis represented in each selection phase of explored databases. For IEEE, from phase 1 to the final phase, there was a paper reduction of 74.44%. In Springer, there was a reduction of 55%, for ScienceDirect, there was a reduction of 46.67%, for IEEExplore, there was a reduction of 50%, for Google Scholar, there was a reduction of 25% and finally, for The Royal Society, Computer Science, SpringerLink, Elsevier, Database, there was a reduction of 50%. From this analysis the databases that contain the most relevant information are IEEE, Springer, IEEEXplore and Google Scholar. Finally, in Fig. 11, we noticed that from 2015 to 2018, almost have the same percentage of the reviewed literature there are approximately 16% .For 2019 this percentage of the reviewed literature low to the 11% approximately and for 2020 the percentage of the reviewed literature increased approximately to 23% It implies that, in this year the research in deep learning and its application in agriculture is increasing and it is expected that in the remainder of this year, more research can be published to improve the agriculture sector.

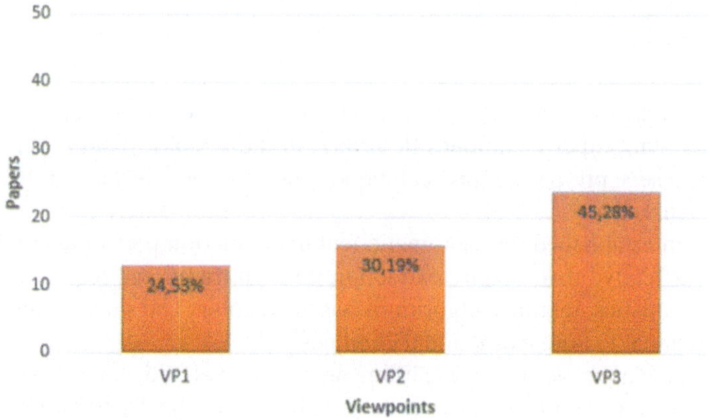

Fig. 9 Selected papers classification

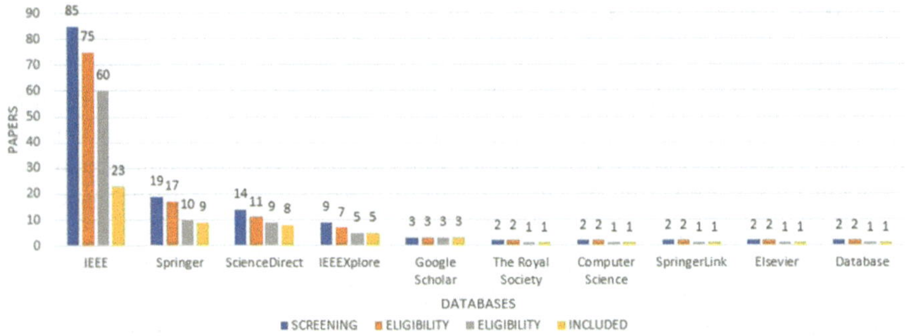

Fig. 10 Phases analysis

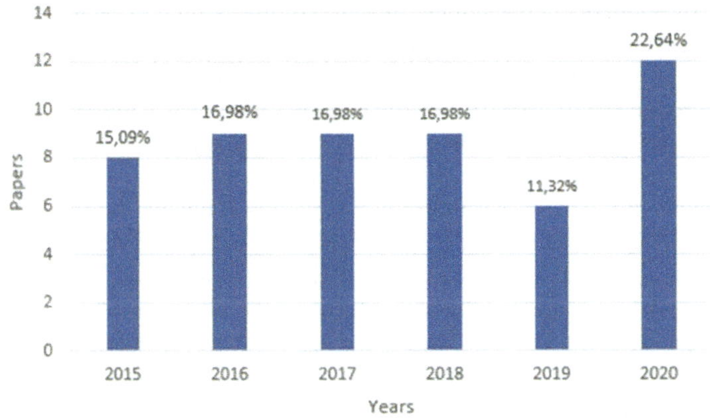

Fig. 11 Years analysis

4 Discussion

The present analysis shows that DL has a superior performance when comparing its performance with other techniques described in the articles. Each article involved different data sets, pre-processing techniques, metrics, models and parameters, so it cannot be generalized.

It has been established that automatic feature extraction performed by DL models is more effective than feature extraction through traditional techniques such as histograms,statistics, textures, algorithms based on colors and shapes, random fields and other manual feature extraction techniques.

Selecting the appropriate deep learning architecture is critical to obtaining robust and reliable results, so understanding the capabilities of the deep learning architecture as well as the characteristics of the input data and research objectives is really important. Apart from the architecture there are many hyper-parameters such as:

the number of hidden units, the number of layers, the learning rate and the learning iterations, which can influence the results [65].

Although most of the articles used traditional architectures to perform the classification, in some articles experimented with more advanced models to solve more complex problems. These approaches constitute a very promising research direction, since the task of identifying the morphological quality of fruits in an image this technique has numerous applications in real life and could solve several agricultural problems.

An application like Multimodal deep learning [66], which leverages information from multiple input sources, is a promising technique for the future of deep learning research. The more parameters of the deep learning model and training data you have, you can improve learning performances, but this increases training time and create the need for accelerated deep learning. To accelerating deep learning can be divided into three groups: advanced optimization algorithms, parallel and distributed computing, and specialized hardware [67].

Currently, field programmable gate array based processors are being developed, neuromorphic chips modeled from the brain are promising technologies [68].

The advantage of the DL is that it shows reduced effort in feature engineering. Manual techniques take considerable time, but DL does it automatically. Also, sometimes manual search for good feature extractors is not an easy and obvious task, DL models seem to generalize well, such as in challenging conditions like image size and orientation, lighting, complex background, different resolution, occlusion, variation, and scale. These models are used for a variety of fruits such as apples, citrus, peach and others.

If different data sets are used for testing, model performance was generally high, with only small reductions in performance compared to performance when using the same data set for training and testing.

DL has the ability to develop simulated data sets to train the model, which could be properly designed to solve real-world problems [69].

Although DL takes longer to train than other approaches, its test time efficiency is quite fast.

The downside in using DL is the need for large data sets, which serve as input for training procedure. At least a few hundred images are actually required, depending on the complexity of the problem of study such as: number of classes, required precision and others.

It has been found that limited and unbalanced data can compromise deep learning performance. Solutions for limited and unbalanced data can be divided into three main groups [70]: data pre-processing, cost-sensitive learning, and algorithmic modification.

Data preprocessing generally provides a better data set through sampling or extraction of basic characteristics.

Basic extraction methods deviate from the concept of deep learning, occasionally being used to show the difficulties of learning from limited and unbalanced data.

Algorithmic modification methods adapt to learning algorithms to increase their suitability for limited and unbalanced data.

Incorporating traditional deep learning architectures is a promising future trend.

Semi-supervised learning and reinforced learning are also receiving attention. Semi-supervised learning explore labeled and unlabeled data, and some algorithms have been proposed.

Reinforcement learning takes advantage of reward outcome cues that result from actions rather than properly labeled data. Since reinforcement learning is more like the way humans actually learn, this approach holds great promise for general artificial intelligence [71].

More complex architectures would appear, combining various DL models and classifiers together, or combining manual features with automatic features extracted using various techniques, merged to improve the overall result.

It is expected that researchers to test their models using more realistic data sets, demonstrating the ability of the models to generalize to various real-world situations.

5 Conclusions

This paper has presented a methodological revision of the literature presented in last 5 years in the field of deep learning applied in agriculture specifically in the morphological analysis of fruits. The latest neural networks has been analyzed establishing a mode to improve the images processing to obtain better results.

In this paper, we provide a review that involves deep learning in terms of input data, research objectives, the characteristics of established deep learning architectures, technical details of the models employed, preprocessing tasks and existing techniques, in performance terms. Our findings indicate that deep learning offers better performance and outperforms than other popular image processing techniques.

Also, we show the limitations of the approach and promising directions for future research. Although deep learning is promising, many potential challenges remain, including limited or unbalanced data, interpreting deep learning results, and selecting an appropriate architecture and hyperparameters.

The overall benefits of deep learning are encouraging for further use towards smarter and more sustainable agriculture, we believe this review will provide valuable information and serve as a starting point for the application of deep learning to advance future research.

References

1. Wan B, Sung Lee Y, Ho J (2019) Computer Vision-Based Bridge Displacement Measurements Using Rotation-Invariant Image Processing Technique, p 12–16
2. Ireri D, Belal E, Okinda C (2019) A computer vision system for defect discrimination and grading in tomatoes using machine learning and image processing, pp 28–29

3. Husen S, Khamitkar S, Bhalchandra P (2020) Modeling groundwater spring potential of selected geographical area using machine learning algorithms. Appl Comput Vision Image Process 1:436–438
4. Arias E, Encalada P, Encalada P, Tigre F, Granizoand C, Gordon C, Garcia MV (2020) A ConvNet-Based Approach Applied to the Gesticulation Control of a Social Robot, pp 11–19
5. Megha P, Lakshmana B (2016) Computer vision based fruit grading system for quality evaluation of tomato in agriculture industry, pp 427–428
6. Dandgawhal T, Momin B (2020) Customer feedback through facial expression recognition using deep neural network. Appl Comput Vision Image Process 1:421–424
7. Patil P, Ranganathan M, Patil H (2020) Ship image classification using deep learning method. Appl Comput Vision Image Process 1:232–234
8. Wiley V, Lucas T (2018) A Paper Review, Computer Vision and Image Processing, pp 28–36
9. Deva J, Lucas T (2019) Computer Vision Based Traffic Sign Sensing for Smart Transport, pp 11–19
10. Annamalai J, Lakshmikanthan C (2019) An optimized computer vision and image processing algorithm for unmarked road edge detection, pp 429–432
11. Patel K, Parmar B (2020) Assistive device using computer vision and image processing for visually impaired; review and current status, vol 1, pp 3–5
12. Barua S, Patil H, Dharmeshkumar P (2020) Deep learning-based smart colored fabric defect detection system. Appl Comput Vision Image Process 1:212–214
13. Garcia MV, Armentia A, Perez F, Marcos M (2019) An approach of load balancers for low-cost CPPSs in software-defined networking architecture. In: 2019 15th IEEE International Workshop on Factory Communication Systems (WFCS), pp 1–4 (2019)
14. Naranjo JE, Urrutia FU, Garcia MV, Gallardo-Cárdenas F, Franklin TO, Lozada-Martínez E (2019) User experience evaluation of an interactive virtual reality-based system for upper limb rehabilitation. In: 2019 Sixth International Conference on eDemocracy eGovernment (ICEDEG), pp 328–333 (2019)
15. Naranjo D, Cordova P, García MV, Saltos T, Robayo A, Altamirano S, et al. (2019) Wearable Telemedicine System for Real-Time Monitoring of Electrocardiographic Signals, pp 69–75. https://doi.org/10.1109/ICEDEG.2019.8734294
16. Garcia CA, Lanas D, Edison AM, Altamirano S, Garcia MV (2018) An approach of cyber-physical production systems architecture for robot control, pp 2847–2852. https://doi.org/10.1109/IECON.2018.8591286
17. Tiwari M, Lamba S, Gupta B (2018) An image processing and computer vision framework for efficient robotic sketching. Procedia Comput Sci 1:284–289
18. Garcia CA, Salazar LF, Mariño C, Villalba WR, Garcia MV (2018) Design of flexible cyber-physical production systems architecture for industrial robot control, pp 1–6. https://doi.org/10.1109/ETCM.2018.8580338
19. Vimos VH, Benalcázar M, Oña AF, Cruz PJ (2020) A novel technique for improving the robustness to sensor rotation in hand gesture recognition using sEMG. In: Advances and Applications in Computer Science, Electronics and Industrial Engineering, pp 226–243. Springer International Publishing, Cham
20. Mantilla-Brito J, Pozo-Espín D, Solórzano S, Morales L (2020) Embedded system for hand gesture recognition using EMG signals: effect of size in the analysis windows. In: Advances and Applications in Computer Science, Electronics and Industrial Engineering, pp 214–225. Springer International Publishing, Cham
21. Hung C, Yang H (2016) A Simple Image-based Object Velocity Estimation Approach, pp 2–6
22. Singh S, Pratap U (2019) Applications of Computer Vision in Plant Pathology: A Survey, pp 1–5
23. Goswami T (2018) Impact of Deep Learning in Image Processing and Computer Vision, pp 1–5
24. Matich DJ (2001) Redes Neuronales: Conceptos Básicos y Aplicaciones. Historia, p 55
25. Alonso R, Chávez A (2017) Short term load forecast method using artificial neural network with artificial immune systems. In: 2017 IEEE URUCON, pp 1–4

26. Strelbitsky M, Katerynchuk I (2019) Model of information flows of the information system at the stage of modernization. In: 2019 3rd International Conference on Advanced Information and Communications Technologies (AICT), pp 200–203
27. Arakeri MP, Lakshmana (2016) Computer vision based fruit grading system for quality evaluation of tomato in agriculture industry. Procedia Computer Science 79:426–433. Proceedings of International Conference on Communication, Computing and Virtualization (ICCCV) 2016. https://doi.org/10.1016/j.procs.2016.03.055
28. Spetale FE, Murillo J, Vazquez DV, Cacchiarelli P, Rodríguez GR, Tapia E (2020) LocAnalyzer: a computer vision method to count locules in tomato fruits. Comput Electron Agric 173. https://doi.org/10.1016/j.compag.2020.105382
29. Kumar SD, Esakkirajan S, Bama S, Keerthiveena B (2020) A microcontroller based machine vision approach for tomato grading and sorting using SVM classifier. Microprocess Microsyst 76. https://doi.org/10.1016/j.micpro.2020.103090
30. Condotta ICFS, Brown-Brandl TM, Pitla SK, Stinn JP, Silva-Miranda KO (2020) Evaluation of low-cost depth cameras for agricultural applications. Comput Electron Agric 173. https://doi.org/10.1016/j.compag.2020.105394
31. Mazare A, Ionescu L, Lita A, Serban G, Ionut M (2015) Mobile system with real time route learning using hardware artificial neural network. In: 2015 7th International Conference on Electronics, Computers and Artificial Intelligence (ECAI), pp P–45–P–48
32. Hornung M, Brand O, Paul O, Baltes H (1997) Ultrasound transducer for distance measurements. In: Proceedings of International Solid State Sensors and Actuators Conference (Transducers 1997), vol 1, pp 441–444
33. Yu X-H (1992) Explicit solutions of the optimum weights of layered neural networks. In: Proceedings 1992 IJCNN International Joint Conference on Neural Networks, vol 1, pp 719–724
34. Go J, Lee C (1999) Analyzing weight distribution of neural networks. In: IJCNN 1999. International Joint Conference on Neural Networks. Proceedings (Cat. No. 99CH36339), vol 2, pp 1154–1157
35. Lippmann R (1987) An introduction to computing with neural nets. IEEE ASSP Mag 4(2):4–22
36. Igarashi S, Kitagawa Y, Fukunaga T, Azumi T (2020) Accurate contention estimate scheduling method using multiple clusters of many-core platform. In: 2020 28th Euromicro International Conference on Parallel, Distributed and Network-Based Processing (PDP), pp 67–71
37. Kitagawa Y, Ishigooka T, Azumi T (2018) DAG scheduling algorithm for a cluster-based many-core architecture. In: 2018 IEEE 16th International Conference on Embedded and Ubiquitous Computing (EUC), pp 150–157
38. Perret Q, Maurere P, Noulard E, Pagetti C, Sainrat P, Triquet B (2016) Mapping hard real-time applications on many-core processors. In: 24th International Conference on Real-Time and Network Systems (RTNS 2016), Brest, France, pp 235–244. https://hal.archives-ouvertes.fr/hal-01692702
39. Lee KY, Cha YT, Park JH (1992) Short-term load forecasting using an artificial neural network. IEEE Trans Power Syst 7(1):124–132
40. Smith CL (2015) Genesis of the large hadron collider. Philos Trans Royal Soc A: Math Phys Eng Sci 373(2032):20140037. https://doi.org/10.1098/rsta.2014.0037
41. Shukla S, Agarwal K, Lakhmani A (2016) Approaches of Artificial Intelligence in Biomedical Image Processing, pp 4–7
42. Li Y, Di X (2016) Image mixed blur classification and parameter identification based on cepstrum peak detection. In: 2016 35th Chinese Control Conference (CCC), pp 4809–4814
43. Liu Y, Zhai G, Liu X, Zhao D (2015) Quality assessment for out-of-focus blurred images. In: 2015 Visual Communications and Image Processing (VCIP), pp 1–4
44. Gu K, Zhai G, Lin W, Yang X, Zhang W (2015) Visual saliency detection with free energy theory. IEEE Sign Process Lett 22(10):1552–1555
45. Motohashi S, Nagata T, Goto T, Aoki R, Chen H (2018) A study on blind image restoration of blurred images using R-map. In: 2018 International Workshop on Advanced Image Technology (IWAIT), pp 1–4

46. Kearney S, Li W, Parsons M, Kim KI, Cosker D (2020) RGBD-Dog: predicting canine pose from RGBD sensors. In: 2020 IEEE/CVF Conference on Computer Vision and Pattern Recognition (CVPR), pp 8333–8342
47. Khaliluzzaman M, Yakub M, Chakraborty N (2018) Comparative analysis of stairways detection based on RGB and RGB-D Image. In: 2018 International Conference on Innovations in Science, Engineering and Technology (ICISET), pp 519–524
48. Phetnuam S, Yingthawornsuk T (2018) Classfication of categorized KMUTT-BKT's landscape images using RGB color feature. In: 2018 14th International Conference on Signal-Image Technology Internet-Based Systems (SITIS), pp 327–331
49. Nishani E, Çiço B (2017) Computer vision approaches based on deep learning and neural networks: deep neural networks for video analysis of human pose estimation. In: 2017 6th Mediterranean Conference on Embedded Computing (MECO), pp 1–4
50. Pfister T (2015) Advancing human pose and gesture recognition
51. Alom MZ, Alam M, Taha TM, Iftekharuddin KM (2017) Object recognition using cellular simultaneous recurrent networks and convolutional neural network. In: 2017 International Joint Conference on Neural Networks (IJCNN), pp 2873–2880
52. Ding S, Wu Q, Chang X, Zhang F (2017) Prediction of laboratory equipment demand based on grey relation analysis and Elman neural network. In: 2017 10th International Symposium on Computational Intelligence and Design (ISCID), vol 1, pp 408–411
53. Yin Z, Yin D, Chen Z, Li Q (2015) A new combination model for short-term wind power prediction. In: 2015 5th International Conference on Electric Utility Deregulation and Restructuring and Power Technologies (DRPT), pp 1869–1873
54. Ay B, Aydın G, Koyun Z, Demir M (2019) A visual similarity recommendation system using generative adversarial networks. In: 2019 International Conference on Deep Learning and Machine Learning in Emerging Applications (Deep-ML), pp 44–48
55. Kiapour MH, Han X, Lazebnik S, Berg AC, Berg TL (2015) Where to buy It: matching street clothing photos in online shops. In: 2015 IEEE International Conference on Computer Vision (ICCV), pp 3343–3351
56. Gonzalez TF (2007) Handbook of approximation algorithms and metaheuristics. Handbook of Approximation Algorithms and Metaheuristics, pp 1–1432. https://doi.org/10.1201/9781420010749
57. Cong G, Bhardwaj O (2017) A hierarchical, bulk-synchronous stochastic gradient descent algorithm for deep-learning applications on GPU clusters. In: 2017 16th IEEE International Conference on Machine Learning and Applications (ICMLA), pp 818–821
58. Kaskavalci HC, Gören S (2019) A deep learning based distributed smart surveillance architecture using edge and cloud computing. In: 2019 International Conference on Deep Learning and Machine Learning in Emerging Applications (Deep-ML), pp 1–6
59. Lee S, Lim S, Choi NY, Bae J, Kwon D, Park B et al (2019) Operation scheme of multi-layer neural networks using NAND flash memory as high-density synaptic devices. IEEE J Electron Devices Soc 7:1085–1093
60. Shiddieqy HA, Hariadi FI, Adiono T (2017) Implementation of deep-learning based image classification on single board computer. In: 2017 International Symposium on Electronics and Smart Devices (ISESD), pp 133–137
61. Clifford G, Liu C, Moody B, Lehman L, Silva I, Li Q et al (2017) AF classification from a short single lead ECG recording: the PhysioNet/computing in cardiology challenge 2017. Comput Cardiol (CinC) 2017:1–4
62. Tsaftaris S, Minervini M, Scharr H (2016) Machine learning for plant phenotyping needs image processing. Trends Plant Sci 12(21):989–991. https://doi.org/10.1016/j.tplants.2016.10.002
63. Wang J, Ma Y, Zhang L, Gao RX, Wu D (2018) Deep learning for smart manufacturing: methods and applications. J Manuf Syst 48:144–156. Special Issue on Smart Manufacturing. https://doi.org/10.1016/j.jmsy.2018.01.003
64. Quispe M, Arroyo J, Kemper G, Soto J (2020) An algorithm oriented to the classification of quinoa grains by color from digital images, pp 237–247
65. Claesen M, De Moor B (2015) Hyperparameter Search in Machine Learning, February 2015

66. Li L, Du B, Wang Y, Qin L, Tan H (2020) Estimation of missing values in heterogeneous traffic data: application of multimodal deep learning model. Knowl-Based Syst 194:105592. https://doi.org/10.1016/j.knosys.2020.105592
67. Zhang T, Li Y, Li Y, Sun S, Gao X (2020) A self-adaptive deep learning algorithm for accelerating multi-component flash calculation. Comput Methods Appl Mech Eng 369:113207. https://doi.org/10.1016/j.cma.2020.113207
68. Ovtcharov K, Ruwase O, Kim JY, Fowers J, Strauss K, Chung E (2015) Accelerating Deep Convolutional Neural Networks Using Specialized Hardware, February 2015
69. Dyrmann M, Jørgensen R, Midtiby H (2017) RoboWeedSupport - detection of weed locations in leaf occluded cereal crops using a fully convolutional neural network. Adv Anim Biosci 07(8):842–847. https://doi.org/10.1017/S2040470017000206
70. Sun Y, Wong AKC, Kamel MS (2009) Classification of imbalanced data: a review. Int J Patt Recogn Artif Intell 23(04):687–719. https://doi.org/10.1142/S0218001409007326
71. Bini SA (2018) Artificial intelligence, machine learning, deep learning, and cognitive computing: what do these terms mean and how will they impact health care? J Arthroplasty 33(8):2358–2361. https://doi.org/10.1016/j.arth.2018.02.067

Ergonomic Working Conditions in Workers Under the Modality of "homeoffice" Due to a Covid-19 Pandemic, in a Bottling Company in Ecuador

Karla Cabrera-Abad, Paola Pinos-Úrgiles, Oswaldo Jara-Diaz, Luis Duque-Córdova, and Kenny Escobar-Segovia

Abstract Due to the pandemic caused by the SARS-CoV-2 virus many companies worldwide had to adapt to work from home. Therefore, the objective of this research was to know the ergonomic working conditions in workers under the "homeoffice or telework" modality due to a pandemic of Covid-19 in a bottling company in Ecuador. The study was descriptive, quantitative, with a cross-sectional cohort. We worked with a sample of 222 workers and applied a self-elaborated questionnaire validated by expert judgment. It was found that 72.5% do not have a specific area within their home to work and 83.3% of the respondents do not perform their homeoffice in an ergonomic chair. Based on the results obtained, it is concluded that most workers are not performing their daily activities with the appropriate equipment and space to avoid the risks derived from homeoffice conditions and thus ensure their labor and productive well-being.

Keywords Working conditions · Covid-19 · SARS-CoV-2 · Bottler

K. Cabrera-Abad · P. Pinos-Úrgiles · L. Duque-Córdova
Universidad Espíritu Santo, Samborondón, Ecuador
e-mail: karlacabrera@uees.edu.ec

P. Pinos-Úrgiles
e-mail: katherinepinos@uees.edu.ec

L. Duque-Córdova
e-mail: luisduque@uees.edu.ec

O. Jara-Diaz
Universidad Internacional SEK, Quito, Ecuador
e-mail: oswaldo.jara@uisek.edu.ec

K. Escobar-Segovia (✉)
Escuela Superior Politécnica del Litoral, Guayaquil, Ecuador
e-mail: kescobar@espol.edu.ec

© The Author(s), under exclusive license to Springer Nature Switzerland AG 2022
M. V. Garcia et al. (eds.), *Advances and Applications in Computer Science, Electronics, and Industrial Engineering*, Lecture Notes in Networks and Systems 433,
https://doi.org/10.1007/978-3-030-97719-1_2

1 Introduction

The pandemic unleashed by the Coronavirus (Covid-19) is an infectious disease caused by a type of coronavirus discovered in 2019, called SARS-CoV-2 [1]. Most people infected with Covid-19 have mild to moderate respiratory symptoms and recover without special treatment. The elderly and those with underlying medical conditions, such as cardiovascular disease, diabetes, chronic respiratory disease, or cancer, are more likely to experience severe disease [2]. Since the world began to talk about Covid-19 and borders were no longer a limit to contain it, a new scenario for humanity arose and social isolation has been the most effective measure so far. As a consequence, public places were closed and the vast majority of the population, especially those with some vulnerability, were forced to acquire the homeoffice modality (teleworking) to fulfill their work activities [3].

On March 11, 2020, the World Health Organization (WHO) through its general director had declared the outbreak of coronavirus as a global pandemic and requested countries to intensify actions to mitigate its spread, protect people and health workers, and save lives; through a "Technical Report for the Declaration of Emergency Covid-19", they sent to the highest authority of the Ministry of Public Health the current status of the Coronavirus Covid-19 in Ecuador [4], suggesting the "declaration of health emergency to the National Health System" [5, 6]. The term homeoffice (teleworking) appeared with the need to define a person who took his work home with him, in the seventies when the crisis forced many companies to reduce consumption and vehicle pollution in the United States, with the oil crisis, began as an incipient and scarce activity due to high costs in terms of technology and on the other hand because hiring people outside the company generated distrust [7]. The International Labor Organization (ILO), through its manual of good practices on teleworking, offers us a concept considering that "teleworking is work at a distance (including work at home) carried out with the help of telecommunication means and/or a computer" [8]. The European Framework Agreement establishes the following: Teleworking is a form of organization and or performance of work using information technologies within the framework of a contract or an employment relationship, in which a work that could be performed in the premises of the company are carried out outside of them on a regular basis [9].

Faced with this harsh reality, it should be noted that Ecuador did not have a legal regulation that regulates telework, only the Labor Code mentions a chapter to regulate home work, but telework was not named; However, in 2016, with the Ministerial Agreement MDT2016-0190 of the Ministry of Labor, the norms that regulate teleworking in the private sector were issued, necessitating a legal regulation of teleworking in Ecuador [10], Likewise, on March 12, 2020, the Ecuadorian government issued a Ministerial Agreement MDT2020-076, where guidelines were established for the application of emergency teleworking during the declaration of health emergency [11].

Teleworking is incorporated into the business activity as an option that includes a wide variety of possibilities, but also risks for the worker's health, among which stand out personal and social isolation, aspects such as fatigue, environmental and ergonomic conditions of the workplace, which makes this form of work organization a matter in need of reflection and research both prospective and retrospective [12]; not counting the possible psychosocial risks in which workers in any area and industry are immersed [13]. Being this our new and current reality due to the pandemic we are going through, the general objective of our research is to know the ergonomic working conditions in workers under the modality of "telework" due to Covid-19 pandemic, in a bottling company in Ecuador, through the application of a validated questionnaire developed by the authors.

2 Methodology

The research work is a quantitative type of cross-sectional cohort. The population of the study consisted of all workers who fulfilled their activities in teleworking mode, the number was 519 people, for our work was calculated a sample size by finite population formula with a 95% confidence level and 5% error, which gave a number of 222 workers (103 men and 119 women).

As inclusion criteria for this research work the following points were considered: i) that they are older than 18 years, ii) that they fulfill their activities under the modality of telework, during the months of July and August of 2020, during the isolation that was carried out by the pandemic by the Covid-19, iii) that they accepted to participate in the present investigation. The variables measured are all the conditions that the worker adopts for the fulfillment of his work. In addition, variables conformed by the sociodemographic data were considered, such as, age, sex, time working in the company, area where work is performed and the presence or not of children in the place of teleworking.

To collect the information, a self-elaborated questionnaire was implemented, based on the technical guide of the INHST now INSST (National Institute of Safety and Health at Work of Spain) on the display screens (PDV), in which some variables are collected about the conditions in which the participant works in the research, in the telework modality, as well as characteristics that may cause some discomfort within their activities. For the questionnaire, a validation was performed using Delphi methodology [14], with the help of experts on the subject and statistical validation using indicators such as Kendall's W concordance coefficient [15–17], the questionnaire was sent to the experts by electronic means, explaining the purpose of the questionnaire, research objectives and evaluation methodology; In order to evaluate certain characteristics such as sufficiency, clarity, coherence and relevance of the questions elaborated [18], it was evaluated with a Likert scale from 1 to 4, with 1 not meeting the criterion and 4 High level of fulfilling the criterion. Once the proposed questionnaire was validated by the experts, it was sent electronically to workers who met the inclusion criteria.

To carry out the following research, the participants were totally volunteers, and per-mission was obtained through consent for the confidential use of the information. The data were stored anonymously, and the SPSS version 26 software was used to carry out the analyses, where means and standard deviations, frequencies and percentages divided by sex, and confidence intervals for proportion with a level of 95% confidence were calculated.

3 Results

For the evaluation by the expert judges, an analysis of the averages of their responses for each question and each criterion was first performed, as shown in Table 1. It was observed that for most of the questions and criteria, the results correspond to levels 3 and 4, which would be between moderate and high levels of compliance.

Table 1 Quantitative analysis (Average) of evaluations by criteria

	Suf.	Clar.	Coh.	Red.
Question 1	4.00	4,00	4,00	4,00
Question 2	4.00	4,00	4,00	4,00
Question 3	4.00	4,00	4,00	3.67
Question 4	3.67	4,00	4,00	4,00
Question 5	3.83	4,00	3.83	3.83
Question 6	3.83	3.67	4,00	4,00
Question 7	3.83	3.67	3.67	4,00
Question 8	4.00	4,00	4,00	4,00
Question 9	3.83	4,00	3.33	3.50
Question 10	4.00	4,00	3.67	4,00
Question 11	3.00	4,00	3.83	3.83
Question 12	4.00	4,00	4,00	4,00
Question 13	4.00	4,00	3.50	4,00
Question 14	4.00	4,00	4,00	4,00
Question 15	4,00	4,00	4,00	4,00
Question 16	4,00	4,00	3.83	4,00
Question 17	3.83	4,00	3.83	4,00
Question 18	4,00	4,00	4,00	4,00
Question 19	4,00	4,00	4,00	4,00
Question 20	4,00	4,00	4,00	4,00
Question 21	4,00	4,00	4,00	4,00
Question 22	4,00	4,00	3.67	4,00
Question 23	4,00	4,00	4,00	4,00
Question 24	4,00	4,00	4,00	4,00
Average	3.91	3.97	3.88	3.95
Stan. dev.	0.21	0.09	0.19	0.13

Suf.: Sufficiency, Clar.: Clarity, Coh.: Coherence
Rel.: Relevance, Stan. Dev: Standard Deviation

Table 2 Degree of concordance between judges

Kendall's W test	Suf.	Clar.	Coh.	Red.
W	0.562	0.357	0.369	0.283
Chi-square	90.487	57.400	59.334	45.561
d.f.	4	4	4	4
p values	0.000*	0.000*	0.000*	0.003*

Suf.: Sufficiency, Clar.: Clarity, Coh.: Coherence, Rel.: Relevance
d.f.: degrees of freedom; agreement between the
judges is considered for significant p values* less than 0.05

Table 3 Sociodemographic variables by Sex

Age group (Years)	Frequency	% (IC 95%)	Frequency	Women % (IC 95%)
Younger than 25	5	4.9 (0.7–9.0)	18	15.1 (8.7–21.6)
From 26 to 35	34	33.0 (23.9–42.1)	55	46.2 (37.3–55.2)
From 36 a 45	33	32.0 (23.0–41.1)	33	27.7 (19.7–35.8)
From 46 a 55*	19	18.4 (11.0–25.9)	7	5.9 (1.7–10.1)
Older than 56	12	11.7 (5.5–17.8)	6	5.0 (1.1–9.0)
Time of work (years)				
Less than 5	34	33.0 (23.9–42.1)	58	48.7 (39.8–57.7)
From 6 to 10	27	26.2 (17.7–34.7)	33	27.7 (19.7–35.8)
From 11 to 15	17	16.5 (9.3–23.7)	14	11.8 (6.0–17.6)
From 16 to 20	10	9.7 (4.0–15.4)	7	5.9 (1.7–10.1)
More than 21	15	14.6 (7.8–21.4)	7	5.9 (1.7–10.1)
Department				
Administration *	51	49.5 (39.9–59.2)	95	79.8 (72.6–87.0)
Logistics	10	9.7 (4.0–15.4)	4	3.4 (0.1–6.6)
Manufacture	4	3.9 (0.2–7.6)	4	3.4 (0.1–6.6)
Transportation and refrigeration *	13	12.6 (6.2–19.0)	1	0.8 (0.0–2.5)
Sales	25	24.3 (16.0–32.6)	15	12.6 (6.6–18.6)
Have kids?				
Yes	62	60.2 (50.7–69.6)	59	49.6 (40.6–58.6)
No	41	39.8 (30.4–49.3)	60	50.4 (41.4–59.4)

*The columns present differences at a significant coefficient of 0.05; %: Percentage; IC 95%: Interval of confidence 95%
Source Prepared by authors

Then a statistical analysis was carried out through Kendall's W concordance test, to see the agreement between the judges, giving the following results in Table 2.

Table 4 Teleworking area according to Sex

	Men		Women	
	Frequency	% (IC 95%)	Frequency	% (IC 95 %)
Part of your home where you telecommute				
Individual study	26	25.2 (16.9–33.6)	35	29.4 (21.2–37.6)
Kitchen	1	0.9 (0.0–2.9)	1	0.8 (0.0–2.5)
Dining room	37	35.9 (26.7–45.2)	31	26.1 (18.2–33.9)
Bedroom	19	18.4 (11.0–25.9)	25	21.0 (13.7–28.3)
Living room	20	19.4 (11.8–27.1)	27	22.7 (15.2–30.2)
Provides privacy and is hassle free				
Yes	53	51.5 (41.8–61.1)	64	53.8 (44.8–62.7)
No	48	46.6 (37.0–56.2)	52	43.7 (34.8–52.6)
Does not apply	2	1.9 (0.0–4.6)	3	2.5 (0.0–5.3)
Enough space to carry out your work				
Yes	78	75.7 (67.4–84.0)	93	78.2 (70.7–85.6)
No	22	21.4 (13.4–29.3)	23	19.3 (12.2–26.4)
Does not apply	3	2.9 (0.0–6.2)	3	2.5 (0.0–5.3)
Connecting cables run through passageways				
Yes	33	32.0 (23.0–41.1)	34	28.6 (20.5–36.7)
No	65	63.1 (53.8–72.4)	79	66.4 (57.9–74.9)
Does not apply	5	4.9 (0.7–9.0)	6	5.0 (1.1–9.0)

*The columns present differences at a significant coefficient of 0.05;%: Percentage; IC 95%: Interval of confidence 95%
Source Prepared by authors

According to the data obtained, we proceeded to work with 222 workers (103 men and 119 women), with an average age of 37.7 years of age and a standard deviation of 10.3 years of age. As can be seen in Table 3, most of the employees are between 26 and 35 years of age; most of the workers (41.4%) have been with the company for less than 5 years, followed by 27% who have been with the company for between 6 and 10 years; the largest number of workers is 65.7% in the administrative area and 55.4% have children.

In Table 4 we can observe that most of the collaborators telework in the dining room (30.6%), followed by an individual study with 27.5%; the majority determine that they have an area that provides them with privacy and free of disturbances (52.7%), as well as sufficient space to develop their work (77.0%).

Table 5 Implements for teleworking according to Sex

	Men		Women	
	Frequency	% (IC 95%)	Frequency	% (IC 95 %)
Computer equipment (higher frequency)				
Cellphone	16	15.5 (8.5–22.5)	11	9.2 (4.0–14.4)
PC Desktop	17	16.5 (9.3–23.7)	19	16.0 (9.4–22.5)
Laptop	70	68.0 (58.9–77.0)	89	74.8 (67.0–82.6)
Table/Surface				
Adequate	80	77.7 (69.6–85.7)	87	73.1 (65.1–81.1)
Inappropriate	23	22.3 (14.3–30.4)	32	26.9 (18.9–34.9)
Chair				
Standard house chair	70	68.0 (58.9–77.0)	87	73.1 (65.1–81.1)
Ergonomic chair	19	18.4 (11.0–25.9)	18	15.1 (8.7–21.6)
Not an ergonomic char	14	13.6 (7.0–20.2)	14	11.8 (6.0–17.6)
MonitorÁdequate position	54	52.4 (42.8–62.1)	66	55.5 (46.5–64.4)
Inadequate position	49	47.6 (37.9–57.2)	53	44.5 (35.6–53.5)
Keyboard and mouse				
Independent in use of the laptop	30	18.5 (12.5–24.5)	33	15.6 (10.7–20.5)
Adjustable	20	12.3 (7.3–17.4)	25	11.8 (7.5–16.2)
Enough space	42	25.9 (19.2–32.7)	57	27.0 (21.0–33.0)
Readable symbols	60	37.0 (29.6–44.5)	73	34.6 (28.2–41.0)
The use of wrist rest	10	6.2 (2.5–9.9)	23	10.9 (6.7–15.1)
Foot rest				
Yes	84	81.6 (74.1–89.0)	98	82.4 (75.5–89.2)
No	19	18.4 (11.0–25.9)	21	17.6 (10.8–24.5)
Document holder				
Transcribe handwritten documents	14	13.6 (7.0–20.2)	18	15.1 (8.7–21.6)
Use document holder	1	0.9 (0.0–2.9)	4	3.4 (0.1–6.6)
Does not apply	88	85.4 (78.6–92.2)	97	81.5 (74.5–88.5)

*The columns present differences at a significant coefficient of 0.05; %: Percentage; IC 95%: Interval of confidence 95%

Source Prepared by authors

Table 6 Disconfort in teleworking according to Sex

	Men		Women	
	Frequency	% (IC 95%)	Frequency	% (IC 95 %)
Acoustic disconfort				
Adequate	74	71.8 (63.2–80.5)	73	61.3 (52.6–70.1)
Inadequate	29	28.2 (19.5–36.8)	46	38.7 (29.9–47.4)
Thermal disconfort				
Adequate	84	81.6 (74.1–89.0)	104	87.4 (81.4–93.4)
Inadequate	19	18.4 (11.0–25.9)	15	12.6 (6.6–18.6)
Light discomfort				
Adequate	82	79.6 (71.8–87.4)	96	80.7 (73.6–87.8)
Inadequate	21	20.4 (12.6–28.2)	23	19.3 (12.2–26.4)

*The columns present differences at a significant coefficient of 0.05; %: Percentage; IC 95%: Interval of confidence 95%
Source Prepared by authors

The equipment most frequently used in teleworking is a laptop or portable computer (71.6%), followed by the desktop computer with 16.2%; As for the conditions that can create some discomfort when working, 24.7% carry out their activities at a table that does not provide the facilities or is not the most appropriate; Likewise, 83.3% work in a chair that is not ergonomic, 13.6% use cushions to regulate their height and 24.1% use pads to support their back; Regarding the monitor, 45.9% work in an in-appropriate position in front of the monitor; 59.9% of employees who use a laptop computer use a separate keyboard or mouse, and 18% cannot fully support their feet (Table 5).

In terms of the noise generated in or near their telework environment, 33.7% say that they perform their activities with an uncomfortable noise. In terms of temperature, 15.5% say that they work in inadequate temperature conditions; and, finally, in terms of lighting, 19.8% say that they work with non-optimal lighting conditions (Table 6).

Table 7 shows that 59.4% know the applicable procedures and standards regarding safe and healthy working conditions, 43.2% do not have training or information on when and how to adjust their workstation to avoid discomfort; 37.3% do not have the habit of taking active breaks during their workday and 56.3% are not trained in first aid in case of emergencies in their teleworking place.

Table 7 Training, organization and emergency in teleworking according to Sex

	Men		Women	
	Frequency	% (IC 95%)	Frequency	% (IC 95 %)
Know the applicable procedures and standards regarding safe and healthy working conditions				
Yes	55	53.4 (43.8–63.0)	77	64.7 (56.1–73.3)
No	43	41.7 (32.2–51.3)	40	33.6 (25.1–42.1)
Does not apply	5	4.9 (0.7–9.0)	2	1.7 (0.0–4.0)
You have received training or information on when and how to adjust your workstation to avoid discomfort				
Yes	46	44.7 (35.1–54.3)	71	59.7 (50.8–68.5)
No	50	48.5 (38.9–58.2)	46	38.7 (29.9–47.4)
Does not apply	7	6.8 (1.9–11.7)	2	1.7 (0.0–4.0)
The times of day and what days are established, you are available and accessible to establish contact with the company				
Yes	58	56.3 (46.7–65.9)	74	62.2 (53.5–70.9)
No	39	37.9 (28.5–47.2)	38	31.9 (23.6–40.3)
Does not apply	6	5.8 (1.3–10.3)	7	5.9 (1.7–10.1)
You have an Internet connection that allows an adequate connection				
Yes	88	85.4 (78.6–92.2)	104	87.4 (81.4–93.4)
No	14	13.6 (7.0–20.2)	14	11.8 (6.0–17.6)
Does not apply	1	0.9 (0.0–2.9)	1	0.8 (0.0–2.5)
Are you in the habit of taking breaks every day?				
Yes	64	62.1 (52.8–71.5)	70	58.8 (50.0–67.7)
No	37	35.9 (26.7–45.2)	46	38.7 (29.9–47.4)
Does not apply	2	1.9 (0.0–4.6)	3	2.5 (0.0–5.3)
Do you know how to request help and advice in the use of ICTs and equipment?				
Yes	73	70.9 (62.1–79.6)	81	68.1 (59.7–76.4)
No	28	27.2 (18.6–35.8)	37	31.1 (22.8–39.4)
Does not apply	2	1.9 (0.0–4.6)	1	0.8 (0.0–2.5)
Are you trained in emergency first aid at your teleworking location?				
Yes	38	36.9 (27.6–46.2)	51	42.9 (34.0–51.7)
No	61	59.2 (49.7–68.7)	64	53.8 (44.8–62.7)
Does not apply	4	3.9 (0.2–7.6)	4	3.4 (0.1–6.6)
Do you know the emergency measures to apply in case of force majeure circumstances (fire, earthquake, etc.)?				
Yes	77	74.8 (66.4–83.1)	92	77.3 (69.8–84.8)
No	24	23.3 (15.1–31.5)	25	21.0 (13.7–28.3)
Does not apply	2	1.9 (0.0–4.6)	2	1.7 (0.0–4.0)

*The columns present differences at a significant coefficient of 0.05; %: Percentage; IC 95%: Interval of confidence 95%

Source Prepared by authors

4 Discussion and Conclusion

Given the compliance of teleworking and once concluded the health emergency in Ecuador despite the pandemic has not ended, according to the study conducted is evident that workers do not have the equipment and adequate space to avoid the risks caused by teleworking conditions and thus ensure their welfare and productive work, which could generate musculoskeletal ailments and / or occupational accidents.

It was also evidenced that there are inadequate conditions in terms of lighting and noise levels in the workstations, additionally there is a significant percentage of workers who mention not having training related to the position under the modality of teleworking, as well as a previous study conducted in the city of Quito in the first survey on safety and health conditions in 2016, with the difference that at that time the survey was conducted to personnel who did not perform teleworking, in our study 33.7% presents discomfort with respect to noise, and in the study indicated 40% [19].

According to the Occupational Health and Risk Prevention Guide in teleworking [20]: Avoid placing furniture or storing objects that may cause falls or bumps. Have enough space under the desk so that the person can mobilize their legs (minimum width 70 cm, preferably more than 85 cm). Locate the electrical and telephone line cables along the walls or behind the desk, well organized, in order to prevent them from being traversed in passageways, from movements of the chair or from making them uncomfortable to carry out work. Identify and control risks due to uneven or damaged surfaces, defective ceilings, stairs, walls, or windows. The floor surface must allow the work chair to move.

There is a risk of suffering consequences from musculoskeletal disorders and or accidents, by not following the recommendations of the guide and thus preventing unwanted effects on workers, the study covers a topic of transcendental importance today, especially since the arrival of the Covid-19 pandemic that prevented many workers from going to their workplaces, finding in teleworking a solution that supports biosafety measures; According to the results, most of them do not find significant differences between the proportions by sex.

It can be said that although the workers were sensitized to the survey, there may be biases at the time of filling it out, in addition to being a current topic we do not have enough studies that allow comparisons, with the aim of obtaining much broader discussions. The present research is intended to lay the foundations for future studies and to be able to propose public health policies on the subject.

Acknowledgements We thank the group of participating experts: MSc. Johanna Gordillo, MSc. Gabriela Zambrano, MSc. Diego Quevedo, MSc. Daniela Guzmán, MSc. Anna Carrozzini.

Annexes

Cuestionario para medir condiciones de trabajo (modalidad teletrabajo) en una embotelladora en Ecuador

- Edad (...)
- Sexo: Hombre (...) Mujer (...)
- Tiempo de labores en la Empresa (en años):
  - 0–5 años (...)
  - 6–10 años (...)
  - 16–20 años (...)
  - + de 20 añor (...)

1. Presencia de niños en casa:
   a) Sí (...)
   b) No (...)
2. En qué área de su casa trabaja con mayor frecuencia?
   a) Dormitorio (...)
   b) Área independiente (estudio individual) (...)
   c) Comedor (...)
   d) Sala (...)
3. Transformación de una parte de la vivienda [El área de trabajo proporciona priva-cidad y es libre de molestias e interrupciones?]
   a) Sí (...)
   b) No aplica (...)
4. Transformación de una parte de la vivienda [El área de trabajo tiene el espacio suficiente para desarrollar su trabajo en condiciones seguras. Se supone una superficie libre de 2 m^2]
   a) Sí (...)
   b) No (...)
   c) No aplica (...)
5. Transformación de una parte de la vivienda [Los cables de conexión eléctrica atra-viesan zonas de paso o acceso a su zona laboral (silla, mesa, etc.)]
   a) Sí (...)
   b) No aplica (...)

6. Equipo informático utilizado con mayor frecuencia para desarrollar sus actividades laborales

    a) Máquina portátil (…)
    b) Desktop (…)
    c) Celular (…)

7. Mesa / Superficie de trabajo (múltiples respuestas)

    a) La mesa NO tiene las dimensiones suficientes para permitir una colocación flexible de la pantalla, teclado, documentos y material accesorio (…)
    b) La mesa tiene dimensiones suficientes para permitir una colocación flexible de la pantalla, teclado, documentos y material accesorio (…)
    c) Es regulable en altura (…)
    d) La mesa tiene dimensiones suficientes para permitir una colocación flexible de la pantalla, teclado, documentos y material accesorio (…)
    e) Sus codos (flexionados a 90°) se encuentran a la altura de la mesa de trabajo? (…)
    f) El espacio bajo la mesa, el espacio para colocar las piernas, es suficiente y sin obs-táculos que impidan su movimiento sin golpearse? (…)
    g) El espacio bajo la mesa, el espacio para colocar las piernas, es suficiente y sin obs-táculos que impidan su movimiento sin golpearse? (…)

8. Silla (múltiples respuestas)

    a) La silla utilizada es la estándar de nuestros hogares (…)
    b) Utiliza un almohadón para soportar su espalda (…)
    c) Su altura y respaldo es ajustable, tiene soporte lumbar, apoyabrazos ajustables, tapi-cería transpirable y cinco puntos de apoyo (…)
    d) Silla de oficina sin algunas de las características mencionadas en el ítem anterior (…)
    e) Utiliza cojines para regular su altura (…)

9. Monitor (múltiples respuestas)

    a) Su borde superior se encuentra por debajo o por encima provocando una flexión o extensión del cuello (…)
    b) Puede ajustar su brillo y contraste (…)
    c) Está ubicado totalmente de manera frontal a usted y el borde superior se encuentra a la altura de sus ojos con el cuello en posición neutral (…)
    d) Tiene regulación de altura e inclinación (…)
    e) La distancia entre los ojos y monitor es por lo menos 40 cm (…)
    f) De usar portátil, tiene disponible un elevador de la pantalla (…)
    g) Si utiliza una tablet, tiene disponible un atril o funda que permite colocar la pantalla vertical (…)

10. Teclado y Ratón (múltiples respuestas)

    a) Los símbolos de las teclas resaltan suficientemente y son legibles desde la posición normal de trabajo (…)
    b) De usar portátil, tiene disponible un teclado independiente y ratón (…)
    c) Existe espacio suficiente entre el teclado y el borde la mesa para apoyar los brazos y las manos (…)
    d) Utiliza reposamuñecas (…)

11. Reposapiés (múltiples respuestas)

    a) Al sentarse puede apoyar los pies completamente en el suelo (…)
    b) Utiliza cualquier objeto como reposapiés para poder equilibrar las alturas (…)
    c) El reposapiés es regulable en inclinación y tiene superficie antideslizante (…)

12. Portadocumentos

    a) Transcribe documentos manuscritos o impresos a formato digital y adopta frecuen-temente la postura graficada (…)
    b) Utiliza un portadocumentos para colocar los documentos en el mismo plano del monitor (…)
    c) No aplica (…)

13. Ruido

    a) Tiene dificultades para concentrarse por el ruido existente en su zona laboral (…)
    b) El ambiente acústico es adecuado, no genera molestias ni desconcentración (…)

14. Temperatura

    a) La temperatura de su zona laboral es confortable y con ventilación adecuada (…)
    b) La temperatura de su zona laboral no es confortable y sin ventilación adecuada (…)

15. Iluminación (múltiples respuestas)

    a) El nivel de iluminación es suficiente para realizar su trabajo sin dificultad ( )
    b) En su zona laboral se producen reflejos o brillos molestos ( )
    c) Se proyectan sombras molestas en la zona laboral en la que realiza su trabajo ( )
    d) Desde su posición habitual de trabajo percibe lámparas muy brillantes, ventanas u otros elementos que molesten a la vista ( )

16. Capacitación [Conoce los procedimientos y normas aplicables en el teletrabajo, en particular, en lo referente a condiciones de trabajo seguras y saludables]

    a) Sí (…)
    b) No (…)

c) No aplica (…)

17. Capacitación [Ha recibido capacitación o información sobre cuándo y cómo ajustar su estación de trabajo para evitar disconfort]

a) Sí (…)
b) No (…)
c) No aplica (…)

18. Organización [Están establecidos los momentos del día y qué días, usted está dispo-nible y accesible para establecer contacto con la empresa]

a) Sí (…)
b) No (…)
c) No aplica (…)

19. Organización [Dispone de una conexión a Internet que permita una adecuada cone-xión]

a) Sí (…)
b) No (…)
c) No aplica (…)

20. Organización [Tiene el hábito de realizar pausas todos los días]

a) Sí (…)
b) No (…)
c) No aplica (…)

21. Si usted realiza pausas, cuántas pausas realiza en promedio en un día habitual de teletrabajo

a) 1 (…)
b) 2 (…)
c) 3 (…)
d) 4 (…)
e) 5 (…)
f) 6 (…)
g) 7 (…)
h) No aplica (…)

21.1 Si usted realiza pausas, cuánto es la duración promedio de cada pausa (minutos)

a) 1–3 (…)
b) 4–7 (…)
c) 8–10 (…)
d) 11–15 (…)
e) 16–20 (…)
f) No aplica (…)

22. Emergencias [Conoce como solicitar ayuda y asesoramiento en el uso de TIC's y equipos?]
    a) Sí (…)
    b) No (…)
    c) No aplica (…)
23. Emergencias [Está capacitado en primeros auxilios frente a emergencias en su lugar de trabajo?]
    a) Sí (…)
    b) No (…)
    c) No aplica (…)
24. Emergencias [Conoce las medidas de emergencia a aplicar en caso de suceder alguna circunstancia de fuerza mayor (incendio, sismo, etc.)?]
    a) Sí (…)
    b) No (…)
    c) No aplica (…)

References

1. Organization WH, et al (2020) Novel Coronavirus (2019-nCoV): situation report 11
2. Organization WH (2020) Coronavirus: Synopsis. https://www.who.int/es/health-topics/coronavirus#tab=tab1, Accessed 8 Sept 2020
3. Díaz Pacífico F, Petinari L, Freán P (2020) Teletrabajo y pandemia en las bibliotecas de Rosario: resultados del primer relevamiento
4. Secretaria Nacional de Gestión de Riesgos y Emergencias (2020) Informe de Situación COVID-19 en Ecuador:Informe 001. Gobierno del Ecuador
5. Ministerio de Salud Pública (2020) Acuerdo Ministerial No. 00126–2020: Declárese el estado de emergencia sanitaria en todos los establecimientos del Sistema Nacional de Salud... Gobierno del Ecuador
6. Mahase E (2020) China coronavirus: WHO declares international emergency as death toll exceeds 200. BMJ Brit Med J 368
7. Rojas Jaramillo RR et al (2016) Análisis de la modalidad de teletrabajo y su incidencia en la productividad laboral de las empresas privadas de la ciudad de Quito. Caso de estudio empresa DIT TELECOM. Universidad Andina Simón Bolívar, Sede Ecuador
8. Di Martino V (2004) El teletrabajo en América latina y el caribe. Proyecto Puesta en Marcha del Teletrabajo
9. Arochena FL, Latas RPR (2015) Una mirada periférica al teletrabajo, el trabajo a domicilio y el trabajo a distancia en el derecho español. In: Trabajo a distancia y teletrabajo: estudios sobre el régimen jurídico en el derecho español y comparado. Aranzadi, pp 31–46
10. Ministerio de Trabajo (2016) Acuerdo Ministerial No. MDT-2016-190: Expedir las normas que regulan el teletrabajo en el sector privado. Gobierno del Ecuador
11. Ministerio de Trabajo (2020) Acuerdo Ministerial No. MDT-2020-076: Expedir las directrices para la aplicación de teletrabajo emergente durante la declaratoria de emergencia sanitaria. Gobierno del Ecuador
12. Uguina JRM (2015) La Ley de Prevención de Riesgos Laborales, veinte años después. Revista de información laboral. 8:21–42

13. Collin-Orellana A, Vásquez-Cassinelli J, Duque-Córdova L, Espinoza-Samaniego C, Arias-Ulloa C, Escobar-Segovia K (2020) Improvement in service quality by analyzing the relationship between psychosocial risks and labor engagement in a vertical transport equipment maintenance company. In: 18th LACCEI International Multi-Conference for Engineering, Education Caribbean Conference for Engineering and Technology: "Engineering, Integration, and Alliances for a Sustainable Development" LACCEI
14. Cruz Ramírez M, Rúa Vásquez JA (2018) Surgimiento y desarrollo del método Delphi: una perspectiva cienciométrica. Biblios 71:90–107
15. Escobar-Pérez J, Cuervo-Martínez Á (2008) Validez de contenido y juicio de expertos: una aproximación a su utilización. Avances en medición. 6(1):27–36
16. García C, Castellanos E, García M (2018) UML-based cyber-physical production systems on low-cost devices under IEC-61499. MDPI AG 6:22. https://doi.org/10.3390/machines6020022
17. García MV, Irisarri E, Pérez F, Estévez E, Marcos M (2018) Arquitectura de Automatización basada en Sistemas Ciberfísicos para la Fabricación Flexible en la Industria de Petróleo y Gas, vol 15. Universitat Politecnica de Valencia, p 156. https://doi.org/10.4995/riai.2017.8823
18. Olvera-Ureña E, Huayamave-Torres R, Arias-Ulloa C, Escobar-Segovia K (2019) Validation of a questionnaire to evaluate the level of knowledge of an Ecuadorian Technical Standard INEN 2251: 2013 regarding the sale of liquid fuels, in service stations of the province of Guayas. In: 17th LACCEI International Multi-Conference for Engineering, Education, and Technology: "Industry, Innovation, and Infrastructure for Sustainable Cities and Communities" LACCEI
19. Gómez García AR, Merino-Salazar P, Silva-Peñaherrera M, Suasnavas Bermúdez PR, Vilaret Serpa A (2019) I Encuesta sobre Condiciones de Seguridad y Salud en el Trabajo para Ecuador. Principales resultados en la ciudad de Quito, 2016. Medicina y Seguridad del Trabajo 65(257):238–251
20. Ministerio de Trabajo y Seguridad Social (2018) Guía de Salud Ocupacional y Prevención de los Riesgos del Teletrabajo. Consejo de Salud Ocupacional, Gobierno de Costa Rica

Optimization of Routes for the Collection of Solid Waste

Hugo Muñoz, Daysi Ortiz, Israel Naranjo, and Alex Pazmiño

Abstract This paper presents the results of the optimization of the current routes of the solid waste collection and transportation process in charge of the Public Cleaning and Environmental Management Company of Latacunga canton (EPAGAL). The objective of this research is to minimize the operating costs of each route in the city, using the algorithm applied to Capacitated Vehicle Routing Problem (CVRP). To improve and add robustness to the model it was complemented with an exact heuristic for Miller Tucker Zemlin (MTZ) sub tour elimination, the model was solved using the optimization software ILOG CPEX Optimization Studio and with Python programming plugins. The result of the optimization was obtained a total average thrift in the distance of all the routes, both in a theoretical average saving "in a straight line" of 30.81% and with real data from the routes, an average saving of 11.51%. Furthermore, this optimization streamlines the work of collecting and transporting the solid waste, reduces overtime and reduces environmental impact.

Keywords CVRP · Distribution · MTZ methodology · Mathematical model · Vehicles routing · Software ILOG CPEX Optimization Studio

1 Introduction

Transport is a primary activity for the good performance of both Small and Medium-sized Enterprises (SMEs) and large companies, the search for more efficient services, creates the need to develop mechanisms and strategies to improve the level of service with the lowest possible cost [1].

H. Muñoz · D. Ortiz (✉) · I. Naranjo · A. Pazmiño
Universidad Técnica de Ambato, Ambato, Tungurahua, Ecuador
e-mail: dm.ortiz@uta.edu.ec

H. Muñoz
e-mail: cmunoz4384@uta.edu.ec

I. Naranjo
e-mail: ie.naranjo@uta.edu.ec

A. Pazmiño
e-mail: apazmino4147@uta.edu.ec

© The Author(s), under exclusive license to Springer Nature Switzerland AG 2022
M. V. Garcia et al. (eds.), *Advances and Applications in Computer Science, Electronics, and Industrial Engineering*, Lecture Notes in Networks and Systems 433,
https://doi.org/10.1007/978-3-030-97719-1_3

For this it is necessary to apply mathematical tools and computing systems in order to make appropriate decisions to distribute resources in a better way, this type of optimization is known as vehicle routing problems, which aims to minimize the transport costs associated with distribution and collection routes [2–4]. Companies dedicated to the transport service continuously innovate their processes with mechanisms that optimize logistics activities; according to a study by the National Planning Department (DNP) of Colombia in 2019, transport represents 35.2% of logistics costs [5].

The Vehicle Routing Problem (VRP) has different variants and applications to real cases: distribution of couriers, packages and materials, routing of school groups, postal mailings and distribution of goods, routes of vendors and road maintenance, transport of disabled people, solid waste collection, street cleaning operation, etc. [6].

The collection and transportation of solid waste in the world and especially in Latin America and the Caribbean (LAC) has become one of the main concerns of the administrative entities of each locality, in accordance with technical work of the Department of the Environment, the World Bank made in the year 2019, about 85% of these residues are collected by systems door-to-door, to be subsequently transferred mostly to landfills [7, 8]. This is the reason why companies dedicated to this activity seek to improve the quality of solid waste collection and transport service, to avoid setbacks with customers and the environmental problems that arise from them [9].

In Ecuador, the problem with solid waste is not an isolated issue, as according to the latest report of the INEN of Statistical Information Environmental Economic Decentralized Autonomous Governments Municipal, 2017, on the Integral Management of Solid Waste, each inhabitant of the urban area produces on average 0.58 kg per day of solid waste, while in the country is collected 12.337,26 tons per day, to be subsequently moves-two in a 45.7% to landfills, 28.8% cells are emerging and the 25.6% in open dumps [10, 11]. For this reason, it is necessary to improve the management of solid waste and optimize the collection and transport of these.

In the canton Latacunga, the percentage of dwellings that sent the garbage truck-garbage collector based on the Development Plan and land 2016–2019 is 95,99% for the urban sector and 38,47% for the rural sector, giving a total coverage of solid waste collection of 61,44% [12]. This information justifies the search for and application of strategies that minimize operating costs in order to expand the coverage of the service, especially to rural sectors. Within this perspective, companies will be able to continue contributing to the improvement of the environmental quality of the sectors under their care. The proposed case study is developed in the Public Company of Cleaning and Environmental Management from Latacunga (EPAGAL) responsible for the collection of solid waste in the canton Latacunga province of Cotopaxi.

According to real-world applications, it has been shown that the use of computerized methods and VRP algorithms generate sustainable decreases in transport costs between 5 and 20% of the final operating expenditure [13]. A study of vehicle routing problems for solid waste collection in the city of Ambato in 2017 based on linear programming methods, got that the number of the truck fleet is reduced from 7 to 4, obtaining an annual savings of $ 15885.59 corresponding to a 30.18% reduction unlike the current system [14]. The objective of our research is to optimize solid waste collection routes using the characteristics, variables and constraints of vehicle routing problems with limited capacity Capacitated Vehicle Routing Problem (CVRP).

The CVRP is one of the variants of the VRP, which considers a similar vehicle fleet capacity. This problem seeks to find the optimal route, whether in cost, distance or time, among others. Each vehicle is loaded with a certain amount of a product without exceeding its capacity, to ensure that each of them leaves the warehouse and returns to it, applying a routing strategy of solid waste collection [15]. The CVRP is modeled as a minimum cost flow problem to meet several requirements, including collecting waste from a set of nodes by a homogeneous fleet of fixed capacity vehicles that cannot be breached leaving and returning each one to the reservoir after collecting the solid waste [16]. Methodology Miller Tucker Zemlin (MTZ) is an exact heuristic model applicable to break the sub-tours into exact mathematical models of vehicle routing problems, the objective function can be to maximize or minimize depending on the study problem [17].

2 Methodology

The methodology used in this study describes the application of a model of vehicle routing problems with limited capacity Capacitated Vehicle Routing Problem (CVRP), which is improved by combining with the MTZ methodology to avoid the formation of cycles within the tour, obtaining more robust solutions of the minimum costs depending on the distance, thus determining the most optimal routes for the collection of solid waste.

2.1 Data Collection

In the first phase of the research, the main data of the system were collected and later considered in the CVRP model, variables such as position of the nodes of the collection centers, capacity of the vehicles, capacity of the collection centers, distances in kilometers of current routes. It considers the exact position of containers, floricultural, industrial companies, hospital waste customers and collection points in general where ordinary solid waste is accumulated.

2.2 Development of the CVRP Model

The implementation of the model CVRP was based on previous research, as well as the application of the methodology of the book, Vehicle Routing Problems, Methods, and Applications of the authors Paolo Toth and Daniele Vigo (2014), which describes how a fleet of vehicles with uniform capacities has to satisfy the demand of a group of customers, through a set of routes that start and end in the same place, all this with the lowest possible cost [18].

CVRP mathematical model using the methodology (MTZ):

$$\min Z = \sum_{i=0}^{n}\sum_{j=0}^{n} x_{ij}d_{ij} \tag{1}$$

$$\sum_{i=0}^{n} x_{ij} = 1 \qquad \forall j \epsilon V \setminus \{0\} \tag{2}$$

$$\sum_{j=0}^{n} x_{ij} = 1 \qquad \forall i \epsilon V \setminus \{0\} \tag{3}$$

$$\sum_{i=0}^{n} x_{ij} - \sum_{j=0}^{n} x_{ji} = 0 \tag{4}$$

$$SI \quad x_{ij} = 1 \Rightarrow u_i + q_j = u_j \qquad i, j \epsilon A : i \neq 0, j \neq 0 \tag{5}$$

$$\Rightarrow q_i \leq u_i \leq Q \qquad i \epsilon V \tag{6}$$

where:

min Z: Expresses the shortest distance (km) between solid waste collection centers.
x_{ij}: equal to 1 when you decide to travel in the arc (i, j) ϵ A; zero otherwise.
d_{ij}: minimum travel distance (km) between solid waste collection center (node) i and j.
u_i: auxiliary flow variable for MTZ, number of solid waste collection centers visited up to collection center i.
q_j: capacity in tons of each collection center.
A: set of arches that represents the paths to go from one collection center to another.
Q: vehicle capacity in tonnes.
V: set of vertices that in this case study are the collection centers where solid waste is stored.

The objective function represented by Eq. (1) minimizes the distance traveled between the collection centers. Restrictions (2), (3) ensure that each customer is visited only once. Restriction (4) ensures the continuity of the route. Restriction (5) removes the subtour and ensures that the vehicle departs and returns to the depot and (6) ensures that the total demand of the route does not exceed the capacity of the vehicle on each journey [19].

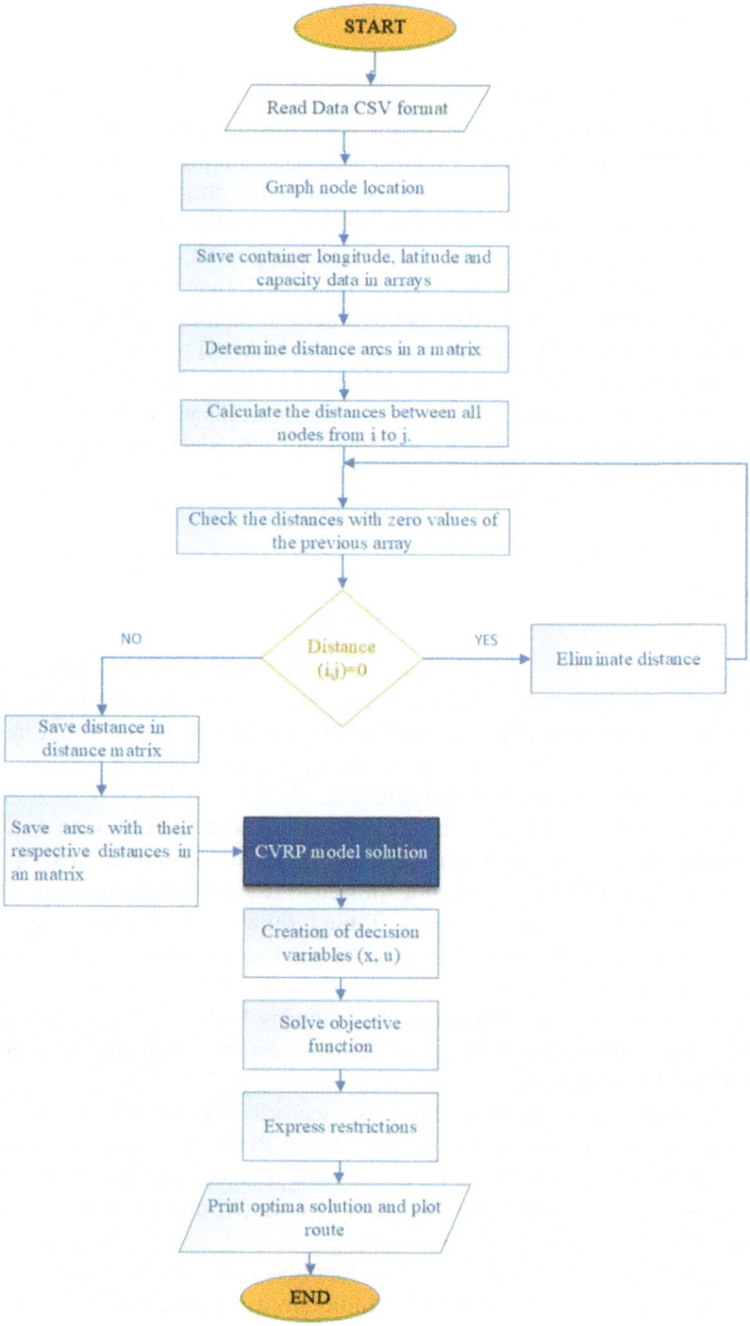

Fig. 1 Flowchart CVRP mathematical model solution

2.3 CVRP Model Solution

The variables that enter the definition of the linear programming model allow the operationalization of the data by applying the software IBM ILOG CPLEX Optimization Studio, which uses mathematical and constraint programming for decision optimization. In addition, it can be combined with other programming languages for ease of use and improved route mapping results [20]. This case was done with the Python code that calculates the optimal routing using a flat coordinate system with matplotlib, to later plot each node and each route in Google Maps [21].

Based on the results obtained, it is possible to establish the best routes for the solid waste collection activities carried out by the EPAGAL company, allowing it to propose improvements according to the needs of the citizens from the Latacunga canton. For a better understanding of how optimization works with IBM ILOG CPLEX Optimization Studio software, a CVRP Mathematical Model solution procedure was developed, as evidenced in Fig. 1.

3 Results

The figures were obtained by collecting information from EPAGAL, where the current situation of solid waste collection routes was established. The respective distances in km of each route, the exact position of containers, floricultural, industrial companies, customers of hospital waste and collection points in general. Table 1 provides a summary of the different routes.

The values of the capacities of the vehicles on their different routes were determined. In the analysis the real average Q was selected only from the first trip, and to keep its fleet in good condition, it is not allowed to reach its maximum capacity, thus avoiding damage to hydraulic systems. Table 2 shows these values, while Table 3 shows the capacities of the containers, the hospital sector does not have containers because the collection is manual. The frequencies of each one of the routes are shown in Table 4, which have been defined by the company EPAGAL, taking into account that each route works differently because it is necessary a collection in both urban and rural parishes of the canton.

As an example, the route of the Industrial sector is detailed, the results obtained from the model determined in the research. In the Python program, the position of the nodes previously obtained from a GPS is first graphed, as shown in Fig. 2.

Then, the program determines the feasible solution and plots it on a plot, as evidenced in Fig. 3.

The program also shows the objective function value, in this case the feasible route for the industrial sector that has a minimum travel distance of 59,605 km, as evidenced in Fig. 4.

Table 1 Initial data of the different routes

Routes	# of nodes	Type of collecting truck
Industrial	14	Front
Underground Islands	29	Back
Western	126	Side
Eastern	138	Side
Route 1	152	Back
Route 7	165	Back
Route 10	166	Back
Hospital	377	Van

Table 2 Vehicle capacities

Vehicles	(Q) Factory rated (tons)	(Q) Average first trip (ton)
Back loading vehiclesles	10	8.08
Side loading vehicles	15	12
Industrial vehicle (Front)	12	10.29
Hospital waste vehicle	1	0.72

Table 3 Container capacities

Type	Nominal volume (L)	Nominal weight
Back loadingng	1300	520 kg = 0.520 tons
Side loading	1800	750 kg = 0.750 tons
Front loading	2400	1000 kg= 1tons

Table 4 Service frequencies of the different routes

Routes	Service frequencies
Industrial sector route	Work tuesday and thursday
Underground Islands route	Work from monday to sunday
Western route	Work from monday to fridays
Eastern route	Work from monday to friday
Route 1	Work monday, wednesday, and friday (frequency 1)
	Work Tuesday and Thursday (frequency 2)
Route 10	Work monday (frequency 1)
	Work tuesday and friday (frequency 2)
	Work wednesday (frequency 3)
	Work thursday (frequency 4)
Hospital sector route	Work monday (frequency 1)
	Work tuesday (frequency 2)
	Work wednesday (frequency 3)
	Work thursday (frequency 4)
	Work friday (frequency 5)

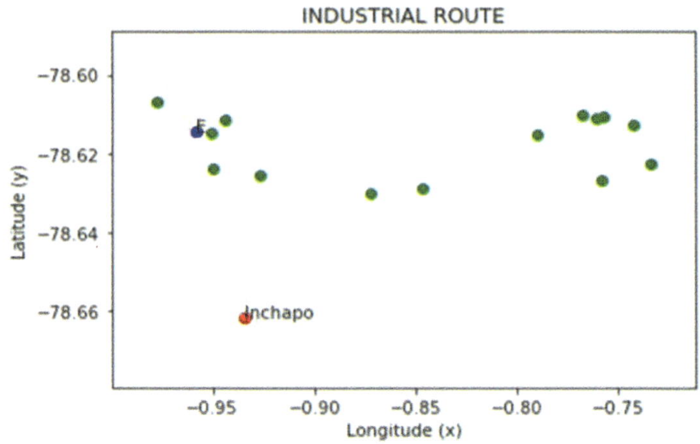

Fig. 2 Industrial route customers location

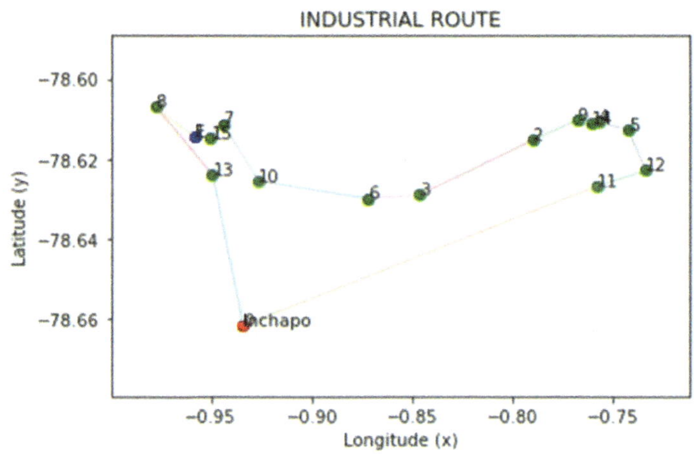

Fig. 3 Optimized feasible route of industrialists

```
Total (root+branch&cut) =  120.13 sec. (5370.07 ticks)
JobSolveStatus.FEASIBLE_SOLUTION
solution for: CVRP
objective: 59.605
```

Fig. 4 Optimized feasible route of the industrial sector

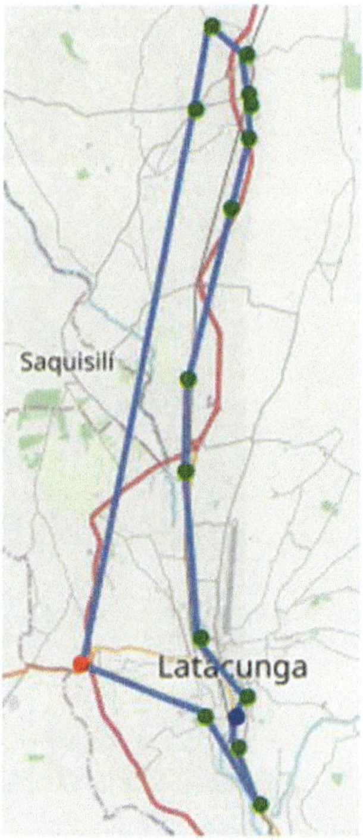

Fig. 5 Optimized route of the industrial sector on a real map

Table 5 Results route of the industrial sector

Previous path route	Feasible route	Real route optimized proposed	Real savings
82.66 km	59.605 km	78.5 km	4.16 km

Later the route visualization is improved by passing it to a real map, with the help of the Python map function, as evidenced in Fig. 5.

Then and with the help of Google Earth the real tour of the route is graphed, as evidenced in Fig. 6.

Finally, and with the help of Python Folium it makes it easy to view data that has been manipulated in Python on an interactive brochure map. A map is created in portable HTML format for the operator and complies with its respective route, as evidenced in Fig. 7. To understand the benefits of the application of the problem of routing vehicles with limited load or CVRP, the solutions obtained are compared with the previous distances determined previously by GPS. Table 5 shows the optimization results for the route of the industrial sector. In all routes the same previous procedure

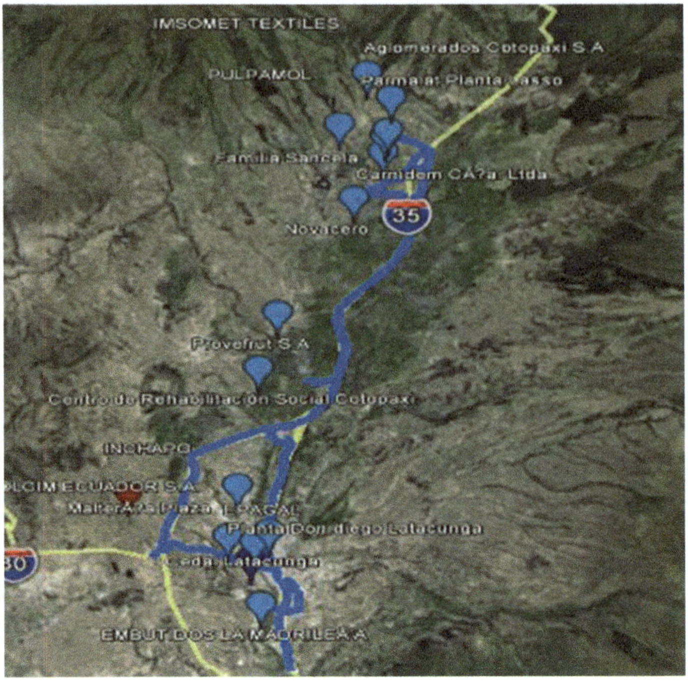

Fig. 6 Optimized route of the industrial sector real distance traveled

Fig. 7 Interactive industrial sector route for the operator

Table 6 Summary of route optimization results

Route	Previous route (km)	Feasible route (km)	Proposed route (km)	Real savings (km)	Optimal savings %	Real savings %
Industrial	82.66	59.61	78.5	4.16	27.89	5.03
Underground Islands	31.98	18.64	26.98	5	41.72	15.63
Western	52.14	48.29	47.02	5.12	7.38	9.82
Eastern	84.68	70.49	73.12	11.56	16.75	13.65
Route 1 MWF	172.24	107.40	140.83	31.41	37.64	18.24
Route 1 TT	178.18	130.68	163.21	14.97	26.66	8.40
Route 7 MWF	105.44	61.44	89.42	16.02	41.73	15.19
Route 7 TT	166.22	107.87	146.19	20.03	35.10	12.05
Route 10 M, Wedn	158.22	124.64	148.7	9.52	21.22	6.02
Hospital mondays	31.78	18.09	27.88	3.9	43.08	12.27
Hospital tuesdays	32.77	22.38	29.11	3.66	31.69	11.17
Hospitable wednesdays	40.35	25.39	35.92	4.43	37.08	10.98
Thursday hospitallers	195.86	159.16	172.56	23.3	18.74	11.90
Hospital fridays	304.7	168.83	271.77	32.93	44.59	10.81
Average					30.81%	11.51%

was performed, it should be noted that the optimal savings is the difference between the previous route and the feasible route, while the real savings is the difference between the previous route and the route with its actual route. Table 6 shows the optimization results for all routes.

4 Discussion

This article presents the optimization of the different routes where the solid waste collection service is offered by the Public Company of Cleaning and Environmental Management from Latacunga (EPAGAL). The problem aims to reduce the operational cost of routes in a framework that guarantees a better service to citizens. A mathematical programming formulation based on the problem of routing of vehicles with limited load or CVRP is applied, which to improve it was complemented with the MTW methodology. In addition, to check the savings obtained from the model, the optimal solutions were transferred to a map to establish the actual routes of the vehicles and then compare them with the current situation of the routes. It was determined that with the optimal solution, an average distance saving of 30.81% and an average real distance saving of 11.51% compared to the previous routes were saved.

All the parameters of the system should be contrasted with a provider of online maps that return actual data of the route, with a theoretical approach "in a straight line" it's totally unrealistic, this ensures the actual efficiency of the route that will carry out the transport, in addition, with the optimization speeds up the work of collection and transportation of solid waste.

5 Conclusions

The process of solid waste collection in the canton Latacunga in charge of the EPA-GAL can be defined as a problem of vehicle routing with load limited (CVRP), in which you get an average savings actual distance of 11.51% with respect to the routes earlier, thus leading to a direct reduction of the operation cost of the routes, since a shorter distance of travel involves a reduced consumption of tires, fuel and overtime that is required to comply with the frequencies and provide a better service to the citizens. Future studies may include in the optimization analysis information regarding the aforementioned costs.

By applying the optimization model, it was possible to determine the reduction in distance of 5.03% in the path of the industrial sector, 15.63% in the path of the islands underground, 9.82% on the western route, 13.65% on the eastern route, 18.24% on route 1 in the frequency of Monday, Wednesday and Friday, 8.40% in the route 1 in the frequency of Tuesday and Thursday, 15.19% on route 7 to the frequency of the Monday, Wednesday and Friday, 12.05% on route 7, the frequency of the Tuesday and Thursday, 6.02% on route 10 on Monday and Wednesday, in the path of hospital Monday 12.27% on Tuesday 11.17%, Wednesday 10.98%, Thursday 11.90% and Friday 10.81% respectively, thus improving the coverage in the collection and transport service of solid waste in the canton of Latacunga.

References

1. Zapata Cortes JA, Vélez Bedoya ÁR, Arango Serna MD (2020) Mejora del proceso de distribución en una empresa de transporte. Investigación Administrativa 49(126). https://doi.org/10.35426/iav49n126.08
2. SimpliRoute. Cómo SimpliRoute resuelve el problema de Ruteo de Vehículos. https://www.simpliroute.com/blog/como-simpliroute-resuelve-el-problema-de-ruteo-de-vehiculos
3. García MV, Irisarri E, Pérez F, Marcos M, Estevez E (2018) From ISA 88/95 meta-models to an OPC UA-based development tool for CPPS under IEC 61499. In: 2018 14th IEEE international workshop on factory communication systems (WFCS), pp 1–9
4. Galleguillos R, Altamirano S, García MV, Pérez F, Marcos M (2017) FAHP decisions developing in low cost CPPs. In: 2017 IEEE 3rd Colombian conference on automatic control (CCAC), pp 1–8
5. Diario La República. Almacenamiento y transporte representan 81, 7% de los costos de logística. https://bit.ly/2YOvaCs
6. Arboleda Zúñiga J, López AX, Lozano YL (2016) El problema de ruteo de vehículos [VRP] y su aplicación en medianas empresas colombianas
7. León J, Plaza P (2017) Análisis de la gestión de residuos sólidos en el cantón Balzar - Provincia del Guayas. Repositorio de la Universidad Católica de Santiago de Guayaquil, p 153
8. Icaza D, Borge-Diez D (2020) Energy supply of a hybrid system of biomass and wind turbines of the pichacay landfill towards an intelligent network for the city of Cuenca-Ecuador. In: Advances and applications in computer science, electronics and industrial engineering. Springer, Cham, pp 287–307
9. World Bank. Convivir con basura: el futuro que no queremos. https://www.bancomundial.org/es/news/feature/2019/03/06/convivir-con-basura-el-futuro-que-no-queremos
10. Instituto Nacional de Estadística y Censos. Según la última estadística de información ambiental: Cada ecuatoriano produce 0,58 kilogramos de residuos sólidos al día. https://bit.ly/3DxvjcK
11. Jurado F, Donoso D, Escobar E, Mayorga T, Bilous A (2021) A prototype electronic toy for the development of mathematical logical reasoning in children from five to seven years old using python. In: Advances and applications in computer science, electronics and industrial engineering. Springer, Singapore, pp 3–18
12. GAD Latacunga (2019) Plan De Desarrollo Del Cantón Latacunga 2016–2019, pp 1–303
13. Álvarez Hernández RJ, et al (2017) Propuesta de solución al problema de ruteo de vehículos en el operador logístico Opperar SA para el transporte y distribución de productos alimenticios secos del Grupo Nutresa SA
14. Lozano ÁGG, Miño GE, Martínez JCC (2017) Optimización del proceso de recolección de desechos sólidos de la ciudad de Ambato mediante el diseño de un modelo de distribución de redes. ECA Sinergia 8(2):158–171
15. Edna Y, Diego E, Fernando J, Miguel A (2020) Algorithms applied to the problem of Routing of Vehicles with Capacity Restriction (CVRP): bibliometric analysis of the solutions 41(29):139–152
16. Qiao Q, Tao F, Wu H, Yu X, Zhang M (2020) Optimization of a capacitated vehicle routing problem for sustainable municipal solid waste collection management using the PSO-TS algorithm. Int J Environ Res Public Health 17(6). https://doi.org/10.3390/ijerph17062163
17. Villegas Flórez J, Zapata C, Gatica G (2017) Una aplicación del método MTZ a la solución del problema del agente viajero. Scientia Et Technica 22(4):341–344. https://doi.org/10.22517/23447214.12751
18. Desaulniers G, Madsen OBG, Ropke S (2014) Chapter 5: The vehicle routing problem with time windows

19. Rodriguez Perez J (2012) Caracterización, modelado y determinación de las rutas de la flota en una empresa de rendering. Universidad de Sevilla, Sevilla
20. IBM. ILOG CPLEX Optimization Studio - Visión general - España|IBM. https://www.ibm.com/es-es/products/ilog-cplex-optimization-studio
21. Medium. Capacitated Vehicle Routing Problem (CVRP) with Python+Pulp and Google Maps AP. https://bit.ly/3oZMOyb

Proposal of a Systemic Model for Integration of Strategic Planning in Corporate Level with Balanced Scorecard

Javier Arturo Gamboa Cruzado, Juan Carlos Chancusig-Chisag, José Vicente Morales Lozada, and Anthony Johan Alfaro Acuña

Abstract A New Model that integrates corporate strategy with the Balanced Scorecard is studied to improve decision-making executives by merging Models Relevant systems, making use of Soft Systems Methodology (SSM), which is based on systems thinking. The research methodology used was action research. This methodology is being applied with very satisfactory results in the generation of new models, as is the case under discussion. Its core is the idea that the researcher does not remain unaware of the research but becomes a participant in the relevant group (diversified firms). In the field, was recorded input and solution models from other authors: the advisor, System Information (SI) staff of corporations, reported investigations and proposals from the maker of researcher? The purpose is to generate a consensus model and confront this model with business reality. The units of analysis considered are international methodologies, knowledge and experience of the investigator, entrepreneurs and researchers, diversified companies, so on. For the testing of the hypothesis was considered: Decision Making of the Head of Admission at the College. The main conclusion was the construction of the New Model, Corporate Strategic Planning and Balanced Scorecard (CSP-BSC), was performed very efficiently from the relevant systems, business requirements and the contributions of the researcher using the SSM, the Philosophical Framework is very important in any investigation because it guides the whole process of research, decreased the percentage of bad decisions, and in-creased the percentage of accuracy of information. It is relevant to corporate and divisional managers, researchers, teachers, and students.

J. A. G. Cruzado
Universidad Nacional Mayor de San Marcos, 15022 Lima, Peru

J. C. Chancusig-Chisag (✉)
Universidad Técnica de Cotopaxi, 05-01491 Latacunga, Ecuador
e-mail: juan.chancusig@utc.edu.ec

J. V. M. Lozada
Universidad Técnica de Ambato, 180207 Ambato, Ecuador
e-mail: jvmorales99@uta.edu.ec

A. J. A. Acuña
Universidad Autónoma del Perú, 15842 Villa el Salvador, Peru
e-mail: a.alfaro@autonoma.pe

© The Author(s), under exclusive license to Springer Nature Switzerland AG 2022

M. V. Garcia et al. (eds.), *Advances and Applications in Computer Science, Electronics, and Industrial Engineering*, Lecture Notes in Networks and Systems 433,
https://doi.org/10.1007/978-3-030-97719-1_4

Keywords Corporate strategy · Indicators · Decision making · Balanced score-card · Business intelligence

1 Introduction

Big companies handle a large amount of information and the challenge is knowing how to manage it. That Information could be used in order to make decision. Understanding the decision-making process [1] is vital for managers to facilitate the process; the tools used in companies are the corporate indicators [2] that should help monitor and control corporate objectives and strategies over time.

Balance Scorecard is a tool highly used in Business Intelligence, which provides reports and indicators to different areas in a company. It is useful at the moment of reducing cost, time and effort, and also it helps to costumers understanding the service or requirements needed without complicated analysis [3–5]. Furthermore, to incorporate Corporate Strategic Planning and Balanced Scorecard (CSP-BSC) will improve the company organization. The CSP-BSC will help the reader understand where to look for new Balanced Scorecard strategies and solutions that address the complex world we face [6]. Many of the stages that are included are based on years of research, teaching and consulting experience. Theories relating to mobile augmented reality applications have been integrated. In addition, the theory about training, learning, heavy machinery [7].

However, In Peru, at national level, there is no methodological proposal that integrates business strategy with decision-making, which allows the development of applications aimed at improving decision-making by managers in organizations. This knowledge and application gap results in executives and business consultants not managing their business work efficiently [8, 9].

The research was carried out as an Action program which is based on the Soft Systems Methodology (SSM), this concept arises in the behavioral sciences and is obviously being applied with very satisfactory results in the generation of models, as in discussion; its core is the idea that the researcher does not remain outside the research field, but becomes a participant in the relevant human group [10] (the diversified company).

The main objective of this research is to generate and apply an Integral Systemic Model of Corporate Strategic Planning with the Balanced Scorecard [11–13] to improve Decision Making in diversified companies. It is proposed a possible solution: the generation and application of a new Integral Systemic Model of Corporate Strategic Planning with the Balanced Scorecard will allow improving the Decision Making in the diversified companies.

2 State of Art

The corporate strategic approach [14] is a process in which an organization defines its long-term mission and the strategies to achieve it based on the analysis of its strengths, weaknesses, opportunities and threats, involves the active participation of the organizational actors, the permanent obtaining of information on their critical success factors, their revision, their monitoring, evaluation and periodic adjustments so that it becomes a style of management that makes the organization a proactive and anticipatory entity.

A large number of companies have more than one business unit and as a consequence, above the business units there is a higher level: the corporate one; In a diversified company, the strategies of each business are not in general independent of each other and neither should they be their Information Technologies and Information Systems (IT/SI) plans [15]. The purpose of this part is to enter into the considerations regarding the corporate level on the business units and their strategies, as well as the corporate organization of the IT/SI, that is, to generate a methodology introducing the role of the corporation [16].

Figure 1 represents the fundamental concepts of the integration between the corporate strategy formulation process and the IT/SI considerations as the element of the IT/SI plans, the central element of the proposed integration is the process of formulating the corporate strategy; each company considers its own procedure according to its management style, its organizational culture or simply its tradition, the chosen process must interact for the different business units.

Within the IT/SI corporate strategic approach it is necessary to develop and operate current and future applications, this means determining the way in which the applications will be developed and how they will acquire, use, control and manage the necessary technological and human resources To meet the needs of the company, all organizations, must have an IT strategy reviewed and approved by the board of directors [17]. The implications of the IT strategy are so important that they should not be left exclusively to the systems director or the technical leader. If you are not sure of the technologies, you can reduce the risks by getting a second opinion from an external consultant.

An IT strategy is just a goal for the entire organization and a map of how to take the company from where it is currently to that goal, along the way there are minor signs or goals along with their scheduled times. The strategy should be reviewed periodically to assess how close it is to the smaller goals [18].

The Balanced Scorecard (BSC) [19] is a powerful instrument for measuring corporate performance and has been shown to be the most effective tool to link vision, mission, and strategy to five performance measures. It also offers a complete view of the organization, being the essential element of the information system that supports the management control system in its mission to improve its level of competitiveness in the long term (See Fig. 2); Although there are 4 perspectives that traditionally identify a BSC, it is not essential that they all be; These perspectives are the most common and can be adapted to the vast majority of companies that are not an

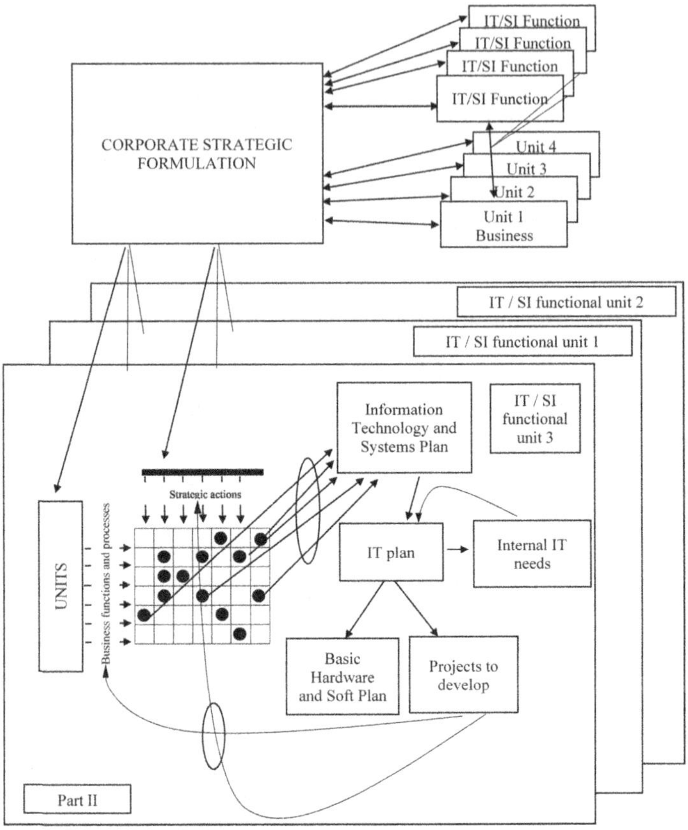

Fig. 1 Integration of corporate strategy and IT/SI plans

indispensable condition for building a business model [20]. The main goal is to Return on Capital Employed (ROCE).

Business intelligence (BI) [21] is the process by which organizations segment, unify, rank and define contexts about information, in order to identify the most relevant and representative information for the organization; the architecture of Business Intelligence is composed of different technologies that are integrated to form a business solution; the components are aimed at transforming the data into timely and reliable information for the end user (See Fig. 3).

Systemic thinking [22] is the attitude of the human being, which is based on the perception of the real world in terms of totalities for analysis, understanding and action, unlike the approach of the scientific method , which only perceives parts of it and in an unconnected way, is integrating, both in the analysis of situations and in the conclusions that arise from there, proposing solutions in which various elements and relationships that make up the structure have to be considered of what is defined as "system", as well as everything that makes up the defined system environment.

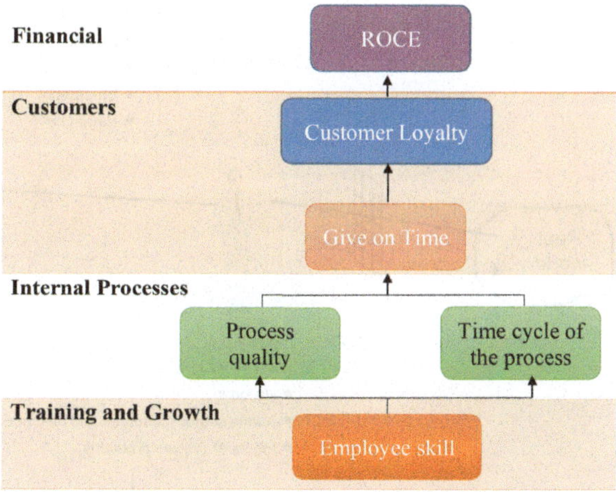

Fig. 2 Integration of corporate strategy and IT/SI plans

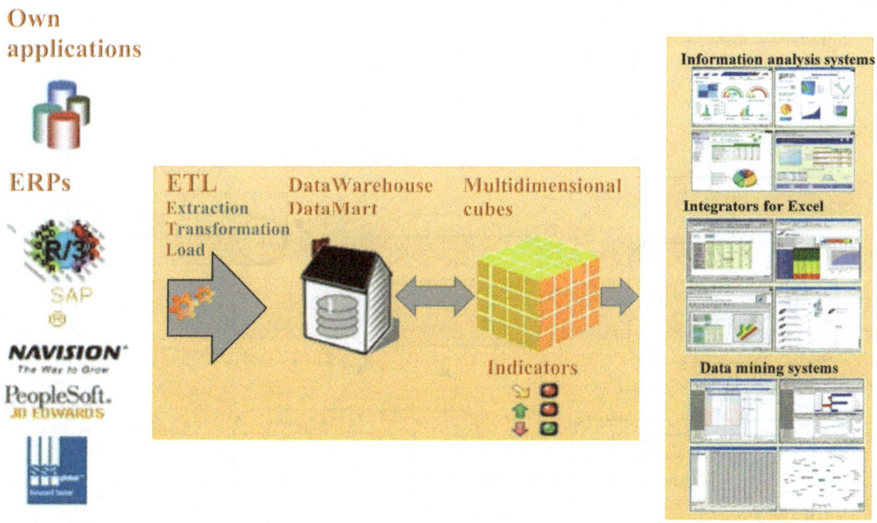

Fig. 3 Cause-effect relationships through the 4 perspectives of the BSC

The philosophical basis that sustains this position is Holism (from the Greek holos = integer) (See Fig. 4).

The Soft Systems Methodology of Peter Checkland [23] is a qualitative technique that can be used to apply structured systems to assistance situations (Fig. 5). It is a way of dealing with situational problems in which there is an activity with a high social, political and human component; it is a process of seven stages or stages of analysis that employ the concept of a Human Activity System (HAS) as a means of getting both "investigate" (diagnose) the situation and "carry out actions" (direct and implement) to improve it [24].

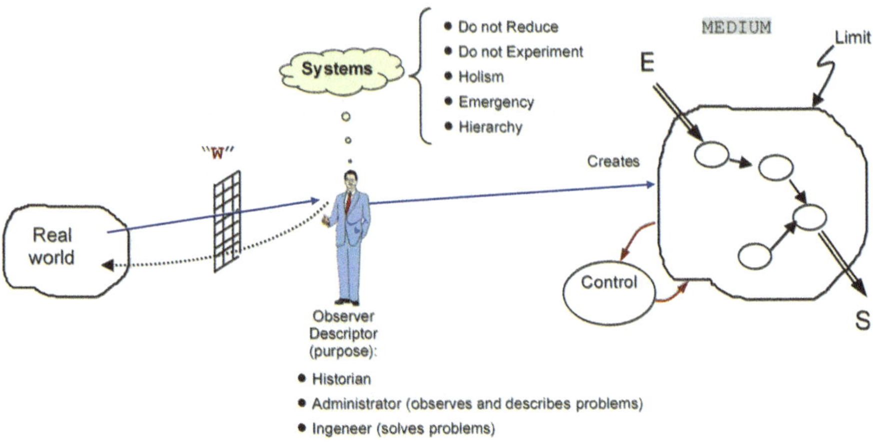

Fig. 4 Actuate the systematic thinker

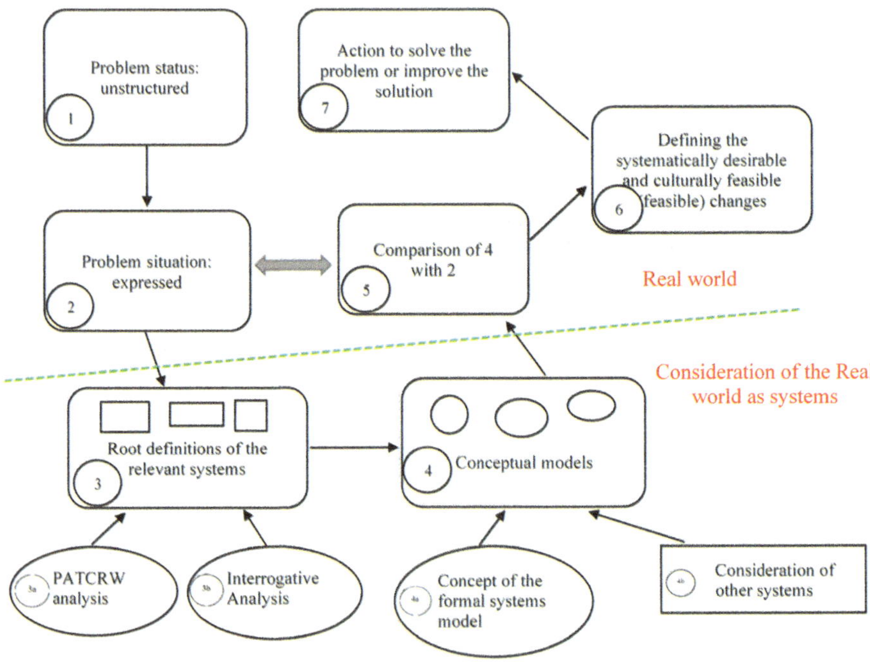

Fig. 5 Stages of the soft systems methodology (SSM)

3 Methods

The investigation was carried out as part of a research program in action. This concept arises in the behavioral sciences and is obviously being applied with very satisfactory results in the generation of new models, as is the case under discussion. Its core is the idea that the researcher does not remain a researcher outside the field but becomes a participant in the relevant human group (the company). The study was carried out as part of an action research program. This concept arises in the behavioral sciences and is being applied with very satisfactory results in the generation of new models, as is the case under discussion. Its core is the idea that the researcher does not continue outside the know field, but rather becomes a participant within the relevant human group, the company (See Fig. 6). In Peru, there are few organizations that dominate and integrate corporate business strategies with the support of the Balanced Scorecard. However, there are some international intellectuals and consultants of recognized prestige who have developed separate proposals on corporate strategies, Balanced Scorecard, and Business Intelligence:

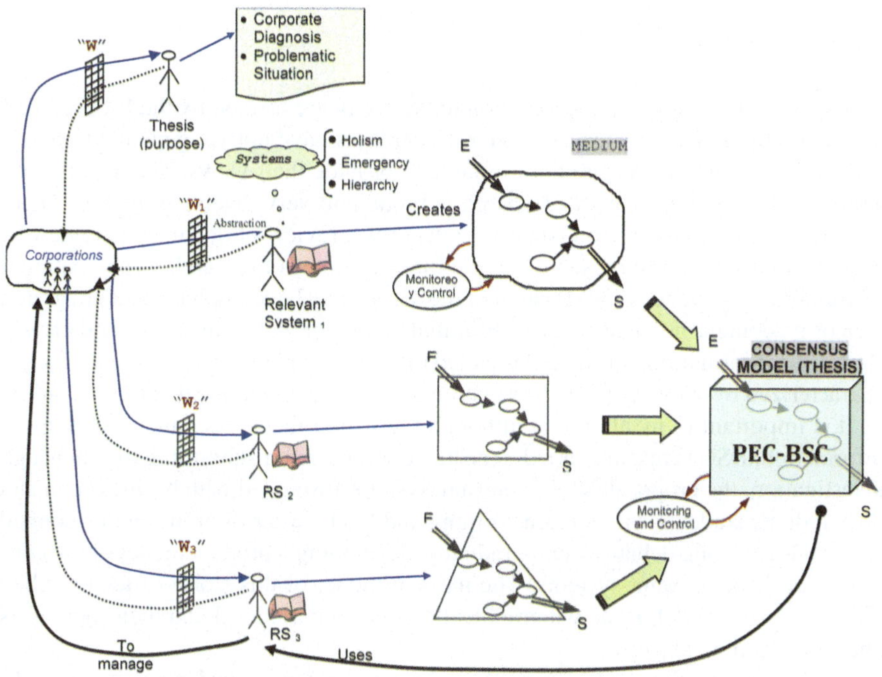

Fig. 6 Actuate the systematic thinker

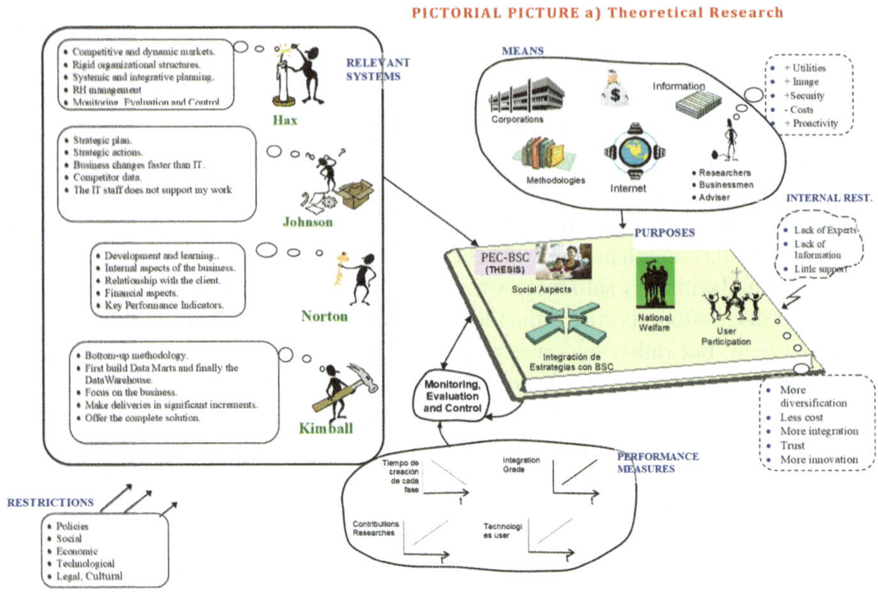

Fig. 7 Pictoral picture

The relevant systems to the situation under study are formed by the Proposals of the International Researchers; Peruvian researchers have not yet generated models and the international ones are not available to provide their views. The information gotten in the previous stage of research is broad and very diverse in nature. There is registered information from surveys, interviews, review of documents, review of business plans, research presented on the Internet, notes of researchers and national consultants, experiences and research of thesis, etc. and have a holistic and integrated idea of its content It is necessary to formulate a model that summarizes and clearly shows this information, this model is called the pictorial picture (See Fig. 7), which is characterized by showing [25], information model on technological communication.

It is important to mention that although there are other tools which offers more information, BSC is relevant as a diagnostic tool [26]. After submitting to a thorough investigation, the proposals of the relevant Systems discussed with businessmen and postgraduate students with a brainstorming and finally after rigorous and prolonged technical-academic debates went, gradually, developing a product model of the consensus of those relevant systems. The nascent model of this storm of ideas, called the Consensus Model, is shown graphically at the generic level and then each of its phases is detailed in Fig. 8.

All decision-making by the Head of the Admission Area at the University throughout the university's operating time is identified as the unit of analysis. Therefore their exact number cannot be known. For this research, a sample of 30 Decision Making from the Head of Admission at the University was taken, since it is an adequate, standard value and several research processes are used. The confidence level of this research will be 95% given the experience of the researcher.

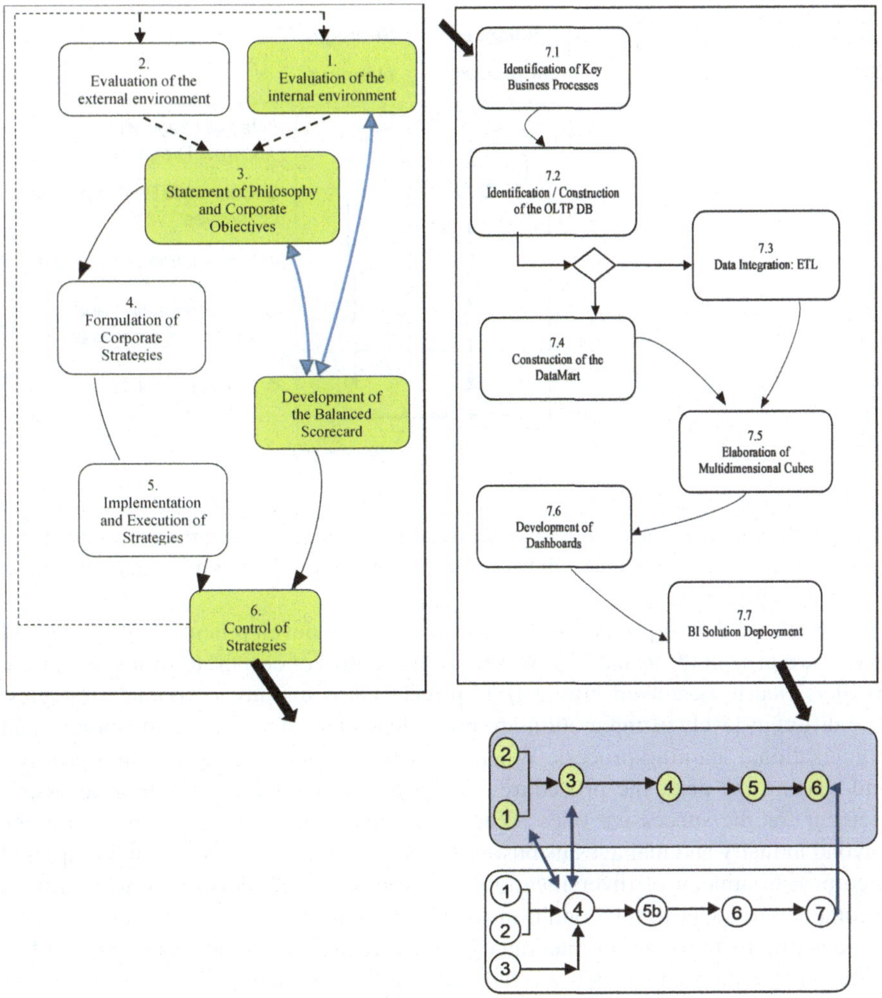

Fig. 8 Consensus model: the new model CSP-BSC

4 An Integrator Model

Figure 9 presents an integrating model of the corporate strategy with the Balanced Scorecard. The integration process begins with the internal and external strategic diagnosis of the corporation (phases 1 and 2). Then the approach of the philosophy and objectives of the organization is carried out (phase 3). This establishes the necessary guidelines for the development of the BSC (1, 2, 3, and 4b). It is necessary to go to the BSC level and formulate their corresponding Key Performance Indicators (KPIs) for the fact table (4b). Next, the corporate strategies are formulated (phase 4),

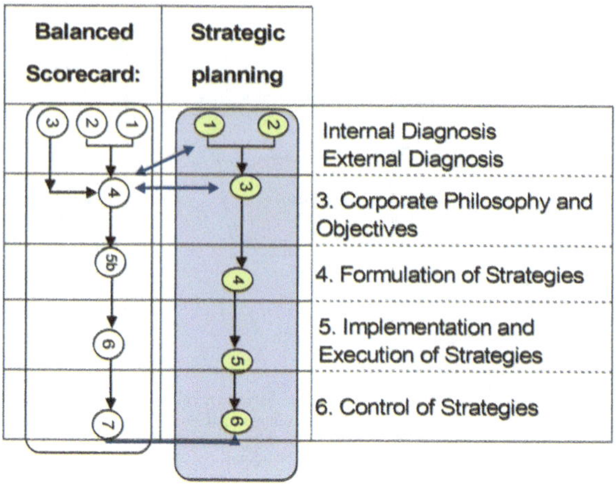

Fig. 9 An integrating model of the corporate strategy with the BSC

which will help meet the corporate objectives proposed in 3. Once the corporate-level strategies are consolidated (4), the corporate strategic implementation and execution take place (phase 5).

The rest of the corporate tasks correspond to the development of the Balanced Scorecard project (5, 6, and 7b), as well as the control of corporate strategies (phase 6). It is clearly perceived how a BSC project helps monitor corporate strategies. The different levels of integration are not independent since they correspond to an organizational learning process. When covering one stage in the learning process, and starting the next, the procedures of the previous stage constitute a necessary element for the successive ones. When a company has diversified its operations beyond industry and manages its business in several industrial sectors, it has applied a corporate strategy of diversification. A corporate diversification strategy allows a company to adapt to the conditions of its external environment. A model can be created which shows, in an integrated manner, which are the internal stakeholders of the corporation and which are the external stakeholders of the corporation. In addition, the interrelation between these stakeholders must be shown. This model gives a very interesting systemic view of the context of the corporation. The Fig. 10 shows the internal and external stakeholders for a study carried out in CONIDA (National Commission of Aerospace Research and Development) of Peru by a group of teachers.

Valuing the Nature of the Environment: Since one of the main problems of strategic management is to overcome uncertainty, it is useful to know how uncertain the environment is and why. The uncertainty of the environment increases the more dynamic and complex are the environmental conditions. The approach to under-standing the uncertainty of the environment depends on its degree of stability or dynamism, as well as its degree of simplicity or complexity. Analysis of Technological Trends: One of the tasks of top management is to perceive the corporation not as it is today but as it may become in the future (See Fig. 11). The vision of what a company can

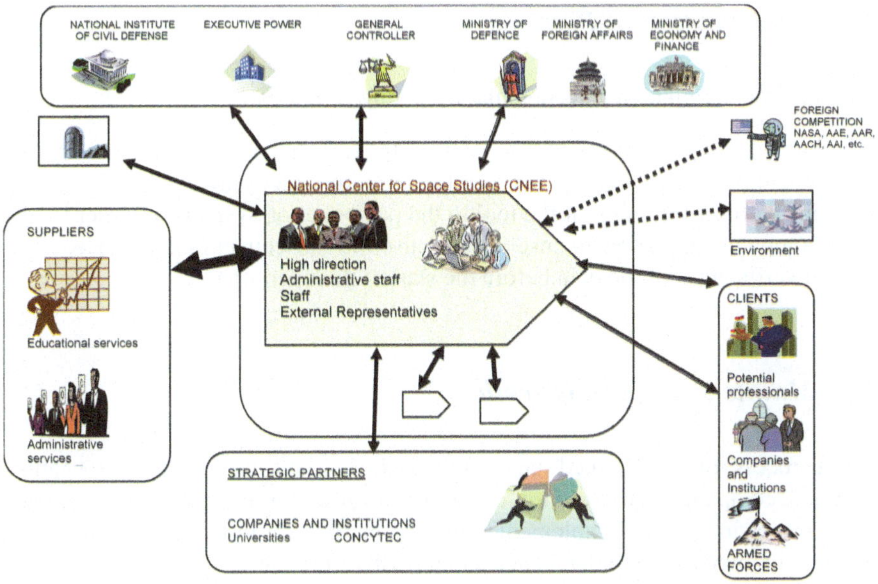

Fig. 10 Map of internal and external Stakeholders for CONIDA

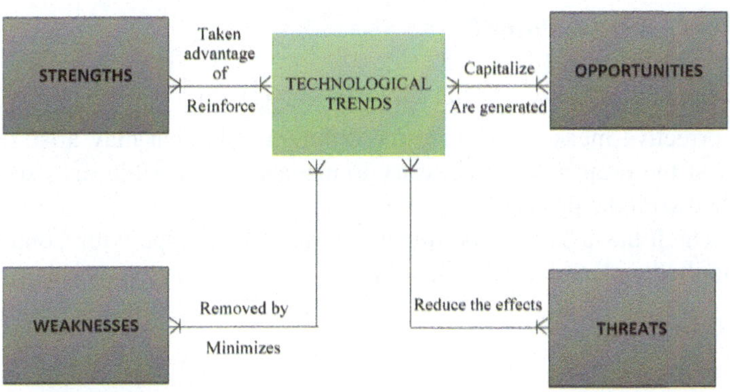

Fig. 11 Technological trends

become is often reflected in a financial model that projects profits and expenses in five or more years in the future and breaks them down in detail.

The implementation represents the convergence of technology, data, and end-user applications accessible from the business user's desktop. There are several extra factors that ensure the proper functioning of all these parts, among them, are training, technical support, and communication. All these tasks must be taken into account before any user can access Business Intelligence.

4.1 Implementation Plan

The implementation plan considers the following activities:

- Training a group of users with the system of the Universidad Alas Peruanas (UAP), these will be able to train the users of the different offices involved.
- Giving technical training of the tool to the personnel of Systems, in order to give the computer support by themselves and also the subsequent maintenance.
- Final verification of the data before the start of the operation.

4.2 Maintenance and Growth

BI is a process (of well-defined stages, with beginning and end, but of a spiral nature) since it accompanies the evolution of the organization throughout its history. Once the BI has been built and implemented there is no time for rest, we must quickly be prepared to manage the maintenance and growth of the same.

4.3 Project Management

This phase allows us to manage the entire life cycle of the project and take preventive and corrective measures to face the different risks that may arise during the exe-cution of the project, it is necessary to manage the possible risks such as user not identified with the project.

This research has a True Experimental design, Panel type with Control Group, which is applied and explained in detail in the testing of the hypothesis.

RGe X O1
RGc – O2

Where:

R: Random choice.

Ge: Experimental Group: Study group to which the stimulus will be applied (Integrated Systemic Model).

Gc: Control Group: Study group to which the stimulus will not be applied (Integrated Systemic Model).

O1: Post-test data for the indicators of the dependent variable: Post-Test measurements of the experimental group.

O2: Post-test data for the indicators of the dependent variable: Post-Test measurements of the control group.

X: (Integrated systemic model): Stimulus or experimental condition.

– : It is the lack of stimulus or experimental condition.

Table 1 Average of the indicators for the Gc and Ge

Indicator	Post-test (Gc) (Media: 1)	Post-test (Ge) (Media: 2)	Commentary
Information accuracy percentage	46.68	85.06	–
Time to report	25.63 min	4.47 min	–
Percentage of bad decisions made	20.33	2,86	–
Number of reports requested per cycle	4 reports/cycle	14 reports/cycle	–
Understanding the reports	–	–	Not contrasted
			Qualitative indicator

4.4 Results

All results in this section were evaluated before and after integrating the new model. Table 1 presents the means of the Indicators for the Post-Test (Gc) and Post-Test (Ge)

- 50.00% of the Percentages of information accuracy in the Post-Test (Ge) were greater than their average accuracy.
- 36.67% of the Percentages of information accuracy in the Post-Test (Ge) were greater than the goal set.
- The 100.00% of the Percentages of information accuracy in the Post-Test (Ge) were greater than the average accuracy in the Post-Test (Gc).

Figure 12 indicates a detailed analysis of the data from KPIs measured.

- The data have close values, which is confirmed by observing that the confidence intervals of the Average and the Median overlap.
- The "average" distance of the individual observations of the Percentages of information accuracy with respect to the average is 3.161%.
- Around 95% of the Percentages of information accuracy are within 2 standard deviations of the mean, that is, between 83.884% and 86.245%.
- The Kurtosis = –1,506 indicates that there are values of time with very low peaks.
- The Asymmetry = –0.037 indicates that the majority of percentages of information accuracy are high.
- The 1st Quartile (Q1) = 81.990% indicates that 25% of the Percentages of information accuracy are less than or equal to this value.
- The 3rd Quartile (Q3) = 88.215% indicates that 75% of the Percentages of information accuracy are less than or equal to this value.

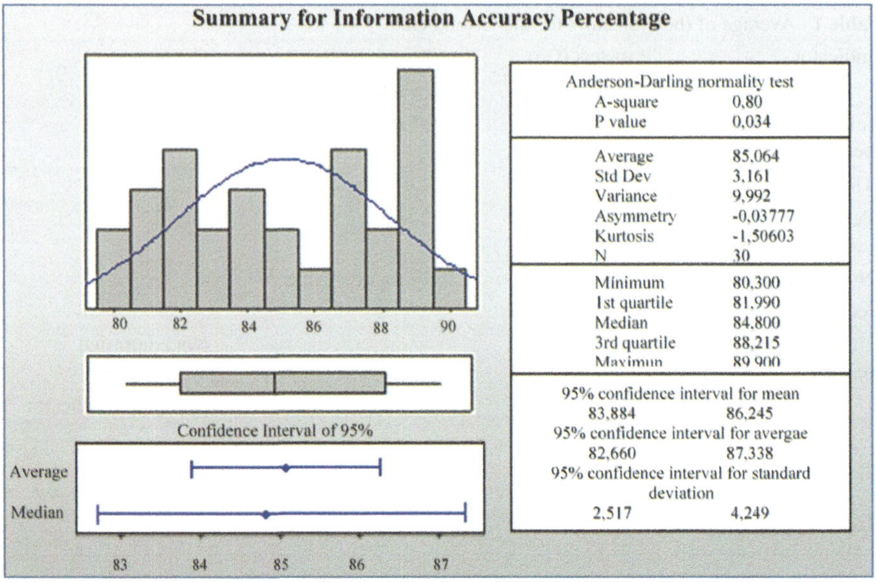

Fig. 12 Summary for information accuracy percentage

11	12	13	14	15	16	17	18	19	20
Eas	Dif	Dif	Nor	Eas	Nor	Nor	Dif	Eas	Eas
21	**22**	**23**	**24**	**25**	**26**	**27**	**28**	**29**	**30**
Nor	Dif	Dif	Dif	Nor	Nor	Nor	Nor	Dif	Eas

Estate	Frequency
Easy	7
Normal	9
Difficult	14

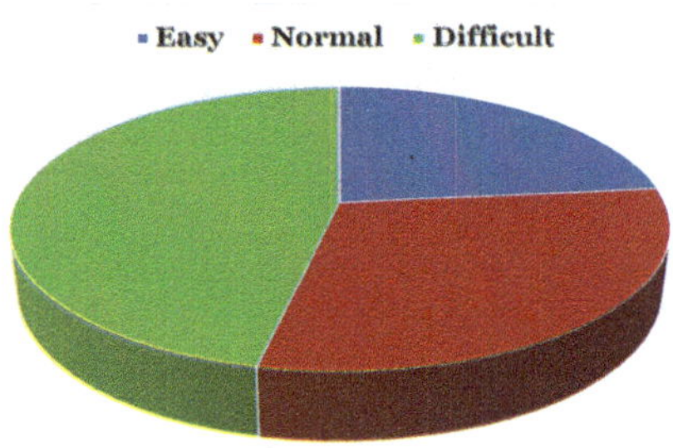

Estate	Frequency
Understandable	7
Not understandable	23

Fig. 13 Pre-test results

4.5 Pre-test

Results showed in Fig. 13 indicates the following information.

- 47% of the time the Understanding of the reports was classified as difficult by the Admission Manager.
- Only 23% of the time the Understanding of the reports was classified as easy by the Admission Manager.

Measurement	1	2	3	4	5	6	7	8	9	10
Value	Nor	Eas	Nor	Nor	Nor	Eas	Nor	Nor	Nor	Eas
	11	12	13	14	15	16	17	18	19	20
	Eas	Eas	Eas	Eas	Nor	Eas	Eas	Eas	Eas	Nor
	21	22	23	24	25	26	27	28	29	30
	Dif	Nor	Eas	Eas	Nor	Eas	Eas	Eas	Dif	Nor

Estate	Frequency
Easy	16
Normal	12
Diffcult	2

Estate	Frequency
Understandable	18
Not understandable	12

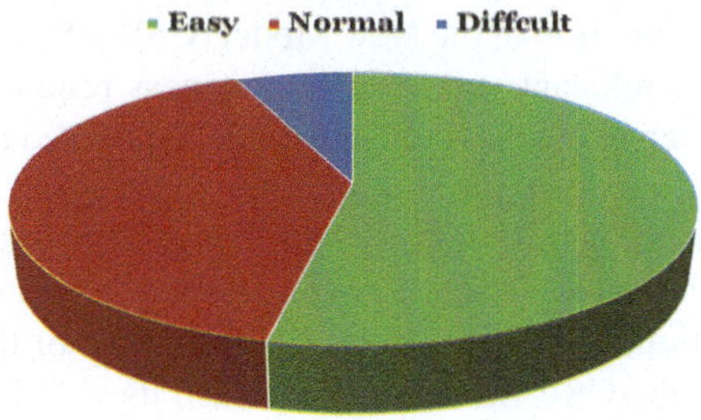

Fig. 14 Post-test results

- It is determined that only 23% of the time the understanding of the reports is understandable.
- It is determined that only 77% of the time the Understanding of the reports is Not Understandable.

4.6 Post-test

Results exposed in Fig. 14 indicates the following items.

- Only 7% of the times the Comprehension of the reports was classified as difficult by the Head of Admissions.
- Now 53% of the times of Comprehension of the reports was classified as easy by the Head of Admissions.
- It is determined that only 60% of the times the Comprehension of the reports is Understandable.
- It is determined that only 40% of the times of Understanding of the reports is Not Understandable.

It is the confrontation of a group Ge intentionally formed by the people who make decisions in the process of Admission in the University, which is applied a systemic integrative model stimulus (X), after which a test is applied after the indicators of the dependent variable (O1). A second group (Gc), also intentionally formed by the people who make decisions in the process of Admission in the UA, which is not given any stimulus, serving only as a control group; Simultaneously a test is applied to the indicators of the dependent variable (O2), it is expected that the O1 values are better than the O2 values.

5 Conclusions

The construction of the New Model (CSP-BSC) was carried out with great efficiency, from the relevant systems, the business requirements and the contributions of the researcher, using the SSM, the model generated is a good support for decision making strategic in diversified companies; the testing of the hypothesis could be carried out efficiently by means of an experiment designed specifically for this purpose, which considers aspects of experimental (true) and non-experimental (Transactional-correlational) designs, therefore the results of the application of the techniques and tools of the CSP-BSC Model confirm its validity.

The Soft Systems Methodology (SSM), based on Systemic Thinking, is an excellent methodological tool for research at postgraduate level, the generation of new methodologies, through consensus is better than the generation from a single point of view; the coordination with business executives and IT specialists was decisive for the integration of the business strategy with IT, the validation of the methodological attempts with the future clients of these enriched said proposal, and even more, gave it the backing of the systemic desirability and cultural feasibility.

It is verified that, having implemented the New Model improved the Decision Making of the Admission Area in a University, it is also observed that the implementation of the New Model helps to minimize the time taken to generate the reports; The implementation of the New Model has made it possible to increase the understanding of the reports and it is appreciated that the implementation of the New Model decreased the percentage of bad decisions made, finally, the successful development of a BSC (BI) project, based on the New Model, increased the percentage of information accuracy.

Finally, theoretical contributions of this research were the in-depth understanding of knowledge on the subject, within the range of areas, themes and subjects that make up the doctorate curriculum. Technical discussions have also been generated with business consultants and corporate executives on the applicability of the new model generated. A new model is presented that integrates the company's strategy at the corporate level with the Balanced scorecard. that allowed to make a contribution, compilation or experimentation of a topic, within the doctorate in Administrative Sciences. Through an own research methodology (in action), which will allow the student to learn to develop this type of work and other similar ones in their future professional performance. Additionally, in the practical sense, some models have been generated to diagnose the current state of some diversified companies and to propose new business models for them.

Acknowledgements Supporting provided by the Technical University of Ambato, Research and Development Directorate (DIDE), and Technical University of Cotopaxi.

References

1. Orazbayev B, Ospanov E, Kissikova N, Mukataev N, Orazbayeva K (2017) Decision-making in the fuzzy environment on the basis of various compromise schemes. Procedia Comput Sci 120:945–952
2. Zeinalova LM (2016) Decision making with combined states under interval uncertainty. Procedia Comput Sci 102:223–230
3. Vega Falcón V, Navarro Cejas M, Cejas Martínez MF, Mendoza Velazco DJ (2020) Balanced scorecard: key tool for strategic learning and strengthening in business organization (Cuadro de mando integral: herramienta clave para el aprendizaje estratégico y el fortalecimiento de la organización empresarial). Acad J Interdisc Stud 9:1–11
4. Alarcon OA, Velastegui HJ, Garcia MV, Gallo CV, Espejo VP, Naranjo JE (2019) Monitoring and control system approach for native threatened species. In: 2019 international conference on information systems and software technologies (ICI2ST), pp 85–91
5. Llamuca ES, Garcia CA, Naranjo JE, Garcia MV (2019) Cyber-physical production systems for industrial shop-floor integration based on AMQP. In: 2019 International Conference on Information Systems and Software Technologies (ICI2ST), pp 48–54
6. Nyonawan M, Utama DN (2018) Evaluation of information technology governance in STMIK mikroskil using COBIT 5 framework. In: 2018 international conference on information management and technology (ICIMTech). IEEE, pp 137–142
7. Brodin M (2017) Security strategies for managing mobile devices in SMEs: a theoretical evaluation. In: 2017 8th international conference on information, intelligence, systems and applications (IISA). IEEE, pp 1–6
8. Pacovilca Alejo GS (2020) El Balance Scorecard y la GestiÓn Educativa EstratÉgica en las Instituciones Educativas del Distrito de Huancavelica 2017. Universidad Nacional de Huancavelica
9. Diaz Arroyo KE (2011) Propuesta del balance Scorecard desde la perspectiva financiera como soporte de información para la toma de decisiones: Caso de estudio: Empresa Cognitio Consulting SAC, Arequipa, 2018. Universidad Nacional de San Agustín de Arequipa
10. Vidal Flores D, Domenge MR (2017) Executive flight simulator as a learning tool in new companies' resource planning based on the balanced scorecard. Contaduría y administración. 62(2):556–576
11. Juiz C, Colomo-Palacios R, Gómez B (2018) Cascading ISO/IEC 38500 based Balanced Score Cards to improve board accountability. Procedia Comput Sci 138:417–424
12. Paredes Larroca F, Saettone Olschewski E, Quino Favero J, Rosales Huamaní J, Castillo Sequera JL (2020) Development of a fuzzy logic-based solar charge controller for charging lead–acid batteries. In: Advances and applications in computer science, electronics and industrial engineering. Springer, Cham, pp 168–183
13. Saeteros M, Paucar W, Molina C, Caiza G (2020) Development and analysis of a PID controller and a fuzzy PID. In: Advances and applications in computer science, electronics and industrial engineering. Springer, Cham, pp 143–154
14. Park SH (2016) A strategic approach to policy tasks for the development of Korea's island areas. J Marine Island Cult 5(1):14–21
15. Winter MG (2016) A strategic approach to debris flow risk reduction on the road network. Procedia Eng 143:759–768
16. Papulova Z, Gazova A (2016) Role of strategic analysis in strategic decision-making. Procedia Econ Finan 39:571–579
17. Volná J, Papula J (2013) Analysis of the behavior of Slovak enterprises in the context of low innovation performance. Procedia-Soc Behav Sci 99:600–608
18. Tsai JM, Chien HH, Shih SC, Lee SC, Tsai LY, Tsay SL (2017) Using balanced scorecard on reducing fall incidents and injuries among elderly cancer patients in a medical center in Taiwan. Int J Gerontol 11(4):253–257
19. Houck M, Speaker PJ, Fleming AS, Riley RA Jr (2012) The balanced scorecard: sustainable performance assessment for forensic laboratories. Sci Just 52(4):209–216

20. Tubis A, Werbińska-Wojciechowska S (2017) Balanced scorecard use in passenger transport companies performing at Polish market. In: Transbaltica. In: Proceedings of the International Scientific Conference, vol 10. Vilnius Gediminas Technical University, Department of Construction Economics, pp 538–547
21. D'Arconte C (2018) Business intelligence applied in small size for profit companies. Procedia Comput Sci 131:45–57
22. DArconte C (2017) CRM, Business intelligence and the impact on companies' performance. In: TIEM 2017, p 77
23. Ibarra D, Ganzarain J, Igartua JI (2018) Business model innovation through industry 4.0: a review. Procedia Manuf 22:4–10
24. Li X, Liu F, Li A, Xu L (2018) Business negotiation based on extenics. Procedia Comput Sci 139:521–528
25. Chancusig JC, Cruzado JAG (2020) Results validation about the adoption model of information and communication technologies applied to the technical University of Cotopaxi. WSEAS Trans Environ Dev 16:158–171
26. Gonzalez-Padron TL, Chabowski BR, Hult GTM, Ketchen DJ Jr (2010) Knowledge management and balanced scorecard outcomes: exploring the importance of interpretation, learning and internationality. Brit J Manag 21(4):967–982

Optimization of Hoeken Mechanism for Walking Prototypes

Vicente Herrera, David Ilvis, Luis Morales, and Marcelo Garcia

Abstract Locomotion prototypes movement shows some lacks in the contact between the links and surfaces, slow speeds, wrong trajectories and variations in the height of the gravity center, that is why the purpose of the present document is to design a locomotion mechanism to minimize this problem. The purposed design in this investigation get inspired by the walker machine "plantigrade" and it is based on a Hoeken's mechanism modification, since it shows several movement advantages. The amendment executed is oriented towards the reduction of links which are part of the plantigrade machine, minimize the friction between the links and reduce the journey time for a stablished distance equal to 220 cm, getting parameters and equations for the design of the proposed mechanism, that owns a numeric value to the driver link equal to 2 cm, rocker, coupler and its extension to 2.5 cm and frame to 2.1 cm. To obtain a straight-line path, a gyroscope MPU 6050 is used to reduce lateral deviations. Finally, a journey time equal to 3.721 s is succeed, a reach of up till 4.5 cm, a maximum speed equal to 308 cm/s, a maximum acceleration of 281 cm/s^2, the average deviation to the left equal to 5.312 cm and an average variation of the center of gravity height equal to 0.6425 cm.

Keywords Mechanism · Plantigrade · Hoeken

V. Herrera · D. Ilvis · L. Morales (✉) · M. Garcia
Universidad Técnica de Ambato, UTA, 180103 Ambato, Ecuador
e-mail: luisamorales@uta.edu.ec

V. Herrera
e-mail: vherrera6509@uta.edu.ec

D. Ilvis
e-mail: dilvis9234@uta.edu.ec

M. Garcia
e-mail: mv.garcia@uta.edu.ec

© The Author(s), under exclusive license to Springer Nature Switzerland AG 2022
M. V. Garcia et al. (eds.), *Advances and Applications in Computer Science, Electronics, and Industrial Engineering*, Lecture Notes in Networks and Systems 433,
https://doi.org/10.1007/978-3-030-97719-1_5

1 Introduction

Globally, robotic focuses in the research and development of robots with ability to walk, that approach is set for an amount of reasons. The principal cause is based on the capacity to adapt to regular or irregular land surfaces that locomotion by limbs has, this kind of locomotion is inspired by the animal kingdom [1, 2]. Another reason is the apparition of new micro controllers and computers which make an easier control of the robot actuator, allowing coordinate and have control of the different mechanisms that conform the limbs [3–5]. The last reason is the researches and developer's necessity for reducing the existing gap between robots and animal in their locomotion forms [6].

In robotic, different locomotion forms are used, not just by limbs that are present in wheel and crawler [7–9]; in the case of wheel; those have been used for a long time, getting results such as: low energy consumption and high speeds, all of this just in flat and regular surfaces. Although when this kind of locomotion is being tested on slippery, muddy or irregular surfaces, robots show a greater effort to get moved, which means a high-power consumption, which often produces the immobilization of those [10]. On the other hand, locomotion by crawler provides better contact with surfaces, giving a better adaptability in irregular terrains, although the resulted speed is lower and shows a high-power consumption.

Finally, the limbs in this kind of locomotion are conformed by mechanisms of four, five, six or more links [11–13]. The use of limbs shows more advantages compared with other forms of locomotion such us: the replication of some animals limbs, this one consists in one single point of support which make each limb being independent, giving a high adaptability on irregular terrains [14, 15]: if obstacles appear in the robots path, the limbs will pass over those, avoiding possible changes in robot trajectory, also the limbs produce less damage on the floor, comparing with wheel and crawler mechanisms [3]; Because the limbs are formed of mechanisms, those will need several actuators to drive the kinematic chain [16], this also have low energy consumption and the facility to manipulate the links dimensions that are part of the mechanism which drive the limbs in this kind of locomotion [10]. The disadvantage more important could be the low speed and low- level instability [7], all of those could be solved with an optimum speed actuator an adding support structures for the limbs [11].

In order to select the appropriate mechanism for the limb locomotion, is important to keep in mind aspects like degrees of freedom, the number of circuits and kinematic chain [17]; taking all of these aspects in consideration, a four-bar linkage is remarkable for its simplicity of design, a good example of that is the Knabe mechanism and the plantigrade machine [17, 18], in the group of the six-bar linkage, Klann and Stephenson are both mechanism with high class structural group, constituted for two circuits [17, 19], those are importantly highlighted, since this classification is possible to find mechanisms conformed by 7 or more links, such us: Theo Jansen, Ghassaei, bipedal robots, trotbot, Wang Jian and NCKU, which are conformed by links in a range from one to eighteen linkages [17, 19, 20], is important to highlight that all these mechanisms have a grade of freedom which indicates that one actuator is needed in order to be activated, nevertheless the number of links and its geometry also should be considered, because the easy control of the mechanism depends on that [19].

Locomotion robots designs most remarkable are based on the plantigrade machine which has four limbs on each side, that is to say eight limbs in total [21], the movement is similar to a bear, rabbit, insect or even some people walks [22]; there are a lot of possible mechanisms for the plantigrade mechanisms, but taking in count the simplicity and the control facility, is important to mention the Hoeken mechanism, better called "Lambda de Chebyshev", which is a kind of Grashof, a crack- rocker mechanism which give a significative advantage; this mechanism has as a principal feature an almost constant speed along the central part of the movement straight to the interest point that describes the walk trajectory, this straight-line movement is a really big advantage because it provides stability in the locomotion, by keeping up the center of gravity of the robot in constant height [23],; additionally, it imitates the plantigrade walk, meaning it has a limb that simulates a flatfoot, all of this improve the stability and reduce the disturbances in the mechanisms [11, 16].

In this investigation, an insect robot prototype (walker) is been developed using the Hoeken mechanism, which originally consists of four links and one extension that allow the simulation of a flatfoot to get the locomotion, so the plantigrade machine is the starting point to get the design, keeping in mind the advantages and disadvantages that the mechanism presents, then a modification is done in it, adding three links that have the goal of creating a new kinematic chain, in this way is possible to provide an optimum advance and an almost constant speed in the robot locomotion; finally a parallel mechanism is added to each limb with the purpose of giving stability in the robot movement.

This article is structured as follows: Sect. 2 presents a series of related studies that encouraged the development of this work. Section 3 provides brief concepts which will allow a better understanding of following sections. Section 4 present the Synthesis of the mechanism by which the dimensions of the Hoeken Linkage are obtained. In the Sect. 5 provides information about the Kinematic Analysis of the Mechanism for two revolutions of the crank. In Sect. 6, the improvement in stability is showed. The construction of the prototype is presented in Sects. 7 and 8. The discussion of results is presented in Sect. 9. Finally, the conclusions and future work are presented in Sect. 10.

2 Related Works

In this section, the works and researches more relevant in the area of development and analysis of walker mechanisms will be exposed. The proposed approach of this document is been described from a technical point of view, which allow to get the relationship of the topic with different studies in the area of machines and mechanisms theory.

A mechanism is the key element inside any machine, that will determinate the success of its operation, given that the mechanism has as a function to transmit the movement and in specific cases, it gives force or torque from a power supply to an output [16, 24]. Because of the biggest capacity of adapting to irregular terrains that walker robots are considered a perfect solution in the robotic field [14, 20].

Several studies and investigations (principal references), were used as a starting point for the elaboration of this paper, this is because all these studies provide some considerations in the analysis and design of prototypes of locomotion mechanisms to walker robots. In the developed work [25], is exposed that the mechanical structure of some prototypes manufactured for locomotion not just include the imitation of living beings limbs, but also is important to take into account the actuators properties such us size, weight and its power, additionally geometric features like the size of the boy and the work space where the limbs will be.

Additionally, in [24] is exposed that since this s about mobile mechanisms, some troubles appeared during the manufacturing and assembly, because fairy large links are dimensioned to the construction which increase the weight, causing that when the kinematic chain gets activated the actuator dos not work properly, because the robots limbs tend to be flexed.

As show in [26], it was possible to verify that the links joints are considered a critic point in the kinematic chain, since if a joint or articulation behave as a bending beam instead of being a gyratory joint, that will have a directly influence in the robots trajectory; the developed prototype of this work could be floundering at the moment of its launching, since it only has four limbs, which means it only touch the surface with two extremities while the other ones are in movement, this trouble could be solved making that each limb has an independent motor, in this way a coordinated movement can be gotten, this will make that three of the four touch the floor at the same time.

On the other hand, in [27], is exposed that the principal problem in a locomotion mechanism with one degree of freedom which is being controlled by a rotative actuator is the limitation in the capacity of locomotion, because the mechanism get stuck following the interest point in the mechanism during its curve path. Although, the advantages that this mechanism offers in locomotion are notable in power consumption and at the moment it walks over obstacles, also in [28]. Lambda de Chebyshev, Klann y Theo Jansen, are being compared in this paper, after researches; the Chebyshev mechanisms has better qualities such its simplicity, an optimal approximation to straight trajectory, a constant speed during the journey and a less power consumption in the actuator source of energy which activate the kinematic chain, additionally in this study is highlighted a critic factor to get a correct development in the synthesis process, since this process is ensurig an excellent trajectory to the locomotion.

As is show in [10], there are several mechanisms configurations which are possible to find while changing the link dimensions, presenting six types of those, been the most remarkable the one with an approximation to the straight line, in the interest point of view trajectory in the kinematic chain, which makes an ideal mechanism to get locomotion, in addition, is exposed the advantage of using a Grashof four-bar mechanism, since won't be troubles in the locomotion forward (when the crank activate the limb); however, in the opposite movement (when the limb activate the crank) is fundamental to verify that there is not blockages or obstructions in some positions.

In this investigation, an insect robot prototype (walker) is been developed using the Hoeken mechanism, which originally consists of four links and one extension that allow the simulation of a flatfoot to get the locomotion, so the plantigrade machine is the starting point to get the design, keeping in mind the advantages and disadvantages that the mechanism presents, then a modification is done in it, adding three links that have the goal of creating a new kinematic chain, in this way is possible to provide an optimum advance and an almost constant speed in the robot locomotion; finally a parallel mechanism is added to each limb with the purpose of giving stability in the robot movement.

This article is structured as follows: Sect. 2 presents a series of related studies that encouraged the development of this work. Section 3 provides brief concepts which will allow a better understanding of following sections. Section 4 present the Synthesis of the mechanism by which the dimensions of the Hoeken Linkage are obtained. In the Sect. 5 provides information about the Kinematic Analysis of the Mechanism for two revolutions of the crank. In Sect. 6, the improvement in stability is showed. The construction of the prototype is presented in Sects. 7 and 8. The discussion of results is presented in Sect. 9. Finally, the conclusions and future work are presented in Sect. 10.

3 State of the Art

Hoeken, Klann y Theo Jansen mechanisms are remarkable in the walking field, all of them have as a feature its mobility, since it is equal to one degree of freedom, which makes easier the adaptability to different actuators and the kinematic chain control. Hoeken mechanisms is a Grashof crank- rocker, which means is a significative practice advantage; it is important to mention that this mechanism has as a feature an almost constant speed along the central part in the straight movement, so this mechanism useful to replicate the walking plantigrade [21] (Fig. 1).

On the other hand, Klann mechanisms is conformed by six links, that has as a purpose to emulate a straight-line trajectory in the interest point, as can be noted in Fig. 2, it just has one contact point with the surface. Generally, competition robots are manufactured based on this kind of locomotion mechanisms, it could include two extremities on either side, producing an unstable walking with lateral deviations during its trajectory.

In respect of Theo Jansen mechanism, this could be considered as a game between art and engineering, this consists in eight links, seven of this eight are mobiles and one is static, since it does not have movement, additionally it has one degree of freedom, which means it just have one drive- link or crank [20], the links length are fixed and those were determined by Theo Jansen is also stressed, the greater the number of links, the greater frictions between them will exist [19] (Fig. 3).

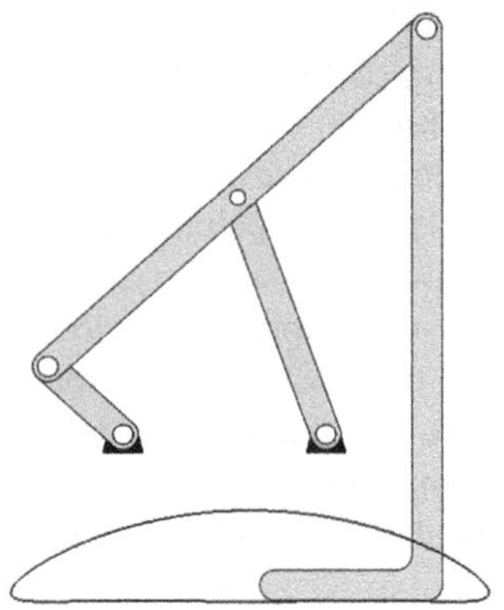

Fig. 1 Hoeken mechanism

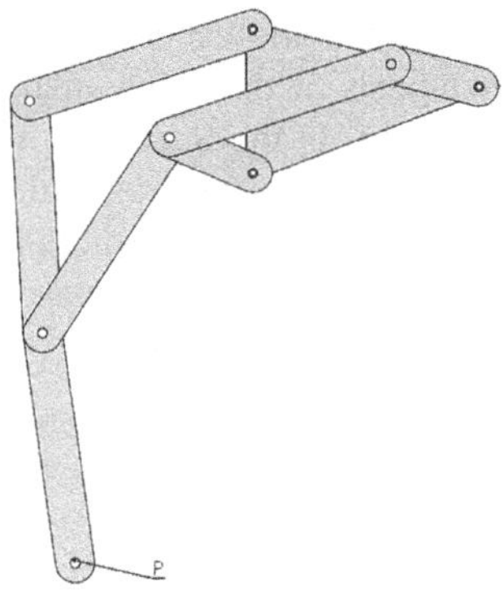

Fig. 2 Klann mechanism

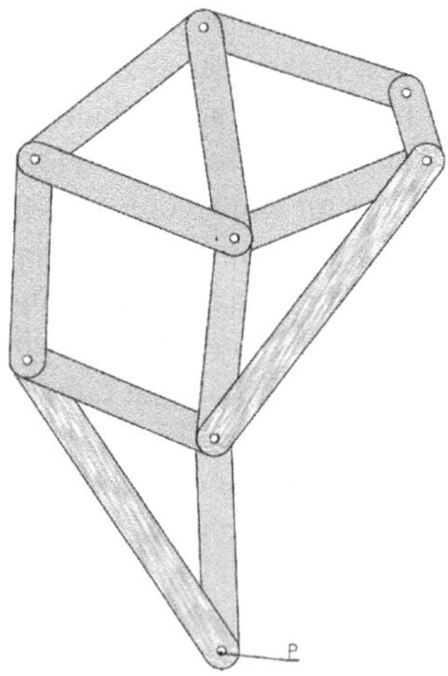

Fig. 3 Theo Jansen mechanism

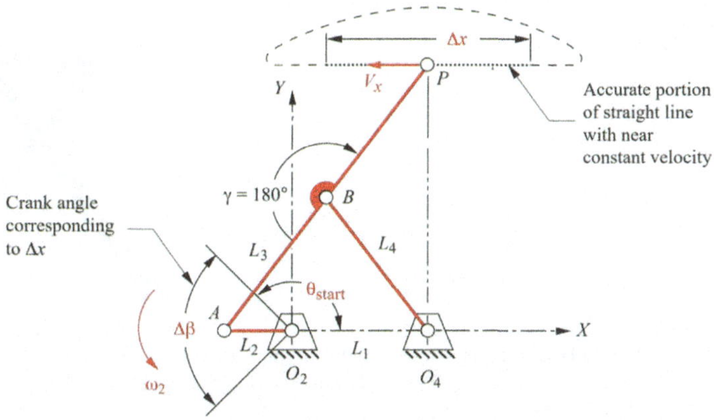

Fig. 4 Hoeken's mechanism geometry [16]

4 Mechanism Synthesis

Several problems in the mechanism design for walking need the creation of a link with movements and speed features. This work establishes as an objective getting a mechanism that will be able to travel a stablished distance in the less time possible, to achieve this, type synthesis is used, since this is a process of definition, choosing the most appropriate kind of mechanisms needed. In [23], the importance of the trajectory in the straight-line of the point of interest is highlighted, this feature could be found in Hoeken mechanism, which transforms a circular movement into a approximated straight - line trajectory in the coupler link, it also provide an almost constant speed during this trajectory, which is well exploited to get a walking movement. Therefore, this mechanism offers an optimum combination between righteousness and speed. since it is a crank- rocker mechanism, it could be activated using a motor. The geometry of it, is showed in Fig. 4.

This is a symmetric four -bars mechanism, because the angle (γ) in the BP segment is described and L_3=L_4=BP and only two links relations are needed to describe the geometry. If the crank L_2 is pushed at constant angular speed ω_2, the lineal speed V_x traveling along the coupler straight- line Δ_x trajectory will be the following to be a constant part in the crack rotation, described by Δ_β. The mechanisms which produce the walking ability in the plantigrade machine are the result of obtain a cognate of the Chebyshev straight - line mechanism, interestingly enough looking that Hoeken and Chebyshev mechanisms are cognate from each other as is showed in Fig. 5.

In order to replicate the linkage dimensions in the plantigrade machine, relations between links, proposed by Hoeken in [16], are used. The Table 1 indicates the values which allow to obtain a four-bar mechanism with the features that were mentioned above, additionally, the sought dimensions wont be excessively large because that could cause deformations on those [24].

Fig. 5 Chebyshev and Hoeken straight line mechanisms

Table 1 Link ratios for a Hoeken type fourbar approximate straight-line mechanism

Range of motion		Optimized for constant velocity		
$\Delta\beta$ (deg)	Θ_{start} (deg)	Link ratios		
		L_1/L_2	L_3/L_2	$\Delta x/L_2$
160	100	1.675	2.013	3.232

Setting the plantigrade machine as a starting point, the link length and the necessity to optimize the constant speed in the mechanism walked, a numeric value equal to 4 cm is assigned to Δ_x, from this parameter is possible to find the next equations:

$$\frac{\Delta_x}{L_2} = 3.232 \tag{1}$$

$$\frac{L_1}{L_2} = 1.675 \tag{2}$$

$$\frac{L_3}{L_2} = 2.013 \tag{3}$$

Based on the Eqs. 1, 2 and 3, the dimensions L_1=2.1, L_2=1, L_3=2.5 are obtained, all of them in centimeters.

Analyzing the plantigrade machine presented in Fig. 6, is notable that there are two Hoeken mechanism in each side, joined by a link connected in the joins A_2 and A_4, this link transmits the movement between mechanisms and coordinates the walked movement, one of the troubles that could be present in this design is the only one actuator which is responsible to activate two mechanisms, this shows that is important to have a high torque in the motor with a high level of energy consumption, other trouble could be the small contact surface that extremities 1, 2, 3 or 4 in the extensions, lastly the several number of links is visible, which could cause frictions between them, giving a high probability to get lateral deviations or even a turnover of the structure.

To reduce the existing troubles, the design presented in Fig. 7, is proposed, here is possible to appreciate the adding of four links to the Hoeken mechanisms, getting

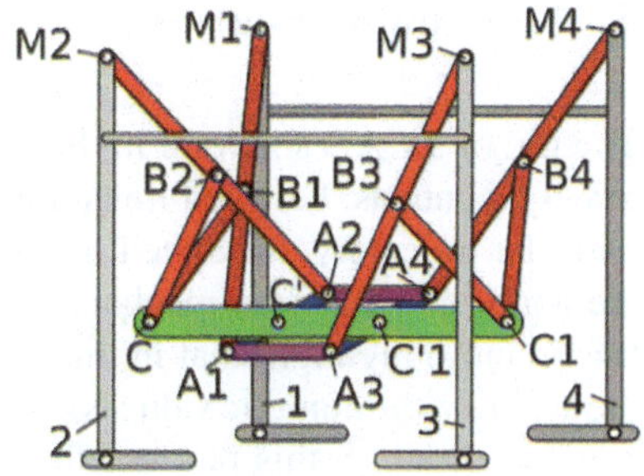

Fig. 6 Plantigrade machine [29]

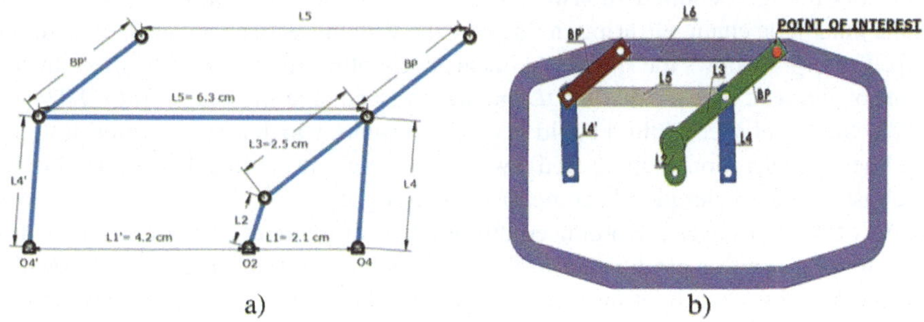

Fig. 7 Modificated mechanism

eight links in total, compared with the thirteen links in the plantigrade machine, which means less effort in the actuator and thereby less energy consumption, links L_5 and L_6 got the function to replicate the Hoeken mechanism movement in the L_4' y BP' that have the same dimensions of L_4 and BP, additionally L_6 link, increase the contact surface in the walking, replacing links 2 and 4 in Fig. 6. The distance from the fixed pivot to L_4' y L_5 dimension is done by an iterative process, which consists in checking that there are no collisions in the links and take care of the mechanism movement in order to be correctly during its journey. It is interesting how connecting the links is not just useful to replicate the movement, but also the kinematic parameters such as: position, speed and acceleration. From the iterative process mentioned above the Eqs. 4 and 5 are gotten.

$$L_1' = 2L_1 \tag{4}$$

$$L_5 = 3L_1 \tag{5}$$

5 Kinematic Analysis

The use of a CAD software in this is section is fundamental, since parameters such gravity, contacts, between links are assigned to get results closer to reality. In this part, the trajectory, the speed and the acceleration in the interest point of interest are analyzed, since the mechanism takes advantage of its trajectory to walk. It is important to highlight that in this analysis is necessary to provide an input value in the actuator, the numeric value is set with a speed equal to 600 rpm and an acceleration equal to 0 rad/s^2, this because the actuator provide constant speed. The presented trajectory in Fig. 8, could demonstrate the additional links replicating the movement in the kinematic chain, ensuring in this way the walking with an advance of 4 cm by step. The Fig. 9, shows the speed and lineal acceleration summary in the mechanism point of interest, that means link L_6 in which is notable the range from 0.02 s to 0.08 s, the lineal speed (blue) could become constant, which can be demonstrate in the lineal acceleration graphic (red); while the range of the lineal speed is getting established the acceleration become lower, near to zero.

All of these behaves are produced during the position interval in the crank, from 100° to 260°, as was established in the mechanism synthesis section, additionally certain variations can be seen in the trajectory while adding contacts between the links and the gravity center.

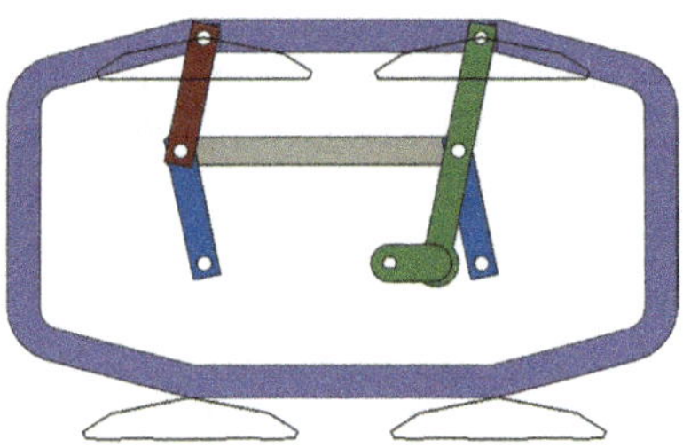

Fig. 8 Point of the interest trajectory

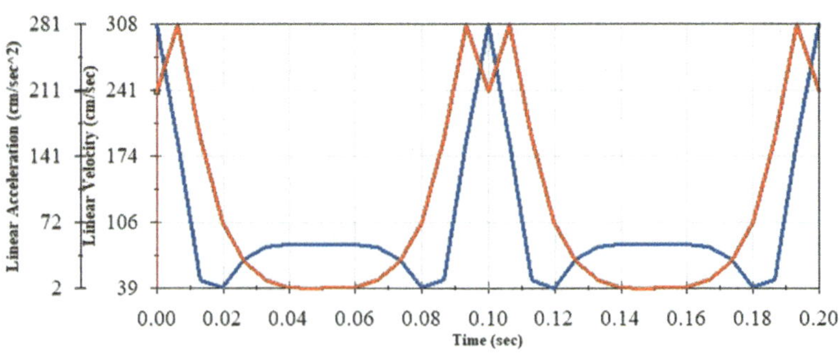

Fig. 9 Sped and lineal acceleration in the point of interest

6 Improvement in Stability

Stability troubles were exposed in [30, 31], there was mentioned that all of these get its origin in walkers robot prototypes with just two extremities or limbs as a support point, that is why the decision of adding one parallel mechanism with the same dimensions appears, this new mechanism needs mover elements to activate the kinematic chain, which are presented in Fig. 10(A) and (B) respectively, the first added link has a function similar to a crankshaft, in this way the link transfers the movement from the crank to L_3 of the parallel mechanism, the second link has the function to be an inverted crank helping to coordinate and balancing the movement in the parallel mechanism, with this modifications is possible to get four extremities or limbs adding parameters such as the walking surface contact, the stability and synchronization during the walk. The final configuration of the mechanism is showed in Fig. 11.

Fig. 10 **A** Crankshaft. **B** Inverted crank

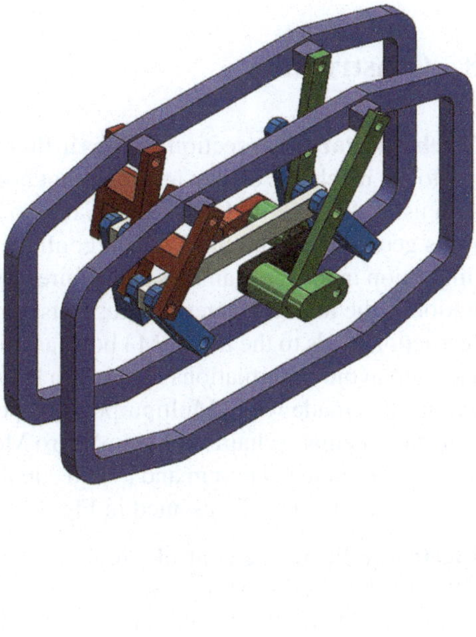

Fig. 11 Locomotion mechanism

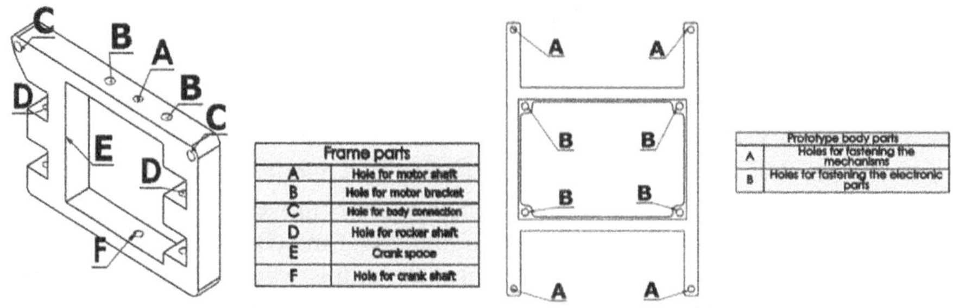

Frame parts	
A	Hole for motor shaft
B	Hole for motor bracket
C	Hole for body connection
D	Hole for rocker shaft
E	Crank space
F	Hole for crank shaft

a)

Prototype body parts	
A	Holes for fastening the mechanisms
B	Holes for fastening the electronic parts

b)

Fig. 12 **A** Frame. **B** Body

7 Prototype's Body

To get the prototype design, first the mechanism mobile links should be joined with the fixed link, that is why the design of the frame was presented in Fig. 12(A), with its parts, the design of it was an iterative process since it should be a warranty of no collisions or unnecessary friction between the links, additionally in this link will be ubicated the actuators and pivot points of the links L_3. Secondly, the mechanisms will be connected using the presented piece in Fig. 12(B) which ubicates the electronic elements that conform the insect robot.

8 Construction

Mechanic Part. This section focuses in the elaboration and assembly of links, according to the mechanic design and the robot body, that were elaborated with 3D impression using PLA (Poly Lactic Acid), due to the impression facility in minute details in its geometry as the links which conform the developed mechanism [20]. Due to dimension in the mechanisms are inspired by the plantigrade machine, there is a limitation in the joint diameter, this restricts the used of bearings because of its size, this restriction leads to the use of M3 bolts as joints. Lubrication in joints is fundamental factor to avoid deformations by friction between the links, therefore, the operational testing was made with "Multipurpose Truper" as the oil lubrication. Finally, to activate the kinematic chains a Pololu Micro Metal Gearmotor HPCB 12V, which offers a torque equal to 6.7 kg-cm and a sped equal to 600 rpm with full load. The assembly in the mechanic part is resented in Fig. 13.

Electronic Part. The control circuit made up of a micro controller Arduino Nano, TB6612FNG Driver Motors, MPU-6050 Accelerometer and Gyroscope and a LIPO Battery 11.1V 1000 mAh, this circuit has as finality control the insect robot trajectory, by reading the navigation angle YAW (obtained with MPU-6050) and the control of actuators by the interaction between the driver and the micro controller, the control

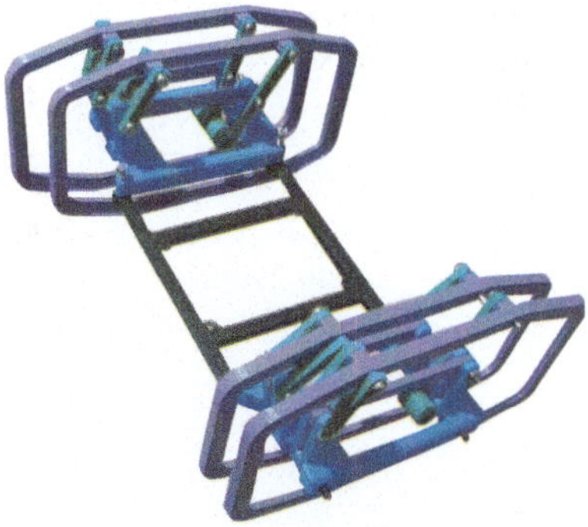

Fig. 13 Mechanic Part

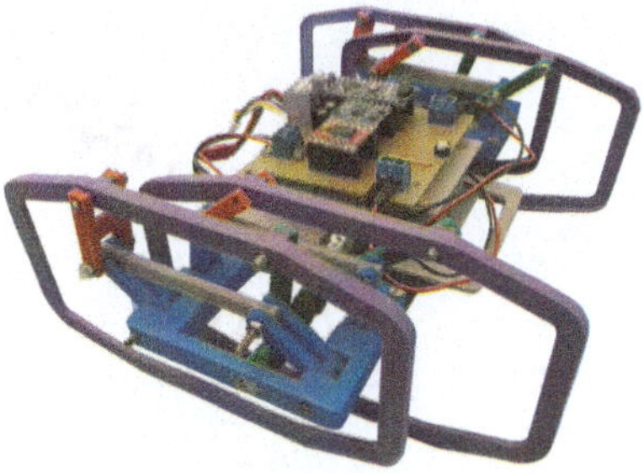

Fig. 14 Electronic part

algorithm detects variations form the angle YAW, and also it controls the actuators sped by the PWM output, to ensure the straight trajectory for the robot, The control circuit and the final assembly is presented in Fig. 14.

9 Discussion of Results

To analyze the results of the trajectory in the point of interest in the mechanisms, the robot journey and the height of the center of gravity of it, Tracker is used as a software for video- analysis.

Prototypes Trajectory. The target path described by the point of interest in the Hoeken mechanism based on the experimental trajectory obtained by video-analysis, shows a variation in the straight-line trajectory, showing an advance equal to 4.5 cm, this variations is credited to existing vibrations in the mechanism joints [24, 30], the real variation is showed in Fig. 15.

Journey and Journey Time in the Prototype. In the operational testing a track is stablished as a surface 25 (width) $\times$ 220 cm (height), it has a left deviation equal to 11.68 cm and a right one equal to 0.389 cm, getting an average deviation equal to 5.312 cm along the track (look Fig. 16(A)); this variations could be biggest in the case of avoiding the developed circuit, additionally this displacement variations exist

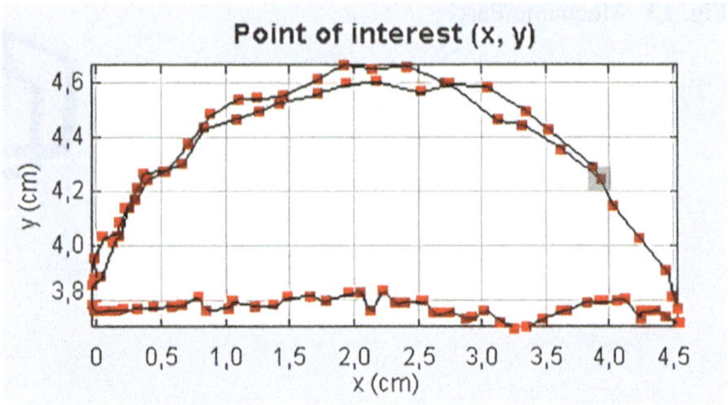

Fig. 15 Trajectory stablished by Tracker software

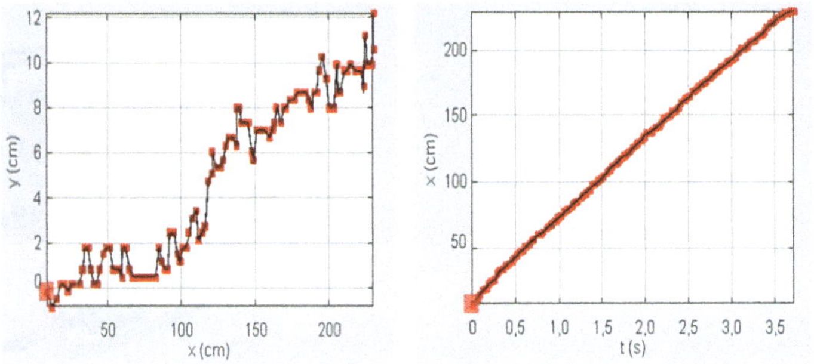

a) b)

Fig. 16 **A** Journey. **B** Journey time in the prototype

because the vibrations in the joints. The time of journey is equal to 3.721 s, this result will be presented in Fig. 16(B).

Height of the Center of Gravity. As it was exposed in [23], the Height of the center of gravity during the locomotion should stays as much as possible in a constant state to improve stability, in the Fig. 17, the real height of the center of gravity variation is presented during the operational testing, setting as a maximum height 0.460 cm and a minimum height −1.713 cm, and an average equal to 0.6425 cm. It is clear that there are several variations due to the existing vibrations in the links joints, high torque and the actuator speed [24, 30]; all of this aspects will directly impact in the straight line journey of the robot, that is why an improvement in the control must be considered.

Advantages of the Optimized Hoeken Mechanism. In this subsection the mechanism advantages are showed about Klann and Theo Jensen mechanisms, that were analyzed in this investigation.

On the one side, implementing an excellent assembly of the Modificated Hoeken mechanism that was presented in this investigation it is possible to reduce the internal friction in the mechanism links, and this allows the robot get a straight trajectory during all the process the actuator (motor) is working, generating a better answer to the loads that could appears in the mechanism structure.

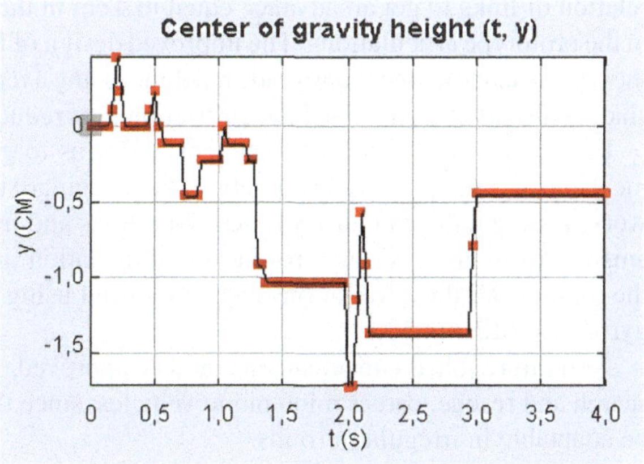

Fig. 17 Results of the analysis of the center of gravity

On the other hand, the optimized Hoeken mechanism configuration depends on a required advance, while Theo Jansen y Klann are stablished to minimize the inertial forces that could appear when the contact points with the surfaces were found on the air: is for this reason that if the links dimensions stablished by the creators of this mechanisms are changed, there is a big probability to get destabilizations and representative changes during its trajectory.

Comparing Theo Jansen y Klann mechanisms and Hoeken in this studio, the last one always will have contact points with surfaces warranting a better stability in the walker robot mechanism. In the same way, optimizing on the Hoeken Mechanism, the adaptability gets better on many surfaces where it can be developed, having as a result, the reduction of variations in the straight trajectory. Besides that, is well-know that four links mechanism can generate straight trajectories not exact, but with an acceptable approximation. In several cases, the search of this straight trajectories could produce vibrations that are the cause to get deviations in the mechanisms work, avoiding the formation of a straight path; all of this problems can get developed mostly in Jansen y Klann, although Hoeken Mechanism tends to straight trajectory really closed where rigidity and the generations of new vibrations are avoided.

The principal advantage of the optimized Hoeken mechanism, is the shift to have constant speed levels which allows the operation and behave of the mechanism. This constant speed permits a successful stability and an increase synchronization in the links.

10 Conclusions

In this document, the design, analysis and construction of an appropriate mechanism has been presented to get locomotion in an insect robot inspired by the plantigrade machine, which consists in a variety of Hoeken straight- line mechanism, with a relation of links to get an advance equal to 4 cm in the straight trajectory, generated in the prototype articulations. The improved design of Hoeken Mechanism optimizes the performance on the plantigrade machine, using a reduction of the number of links, since with this tool, the mechanism friction gets reduced during the movement.

Equations 4 and 5 in this document, allow us to get a better adaptability in the mechanism, taking into consideration, the irregular surfaces where the robot could work, getting reduced journey times. Vibrations and frictions presented in the mechanism movement, give as a result lateral deviation and no constant speeds during the journey, all these variations can be reduced using a control circuit based on the gyroscope MPU 6050.

As a future job, the mechanism could be approved in other fields like; agriculture, search and rescue, career mine mono vehicles, since the locomotion by limbs could be adaptable in irregular terrains.

Acknowledgment This work was financed by Universidad Tecnica de Ambato (UTA) and their Research and Development Department (DIDE) under project CONIN-P-0167-2017.

References

1. Gorrostieta E, Vargas E (2002) Diseñando un robot caminante de seis patas. In: Segundo Congreso Nacional de Robótica. Asociación Mexicana de Robótica e Instituto Tecnológico de Toluca, pp 181–186
2. Demirel M, Carbone G, Ceccarelli M, Kiper G (2018) Design and simulation of a novel hybrid leg mechanism for walking machines. In: New advances in mechanism and machine science. Springer, Heidelberg, pp 283–290
3. Todd DJ (2013) Walking machines: an introduction to legged robots. Springer, Heidelberg
4. Nuñez AV, Encalada P, Garcia CA, Guilcapi JR, Garcia MV (2019) Web-platform for developing man-machine interfaces based on OPC UA. In: 2019 IEEE fourth ecuador technical chapters meeting (ETCM), pp 1–6
5. Caiza G, Torres FI, Garcia MV (2021) Identification of patterns in the involvement of novice software developers in software testing processes. In: 2021 IEEE international conference on artificial intelligence and computer applications (ICAICA), pp 378–382
6. Hong M, Yun-xia L (2018) Walking mechanism and gait design of a novel compound quadruped robot. In: 2018 3rd international conference on robotics and automation engineering (ICRAE). IEEE, pp 10–13
7. Adrian R, García G (2016) Prototipo virtual de un robot móvil multi-terreno para aplicaciones de búsqueda y rescate. ResearchGate
8. Chiliquinga S, Manzano S, Córdova P, Garcia MV (2020) An approach of low-cost software-defined network (SDN) based internet of things, pp 70–74. https://doi.org/10.1109/Incodtrin51881.2020.00025
9. Caiza G, Chiliquinga S, Manzano S, Garcia MV (2020) Software-defined network (SDN) based internet of things within the context of low-cost automation, vol 1, pp 587–591. https://doi.org/10.1109/INDIN45582.2020.9442180
10. Desai SG, Annigeri AR, TimmanaGouda A (2019) Analysis of a new single degree-of-freedom eight link leg mechanism for walking machine. Mech Mach Theory 140:747–764
11. Saab W, Ben-Tzvi P (2016) Design and analysis of a robotic modular leg mechanism. In: International design engineering technical conferences and computers and information in engineering conference, vol 50152. American Society of Mechanical Engineers, p V05AT07A062

12. Lozada-Martinez E, Naranjo JE, Garcia CA, Soria DM, Toscano OR, Garcia MV (2019) SCRUM and extreme programming agile model approach for virtual training environment design, pp 1–5. https://doi.org/10.1109/ETCM48019.2019.9014882
13. Varela-Aldás J, Miranda-Quintana O, Guevara C, Castillo F, Palacios-Navarro G (2020) Educational robot using lego mindstorms and mobile device. In: Advances and applications in computer science, electronics and industrial engineering. Springer, Cham, pp 71–82
14. Kavathia D, Dave J (2021) Kinematic modeling of walking mechanism. In: Mechanism and machine science. Springer, Heidelberg, pp 229–237
15. Villa Tello K (2021) Selection of optimal lodging site in the city of Baños, Ecuador. In: Advances and applications in computer science, electronics and industrial engineering. Springer, Singapore, pp 53–65
16. Robert LN (2008) Design of Machinery. Mc Graw Hill, New York
17. Hoeltgebaum T, da Silva ES, Barreto RLP, Maletz ER, Morlin FV, Carboni AP, et al (2021) Walking mechanisms-a state of the art survey and new developments opportunities. In: International symposium on multibody systems and mechatronics. Springer, Heidelberg, pp 35–43
18. Zielinska T (2004) Development of walking machines; historical perspective. In: International symposium on history of machines and mechanisms. Springer, Heidelberg, pp 357–370
19. Onieva J (2019) Diseñn y construcción de un mecanismo andante. Universidad Politécnica de Madrid, Madrid. http://oa.upm.es/54412/1/TFG_JORGE_GONZALEZ_ONIEVA_JOHANSSON.pdf
20. Cuadrado Mazón KC (2018) Análisis cinemático y cinético del mecanismo Theo Jansen. Diseño y construcción de juguete prototipo [B.S. thesis], Quito, 2018
21. Artobolevsky II (1975) Mechanisms in modern engineering design. Mir Publishers 1:454
22. Çıklaçandır S, Karabiber Ö, Kalafat Y, Can F (2017) Design a robot insect inspired by the camouflage characteristics of the leaf insect
23. Jansen T The Legsystem. https://www.youtube.com/watch?v=FFS-2axFo1Y&t=53s&ab_channel=theojansen
24. Manickavelan K, Singh B, Sellappan N (2014) Design, fabrication and analysis of four bar walking machine based on Chebyshev's parallel motion mechanism. Eur Int J Sci Technol 3(8):56–73
25. Zielinska T, Heng J (2002) Development of a walking machine: mechanical design and control problems. Mechatronics 12(5):737–754
26. Sun J, Zhao J (2019) An adaptive walking robot with reconfigurable mechanisms using shape morphing joints. IEEE Rob Autom Lett 4(2):724–731
27. Vujošević V, Mumović M, Tomović A, Tomović R (2018) Robot based on walking Jansen mechanism. In: IOP Conference series: materials science and engineering, vol. 393. IOP Publishing, p 012109
28. Komoda K, Wagatsuma H (2017) Energy-efficacy comparisons and multibody dynamics analyses of legged robots with different closed-loop mechanisms. Multibody Syst Dyn 40(2):123–153
29. MohdIsharudden F, Mohamed H, Rafaai ZM, Ho TYW, Kamarudin M (2020) Design and prototyping of a motorized legged robot with Klann linkage mechanism. Int J Emerg Trends Eng Res 8(5):1941–1945
30. Bustamante Téllez J (2016) Diseño de mecanismo de locomoción andante con cambio de dirección
31. Wikimedia Commons: Plantigrade machine - Drawing.svg - Wikimedia Commons. https://commons.wikimedia.org/wiki/File:Tchebyshevs$_$plantigrade$_$machine$_$-$_$Drawing.svg

Management by Integrated Processes with Biosafety Parameters. Case Study SMEs Manufacturing Rest Footwear in the Province of Tungurahua

Sebastián López, Freddy Lema, César Rosero, Carlos Sánchez, Jéssica López, and Franklin Tigre

Abstract This study aimed to provide an administrative management methodology through a process management approach integrated with biosafety applications and information technology integration. The operative processes of a manufacturing SMEs (Small and medium-sized enterprises) of rest footwear (slippers) in the province of Tungurahua were established as a case study. Through bibliographic research, knowledge was determined and consolidated regarding preventive measures against COVID-19, management by processes and actions for the safe return to work centres, as well as a model for integrating IT (Information technology) into the organization's processes. A model for the integration of new technologies into the processes of the companies was established, together an analysis of the situation of the company was carried out, the survey of the processes to know the organizational structure and the identification of all the productive activities of each one of the operative processes. Finally, safe work practices, good biosafety practices, hygiene measures in the workplace were established, ensuring biosafety conditions and work standards in safer environments.

S. López · F. Lema · C. Rosero · C. Sánchez · J. López · F. Tigre (✉)
Technical University of Ambato, Ambato, Ecuador
e-mail: fg.tigre@uta.edu.ec

S. López
e-mail: jlopez7891@uta.edu.ec

F. Lema
e-mail: fr.lema@uta.edu.ec

C. Rosero
e-mail: cesararosero@uta.edu.ec; cesar.rosero@epn.edu.ec

C. Sánchez
e-mail: carloshsanchez@uta.edu.ec

J. López
e-mail: jp.lopez@uta.edu.ec

C. Rosero
National Polytechnic School, Quito, Ecuador

© The Author(s), under exclusive license to Springer Nature Switzerland AG 2022
M. V. Garcia et al. (eds.), *Advances and Applications in Computer Science, Electronics, and Industrial Engineering*, Lecture Notes in Networks and Systems 433,
https://doi.org/10.1007/978-3-030-97719-1_6

Keywords Covid-19 · Manufacturing processes · Work standard · Process management · Information technology integration

1 Introduction

SMEs represent an important source for the influence of national economies in Latin America, because they generate jobs and contribute significantly to the GDP of the countries [1]. Within this context, the situation in Ecuador is not very different, where 80% of the companies correspond to SMEs [2], and these generate 75% of the jobs, being the basis of the socioeconomic development of the country mainly producing products or adding value to them, which is why it constitutes a fundamental actor in national progress, promoting economic growth [3–6]. The SMEs in charge of the manufacturing processes correspond to a very active area at the national level, having a 12% participation in their commercial activity, the provinces of Tungurahua, Chimborazo and Cotopaxi represent the largest participation within the country, having a total production of USD 11,878,565.03, which represents 7% of the national production, where there are manufacturing companies dedicated to the handling of textiles and clothing, leather and footwear, processed foods and beverages. Tungurahua reflects a majority participation of this type of companies, in 2016 there were 41,790, according to the INEC, which represents 42% of the total in the area, contributing to the growth of the percentage of employment in the country [7].

However, according to the Ambato Economic Observatory in trade and manufacturing, 18000 jobs were lost in the capital of Tungurahua [8] and this is due to the fact that on March 11, 2020 the World Health Organization declared COVID-19 a global pandemic, causing a health crisis, which has caused a strong socioeconomic impact on industry [9], especially manufacturing, whose sales fell due to a lack of demand, because they depend on the national market. Commerce was reduced by 36.7% [8, 10–12], in addition to facing problems such as a stoppage of the supply chain or working from home, generating massive lay-offs and company closures [7].

This situation places SMEs on the edge of the abyss, facing problems such as the drastic reduction in sales, strict confinements, low productivity or little job creation, all this added to problems that the sector has faced over the years, which means a slow recovery and this is due both to the ravages that the health emergency is leaving, and to the mismanagement that has been handled in this [13].

In this vein, the inclusion of business management measures, a correct application of health and safety measures and proper support from governments should be applied urgently in SMEs to identify the problems and shortcomings they have, in addition to redesign their actions, objectives and processes to adapt to the new economic and social reality they face [14].

The reluctance to innovate is the cause of the failure of SMEs, remaining only those with the capacity to have a progressive regeneration and flexibility to adapt to the changing competitive environment, this is achieved taking into account multiple factors such as the analysis of the competitive environment, knowledge management, communication, information, foreign trade, financial management, operations management, human management, health and safety management, environmental management, organizational culture, social responsibility, strategic direction and administrative management [15–17], an adequate management, in addition to allowing the complete analysis of what has already been described above, allows establishing a continuous improvement factor in its activities.

An organization whose management is focused on processes allows integrating all the parts that compose it through continuous improvement, allowing controls, monitoring and forms of operation, all with the objectives of knowing the companies systematically developing a horizontal vision of them, and thus favour the interaction of work teams for the timely exchange of information and encourage the participation of all their human resources, regardless of their hierarchical position, this to build a new administrative structure focused on the knowledge of the processes having a positive impact on both internal and external customers and users, with suppliers, intermediaries and the environment, taking into account the strategic goals and objectives of the organization [18].

The analysis by processes can be defined as an initial step to include more improvement strategies, because, by knowing the processes and their interrelation with each other, in addition to their actors and responsibilities, the global vision of the organization with the inclusion of having a perspective to continuous improvement by identifying shortcomings or waste with a customer focus [19]. Therefore, the importance of this type of strategies for administrative management in SMEs is a promoter of a measure of change and innovation for the management of the company to achieve business development, having a clear vision to identify all important activities together with the organization's resources [20].

The integration between this management strategy together with biosafety parameters are ideas that are currently necessary for their implementation in sectors related to the production of goods, who are threatened with the closure of their facilities due to poor organizational and logistical management. , by not having a clearly defined organization and responsibilities and above all the lack of basic protection measures against Covid-19 [21].

To face all these problems, thanks to a more interconnected and multifaceted world, it is necessary to adopt first-line measures to face any outbreak of COVID-19 in addition to establishing a robust and flexible organization in companies [22], process management is that tool that allows this organizational structure.

This document is an analysis with a focus on process management as an administrative management strategy with biosafety applications for operational processes for a textile SME manufacturing rest footwear (slippers) in the province of Tungurahua, where this type is carried out. management for the identification of all productive activities, in addition to those responsible, and what measures must be taken to establish biosafety standards.

2 Methodology

The use of an applied type research allowed to identify a deficient standardization, documentation and lack of implementation of biosafety measures within their operational processes, taking into account the above, a bibliographic investigation was used to search and consolidate knowledge about preventive measures against COVID-19, management by processes and action measures for the safe return to the work centres, for this, with the application of the Fink methodology, a total of 61 documents were obtained between scientific articles, books and news collected from different database, from which 13 documents were selected.

All this selection made it possible to establish parameters for safe work in jobs and the correct execution for the development of the proposal for a process management system, in addition to establish a clear picture of the situation of manufacturing SMEs in the current socioeconomic environment in Ecuador, to have a context of the difficulties faced by the company analysed in the case study. The bibliographic search was based on three thematic axes to delimit and organize the information found, the first line of work was with respect to the problems that COVID-19 has generated at an industrial level, and what measures have been taken to face it, there-fore that keywords such as SARSCoV2, industrial production, contagion mitigation, industry, control of work areas, health and safety were used.

The second thematic axis had an approach towards management by processes and strategies for its application, so the search criteria were administrative management, operations management, management by processes, safety, and occupational health. Finally, manufacturing SMEs were analysed in a current context to identify the socioeconomic conditions in Latin America, especially in Ecuador, the development of the proposal focused on the use of the Venkatraman model allowing the integration of a process management system with the application of information technologies.

In addition, the development of management by processes was generated based on the first three phases of the five established in a management by processes (DiANA), which allowed the identification of the processes with the greatest impact on meeting the strategic objectives of the company. A summary of this methodology is presented in Fig. 1 below.

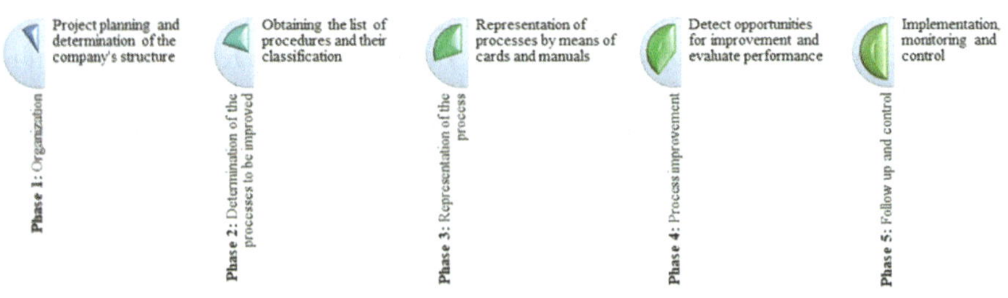

Fig. 1 Process management procedures [23]

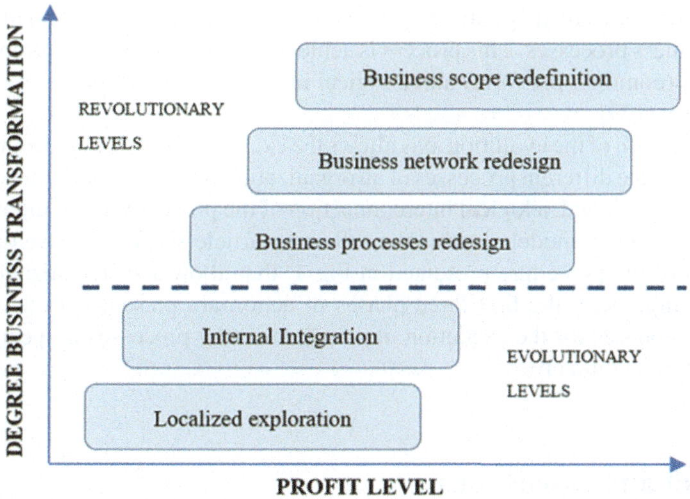

Fig. 2 Venkatraman model [23]

It is important to consider that for the application of a process management model from scratch, there must be a total transformation from a vertical organizational system to a horizontal one with an important focus and inclusion of new information technologies (IT's). The Venkatraman model proposes a method of alignment between processes and IT's allowing that, thanks to this strategy, the business objectives of the company are ideally met, improving the performance of processes, and achieving a competitive advantage in companies.

The Venkatraman model has a focus on five levels of transformation of the company for the easy inclusion of IT's where the first two levels make up an axis called evolutionary change which represents the identification and minimal changes in the processes, the remaining three levels make up an axis called revolutionary change because they require a fundamental change in the routines or business processes of the organization, in Fig. 2 the levels proposed by the author are presented.

The model suggests, in the context of SMEs, to stay at level two of transformation because, by not having an IT application, identifying the processes would be difficult to make large and fundamental changes to the rules' organization tubs [23].

Localized Exploration: This is the initial integration of IT resources for individual activities or as a control tool, this level has minimal changes in the business and the information systems are used for individual tasks and as reinforcement tools to perform them more efficiently [23].

Internal Integration: It is a more systematic attempt to harness the capabilities of IT's in business processes. This process is achieved through a logical interconnection of systems (common platform) and a logical interconnection of processes (process management) [23].

The execution of the evolution axis allows the exchange of information and knowledge between the different processes of an organization, which allows better customer service. As mentioned, a logical interconnection of the processes is an important part of the Venkatraman model, for this the following article shows the development of the phases of the procedure presented in Fig. 1. that allow a systematization of the way of acting. Next, the first three phases of action are presented in detail, since the last two phases for the execution of management by processes is specific to the conditions of a company.

3 Result and Discussions

The development of the case study begins with the survey of the company's processes, which were obtained in three applied phases and two proposed phases, the first three correspond to the collection, classification and documentation of the company's processes, and the two subsequent phases to an analysis and continuous improvement once applied. In the case study, the first two phases were carried out, with suggestions on biosafety issues, in addition to the suggestion of the application of the Venkatraman model for the inclusion of IT in SMEs.

Phase 1: Organization: For the case study in the textile SME, the organizational structure was identified. The company had a total of thirty-eight employees, twenty-nine in the production area, three in the administrative area, four in the managerial area and two in marketing, structuring the company as shown in the Fig. 3.

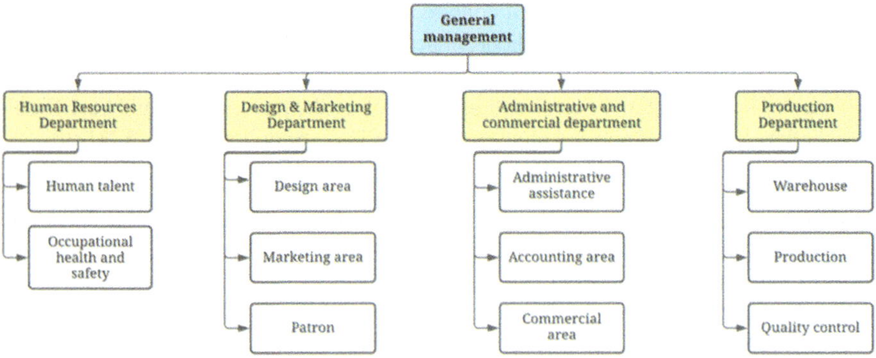

Fig. 3 Structural organization chart of the company

Table 1 Strategic processes

Code	Process	Code	Sub process
E1	Management general	E1.1	Approval of the hiring of personnel authorization
		E1.2	Authorization of the documents for final decisions
		E1.3	Analysis of the monthly reports for the fulfillment of goals
E2	Management resources humans	E2.1	Payment of salaries
		E2.2	Liquidations
		E2.3	Selection and contract of personnel
		E2.4	Issuance and approval of permits and certificates
		E2.5	Establishment of incentive programs
E3	Management administrative	E3.1	Establishment of provisions and policies to be followed in the company
		E3.2	Administration of the physical resources of the company
		E3.3	Analysis of the fulfillment of goals and indicators
		E2.4	Representation of the company in events

Phase 2: Determination of the Processes: The processes are classified by determining strategic, operational and support processes, in addition to the sub-process group that encompasses each of them. The administrative processes and sub-processes are presented in Table 1. These were taken from the case study company through observation and directly from the information provided by the employees.

The operational processes and sub-processes required to produce the slipper are detailed in Table 2. For this case, different models of slippers were taken to determine which processes corresponded to a general development.

Finally, Table 3 describes the support processes and sub-processes, which coordinate and collaborate with the development of the activities of the operational and strategic processes. This information was collected through direct observation and provided by each of those involved in the support processes.

Figure 4 below shows the construction of the process map, detailing the interaction and sequence of the processes that define the organization. As a result, there are three strategic, eight operational and eleven support processes.

Phase 3: Survey and Representation of the Processes: Performing the representation of the processes involves a series of steps that includes the creation of process files (together with their indicators), defining the owner of the process, identifying activities and execution times, as well as generating diagrams that show the logical interconnection.

Table 2 Operative processes

Code	Process	Code	Sub process
O1	Raw material preparation	O1.1	Raw material preparation
O2	External court	O2.1	Upper cut
		O2.2	Strip cutting
		O2.3	Plant cut
		O2.4	Cut of linings
		O2.5	Laser cut
O3	Fillings preparation	O3.1	Cut eva insoles
		O3.2	Eva heel cut
		O3.3	Cut foam insoles
		O3.4	Cut foam heel
		O3.5	Gummed (bagged)
		O3.6	Gummed (side seam)
O4	Finishes	O4.1	Embroidered
		O4.2	Sublimate
O5	Assembly	O5.1	Assembly (bagged)
		O5.2	Gummed (Side seam)
O6	Bagging	O6.1	Bagged
O7	Side seam	O7.1	Side seam
O8	Finishes	O8.1	Finishing and labelling
		O8.2	Packaging

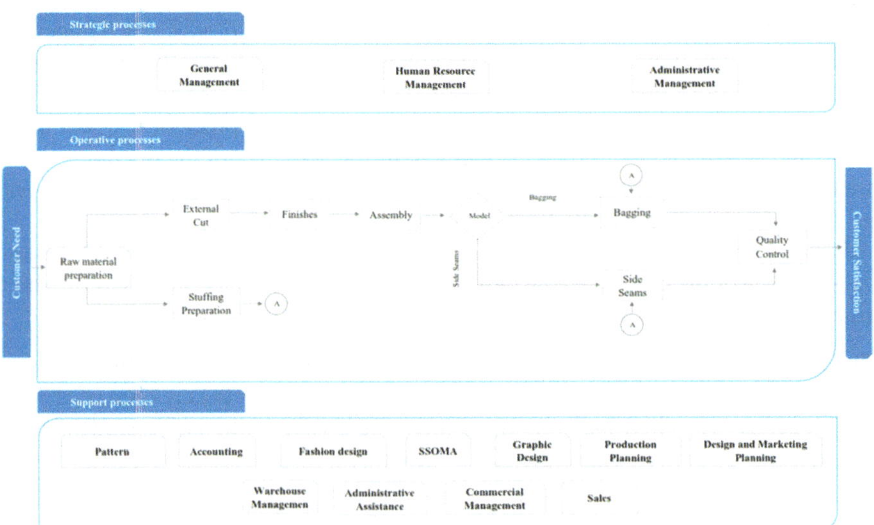

Fig. 4 Company process map

Table 3 Support processes

Code	Process	Code	Sub process
A1	Pattern	A1.1	Generation of patterns required by the designer
		A1.2	Generation of assembly and fitting tests of the slipper
		A1.3	Generation of ergonomics tests on the slipper
		A1.4	Establish raw materials and inputs necessary for production
A2	Fashion design	A2.1	Creation of new designs
		A2.2	Materialization of visual concepts
		A2.3	Search for fashionable materials
		A2.4	Coordinate the assembly of the physical sample
		A2.5	Carry out the weekly report of the activities carried out
A3	Graphic design	A3.1	Carry out the design of the labels and presentation of the product on hangers
		A3.2	Management of social networks and website
		A3.3	Retouching photos of the simple made
		A3.4	Make embroidery, stamping and sublimated designs
A4	Accounting	A4.1	Interpret and record financial information
		A4.2	Coding the company's accounts
		A4.3	Prepare reports on the financial situation of the company
		A4.4	Make bank transfers and conciliations
		A4.5	Order from suppliers
		A4.6	Make withholdings, advances, and invoice payments
		A4.7	Record sales processes
		A4.8	Control customer record
		A4.9	Employee pay control
A5	Administrative assistant	A5.1	Registration and control of financial resources
		A5.2	Petty Cash Control
		A5.3	Make the accounting records of the company
		A5.4	Manage receipts and invoices for company expenses
		A5.5	Payment control
		A5.6	Manage purchasing processes
		A5.7	Update accounting books

(continued)

Table 3 (*continued*)

Code	Process	Code	Sub process
A6	Warehouse management	A6.1	Manage storage of company supplies
		A6.2	Receive invoices and receipts
		A6.3	Stock control of supplies and materials
		A6.4	Management of materials to different areas of the company
		A6.5	Carry out the return of defective supplies
		A6.6	Schedule the receipt of supplies
		A6.7	Verify the quality of the company's supplies
		A6.8	Inventory management
		A6.9	Fabric laminate management
A7	Sales	A7.1	Retail product sales
		A7.2	Manage customer portfolio
		A7.3	Offer products to customers
		A7.4	Make a sample catalogue
		A7.5	Meet customer requirements
A8	Commercial management	A8.1	Billing and Dispatch of the finished product
		A8.2	Customer delivery logistics
		A8.3	Control of customer collection dates
		A8.4	Acquisition of supplies for the company
		A8.5	Payment of invoices
A9	Safety, occupational health and environment	A9.1	Plan and execute training program
		A9.2	Manage the emergency brigade
		A9.3	Manage the EPPS
		A9.4	Plan and execute occupational and environmental health programs
		A9.5	Risks evaluation
		A9.6	Emergency drills
		A9.7	Evaluate jobs
		A9.8	Review of the health of the employees
A10	Design planning and marketing	A10.1	Plan activities of the design area
		A10.2	Analyse the client's proposals
		A10.3	Review history for the creation of molds
		A10.4	Document activities, samples and formats
		A10.5	Establish marketing strategies
A11	Production planning	A11.1	Plan production according to productive capacity

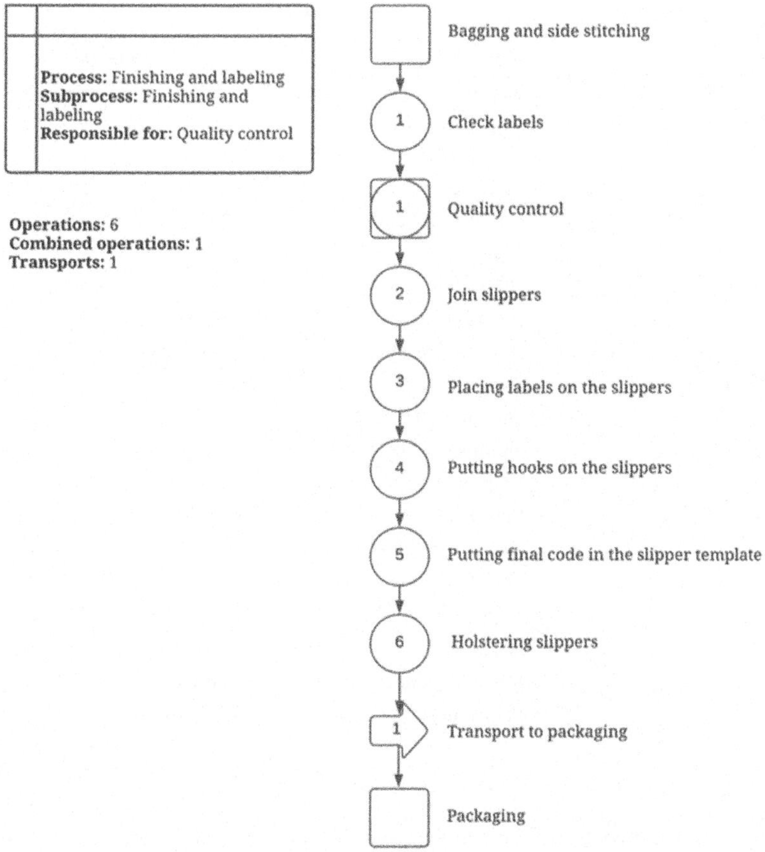

Fig. 5 Structural organization chart of the company

To do this, in the first place, the activities of each of the sub-processes under study were established and identified, together the standard time of each of them was determined. For the identification, the use of an operations process diagram (DPO) is useful where the logical sequence of the execution of activities was detailed. An established format unique to the case study is presented in Fig. 5 as an application example. The format shows operations and inspections, as well as subsequent and preceding threads and processes.

For the knowledge and identification of improvements in the processes, a time study was carried out in each of the operative processes. Figure 4 shows the results obtained.

Having total knowledge of the processes, we proceed to the creation of process sheets and manuals that have the necessary utility to guarantee the development of the work correctly, but above all the ease of communication, knowledge of interrelation

Table 4 Standard time of each operating process

Process	Sub process	Standard time (s)
Raw material preparation	Raw material preparation	3.06
External court	Upper cut	10.71
	Strip cutting	27.94
	Plant cut	12.45
	Cut of linings	11.93
	Laser cut	54.95
Fillings preparation	Cut eva insoles	6.89
	Eva heel cut	12.94
	Cut foam insoles	5.47
	Cut foam heel	3.95
	Gummed (bagged)	45.73
	Gummed (side seam)	43.79
Finishes	Embroidered	1543.13
	Sublimate	29.80
Assembly	Assembly (bagged)	29.80
	Gummed (side seam)	448.53
Bagging	Bagged	135.54
Side seam	Side seam	45.46
Finishes	Finishing and labelling	100.56
	Packaging	18.58

and connection of processes. This tool allows the perpetuation of good work practices, as well as an initial analysis to establish continuous improvement.

Finally, to improve the understanding of the relationship between the processes, a flow diagram was used to represent the sequence of execution of the operational processes to produce the slippers. For this purpose, two diagrams were used, one representing the production of the fillings, shown in Fig. 6, which shows the cutting of the inputs up to delivery in bags and side sewing, and the other, the manufacture of the slippers, shown in Fig. 7, which shows each of the operating processes involved in their development and their relationship with each other, from production planning to dispatch.

Considering that the processes take place in a closed environment, it is necessary that in terms of biosecurity the processes only get involved with those that have an interrelation with each other, in addition, the following actions are proposed to carry out the activities in a safe work environment based on to suggestions from different health organizations.

Safe Work Practices: Only production or related personnel may enter the area or any person who has authorization from the production manager for free circulation and must use all means of protection appropriate to the risks that may arise under

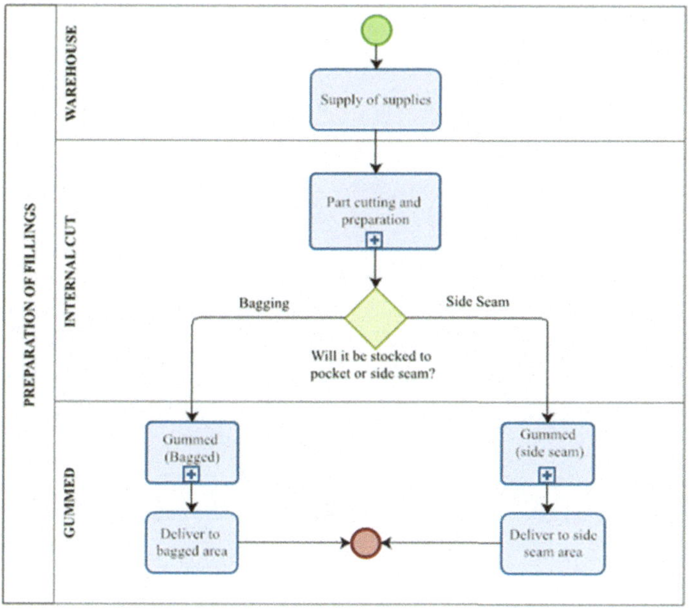

Fig. 6 Flow chart to produce fillings

the knowledge of the company and must also use the circulation areas delimited on the floor and ensure that all tools are in good condition. Finally, the work area must be clean and orderly, respecting the signage found in the work area, and it is the responsibility of each worker to attend the training sessions planned by the company.

Good Biosafety Practices: Comply with all preventive measures indicated by the Occupational Health and Safety process, maintaining interpersonal distance and avoiding any type of physical contact at the time of greeting. In addition, the use of equipment and devices belonging to other workers is prohibited. Company members must wash their hands when entering the facilities, cover their nose and mouth with a handkerchief when sneezing and dispose of it in the corresponding container, avoid contact between hands and face, constantly use masks inside the facilities, and dispose of any personal hygiene waste in the corresponding containers. In case of symptoms, inform the person in charge of the Occupational Health and Safety process.

Hygiene Socks in the Workplace: The members of the company must ensure a high level of ventilation to renew the air on a regular basis, perform daily cleaning of the entire work area, especially in areas that are most frequently touched. In addition, all activities must be carried out in single use masks and gloves, and waste management is handled with separation protocols, i.e., waste bins are labelled for waste disposal. The rubbish cans must be covered, and pedal operated (See Table 5).

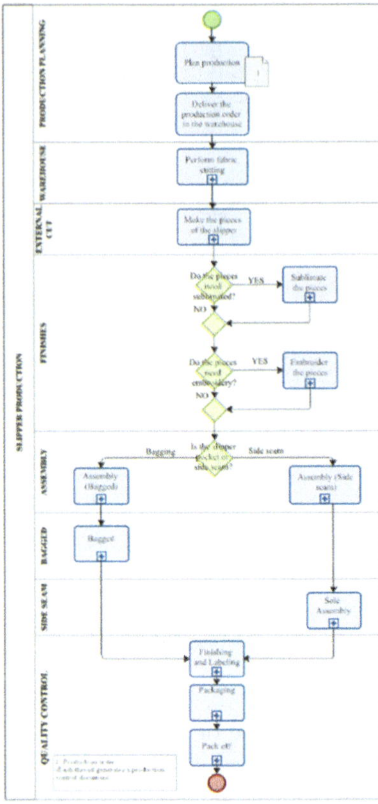

Fig. 7 Flow chart to produce slippers

Access Control Measures and Permanence in the Facilities: It is mandatory that the permanence in the company's facilities is exclusively for those belonging to the operative, administrative, and temporary personnel, and access will be limited to persons outside the company, and their visit will be exclusive to the area where their presence is required.

Transfer of Personnel to Workplaces: Personnel must travel from their homes to the company and vice versa by their own means, so the use of face masks is mandatory on public roads, in transportation and in businesses. It is recommended not to use the public transportation system as much as possible, preferring other methods of mobility not massive and if public transportation is used, it should be sanitized with disinfectant solution, avoid touching handrails, windows, and seats with the hand and if possible, sit away from another person. Additionally, necessary protective equipment was established for the personnel involved in the production area.

Once the processes in the organization have been delimited, and to finalize the Venkatraman model, all the documents present in the company must be digitized, in addition to promoting the use of IT tools for this. Also establish software that facilitates BPM management.

Table 5 Protection equipment

Area	Process	Personal protection equipment
Storage	Raw material preparation	Particulate respirator Nitrile Disposable strip cap Normal safety shoes
Production	Production assistant	Earplugs Particulate respirator Disposable strip cap
Cutting	Laser cut	Half-face mask with filters Glasses Security shoes
Cutting	External cut (Die cut)	Earplugs Particulate respirator Glasses Nitrile Disposable strip cap Normal safety shoes
Cutting	Filling preparation	Half-face mask with filters Particulate respirator
Finishes	Finishes	Earplugs Particulate respirator Disposable strip cap
Assembly	Assembly	Particulate respirator Disposable strip cap
Bagged and side seam	Bagged and side seam	Particulate Respirator Nitrile glove Disposable strip cap
Quality control	Finishing and labelling, packaging	Earplugs Particulate respirator Disposable strip cap

4 Conclusion

The Venkatraman model, together with process management and biosafety measures, allow proper administration of a company's processes, and IT integration contributes to continuous improvement. The application of the evolutionary axis of the Venkatraman model in SMEs allows a transition from the traditional to an inclusive one where IT is applied, including the organization process, according to the technological capabilities, the capabilities of the human resource and the management of the value of the company.

The process management model improves processes to a greater extent to the efficiency and effectiveness of the organization, so that the selection of processes through principles of valid criteria facilitates this area, in addition to the fulfillment of the strategic objectives of the company with a customer focus. The change in the company's processes as well as in the integration of biosafety measures must be carried out jointly through training and awareness programs, ensuring the commitment and involvement of each one of the company's participants.

In future research work, a total execution of the five levels that Venkatraman proposes for the inclusion of information technologies can be established, in addition to monitoring and controlling which correspond to the last two steps of a management by processes and how it influences the integration of computer systems.

References

1. Mendoza MF, Bayón ML (2019) El control interno y las pymes. SINAPSIS (Revista de Investigación de la Universidad IAM) 11(2):19–30
2. Solis Granda LE, Robalino Muñiz RC El papel de las PYMES en las sociedades y su problemática empresarial. https://revistas.uide.edu.ec/index.php/innova/article/view/949/1549
3. Carranco Gudiño R (2017) La aportación de las pequeñas y medianas empresas (pymes) en la economía ecuatoriana
4. Loo N, Hernandez C, Mauricio D (2021) Decision support system for the location of retail business stores. In: Advances and applications in computer science, electronics and industrial engineering. Springer, Singapore, pp 67–78
5. Garcia CA, Castellanos EX, Rosero C, Sanchez C, Garcia MV (2017) Designing automation distributed systems based on IEC-61499 and UML. In: 2017 5th international conference in software engineering research and innovation (CONISOFT), pp 61–68
6. Vaca FC, Reyes-Bozo L, Valdés-González H, Guerra CF, Galleguillos-Pozo R, Garcia MV, et al () Controlled high pressure grinding roll by model predictive control. In: 2017 IEEE 3rd Colombian conference on 2017automatic control (CCAC), pp 1–6
7. Secretaría Nacional de Planificación y Desarrollo (2021) Agendas Zonales Zona3-Centro Cotopaxi, Tungurahua, Chimborazo y Pastaza, p 22
8. Modesto M Leve reactivación del comercio y la manufactura de Ambato | El Comercio. https://www.elcomercio.com/actualidad/ecuador/reactivacion-comercio-manufactura-industrias-ambato.html
9. Ibn-Mohammed T, Mustapha KB, Godsell J, Adamu Z, Babatunde KA, Akintade DD, et al (2021) A critical review of the impacts of COVID-19 on the global economy and ecosystems and opportunities for circular economy strategies. Resour Conserv Recycl 164:105169. https://doi.org/10.1016/j.resconrec.2020.105169
10. Galleguillos R, Altamirano S, García MV, Pérez F, Marcos M (2017) Low cost CPPs for industrial control under FAHP algorithm, pp 1–4. https://doi.org/10.1109/ETFA.2017.8247775
11. Castellanos EX, Garcia CA, Rosero C, Sanchez C, Garcia MV (2017) Enabling an automation architecture of CPPs based on UML combined with IEC-61499, pp 471–476. https://doi.org/10.23919/ICCAS.2017.8204485
12. García CA, Castellanos EX, Buele J, Espinoza J, Beltrán C, Pilatasig M, et al (2017) Fuzzy control implementation in low cost CPPS devices, pp 162–167. https://doi.org/10.1109/MFI.2017.8170423
13. Informe Especial COVID-19 No 4 (2020) Sectores y empresas frente al COVID-19: emergencia y reactivación. https://repositorio.cepal.org//handle/11362/45734

14. San Andrés Reyes PR, Parra Trelles LR, Paredes Chévez IE (2020) Planificación Estratégica Administrativa para Pymes en Tiempos Post Covid. Caso de Estudio de Empresa Textil para la toma de decisiones 2019-2020. INNOVA Res J 5(3.1):185–200. https://doi.org/10.33890/innova.v5.n3.1.2020.1541
15. Carrión L, Zula J, Castillo L (2017) Analysis of model management in small and medium enterprises and its application in catering industry in Ecuador
16. Pusdá-Chulde MR, Salazar-Fierro FA, Sandoval-Pillajo L, Herrera-Granda EP, García-Santillán ID, De Giusti A (2020) Image analysis based on heterogeneous architectures for precision agriculture: a systematic literature review. In: Advances and applications in computer science, electronics and industrial engineering. Springer, Cham, pp 51–70
17. Orellana J, Peña M, Llivisaca J (2021) Assessment of supply chain performance in an assembly company: evaluation of evolutionary algorithms. In: Advances and applications in computer science, electronics and industrial engineering. Springer, Singapore, pp 167–183
18. Hernández Palma H, Martínez Sierra D, Cardona Arbeláez D (2016) Enfoque basado en procesos como estrategia de dirección para las empresas de transformación. Saber, Ciencia y Libertad 11(1):141–150. https://doi.org/10.18041/2382-3240/saber.2016v11n1.499
19. Medina León A, Nogueira Rivera D, Hernández-Nariño A (2019) Procedimiento para la gestión por procesos: métodos y herramientas de apoyo Procedure for process management: methods and support tools. Revista chilena de ingeniería. 27(2):328–342
20. Delfín Pozos FL, Acosta Márquez MP (2016) Analysis and relevance in business development. Revista científica Pensamiento y Gestión (40):184–202. https://doi.org/10.14482/pege.40.8810
21. Gibson C, Carr C, Lyons C, Taksa L, Warren A (2021) COVID-19 and the shifting industrial landscape. Geograph Res 59(2):196–205. https://doi.org/10.1111/1745-5871.12462
22. Lee KW, Saeed M, Warren MF, Hafez hafez HM, Youssef Attia vbA, Hafez HM, et al (2020) Challenges to the poultry industry: current perspectives and strategic future after the COVID-19 outbreak. Front Veter Sci 1:516. https://doi.org/10.3389/fvets.2020.00516, https://www.frontiersin.org/
23. Martínez AM (2014) Gestión por procesos de negocio: Organización horizontal. Ecobook

Computer Science

Computer Vision Technique to Improve the Color Ratio in Estimating the Concentration of Free Chlorine

José Alonso Ruiz-Navarro, Félix Melchor Santos-López, Jhon Manuel Portella-Delgado, and Eulogio Guillermo Santos-de-la-Cruz

Abstract Access to drinkable water is a constitutional right in Peru. However, quality water is not universally accessible, particularly in rural areas where neither water disinfection nor quality control are undertaken. It is therefore crucial to develop methods that allow estimates of water quality characteristics, such as free chlorine concentration. This publication proposes a sensing system that estimates free chlorine concentration using a traditional colorimetric method of chemical reaction with N, N-diethyl-p-phenylenediamine and a stationary digital camera. Using image processing, a subset of computer vision, the algorithm quantifies the color of the water sample after chemical reaction with the reagent and, on the basis of that quantification, tries to estimate the free chlorine concentration of the sample. After the quantification, a model is determined using a best-fit polynomial. The resulting polynomial model is then implemented using serverless cloud computing resources from Amazon Web Services, and its results are also hosted using this platform. The model error has a standard deviation of 0.0129 ppm.

Keywords Free chlorine · Image processing · Model · Cloud · Serverless

J. A. Ruiz-Navarro (✉)
School of Science and Engineering, Pontifical Catholic University of Peru, Lima, Peru
e-mail: ruizn.ja@pucp.edu.pe

F. M. Santos-López · J. M. Portella-Delgado
Department of Engineering, Pontifical Catholic University of Peru, Lima, Peru
e-mail: fsantos@pucp.edu.pe

J. M. Portella-Delgado
e-mail: jportella@pucp.edu.pe

E. G. Santos-de-la-Cruz
Faculty of Industrial Engineering, National University of San Marcos, Lima, Peru
e-mail: esantosd@unmsm.edu.pe

© The Author(s), under exclusive license to Springer Nature Switzerland AG 2022

M. V. Garcia et al. (eds.), *Advances and Applications in Computer Science, Electronics, and Industrial Engineering*, Lecture Notes in Networks and Systems 433,
https://doi.org/10.1007/978-3-030-97719-1_7

1 Introduction

To ensure that water is free of pathogens, chlorine is a commonly used solution to clean water and make it drinkable due to its ease of use. To achieve satisfactory disinfection of the water, it is recommended to maintain a certain concentration of "free chlorine" in order to keep it clean even if new pathogens enter it. There are many methods to measure free chlorine concentration, including field methods such as the traditional colorimetric method, which can be subjective, electronic sensors, and digital colorimetry, which in most cases is expensive and, therefore, harder to implement in the rural areas that suffer from poor water quality. This paper presents a sensor system that allows estimation of free chlorine concentration based on the color of a water sample after chemical reaction with N, N-diethyl-p-phenylenediamine (DPD). This color change is quantified using image processing (computer vision) and then used to estimate the concentration in the sample. Cloud computing resources are used to implement the solution in a serverless architecture. Cloud computing is shown as a viable means of deploying solutions related to environmental parameters, similar to the approach proposed by Murata et al. [9].

This publication is divided into several distinct sections. We begin with a literature review, where we synthesize the findings of various previous studies and summarize relevant data from each. We then outline a system proposal, where a system and its requirements are described. We describe our experiments to show the main parts of the process to implement the previously proposed system, in addition to a description of the results of these experiments in a subsequent section. Finally, we present the conclusions of our study.

2 Literature

2.1 Current Measurement Methods

Traditional methods of chlorine estimate include chemical reactions with reagents that cause a change in color (colorimetric method) of a water sample; the new color of the sample is used as an indicator of the free chlorine concentration in the sample. Different reagents generate different pigmentation in the water sample, the most common of these being pink (DPD), blue (potassium starch), and yellow (toluidine). These colorimetric method results are traditionally evaluated by the user using the naked eye. Interpretation of the results is therefore potentially susceptible to factors such as ambient light intensity variations and the visual health of the user.

Other measurement methods include commercial amperometric sensors, which generate an electric current and relate the change in current to the free chlorine concentration in a system. Some examples of manufacturers of this type of sensors include Bürkert [10, 13, 13], Endress+Hauser [12] and Sensorex [17]. These types of sensors are in most cases expensive and require additional equipment to operate. An alternate approach is digital colorimetry, which is related to the methods described previously. In this method, a digital device is used to generate certain wavelengths of light (based on the reagent used) and aims to measure the concentration of free chlorine.

In recent years, there have been a number of publications where different solutions to chlorine estimation have been proposed. These include a sensing system based on doping a carbon nanotube [14]and sensors based on fiber optics and the absorbance of different light wavelengths in a liquid medium [15, 16, 22]. Other approaches that use absorbance are proposed in [18] and [11], using a laser or flashlight as a light source respectively. These studies use a long glass tube or a triangular mirror in order to increase the optical path, thereby increasing sensitivity. Previous studies have also proposed new types of digital colorimetry. Using cameras and traditional image processing, algorithms have been developed to estimate free chlorine concentration by quantifying the change in color of a sample. A first example is described in [21], where the estimation of iron and chlorine concentration in samples was explored. Authors such as Sumriddetchkajorn et al. explore the estimation of free chlorine concentration by applying image processing and using reagents like potassium starch or o-Toluidine [19, 20]. A structured review of the main details of the above literature is summarized in Table 1.

Table 1 Literature review

Ref. N°	Paper title	Described range	Accuracy/Correlation or resolution
[14]	A carbon nanotube based resettable sensor for measuring free chlorine in drinking water	0.06– mg/L	$R^2 = 0.9760$
[11]	Field analysis free chlorine in water samples by a smartphone-based colorimetric device with improved sensitivity and accuracy	0.8–14 μM	$R^2 = 0.9991$
[21]	Tristimulus Colorimetry Using a Digital Still Camera and Its Application to Determination of Iron and Residual Chlorine in Water Samples	0.3–0.7 mg/L	r = 0.93
[22]	Fiber Optic Sensor for Detection of Chlorine Level in Waters	0–4 g/L	$R^2 = 0.9974$ Resolution = 2.7 mg/L
[19]	Mobile device-based self-referencing colorimeter for monitoring chlorine concentration water	0.3–1.0 mg/L	$R^2 = 0.9976$ Error $\leq 7\%$
[18]	On site determination of free chlorine in water samples by a smartphone-based colorimetric device with improvement sensitivity and reliability	0–80 μg/L	Error = ±2.2 μg/L
[20]	Mobile-platform based colorimeter for monitoring chlorine concentration in water	0.6–2.0 mg/L	$R^2 = 0.9983$ $\sigma = 0.02$ ppm

Where r = correlation coefficient and R^2 is the coefficient of determination

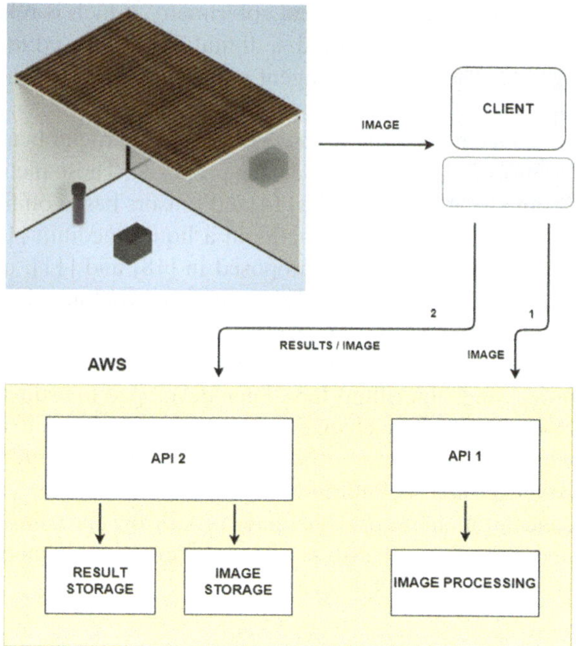

Fig. 1 Simple schema of the proposed system

3 Proposed System

In this study, we propose the design of a sensing system using image processing and a stationary camera, attempting to use the simplest physical set-up possible. This system includes the implementation of the estimation algorithm using cloud computing services from Amazon Web Services (AWS) (Fig. 1). The sensing system's working range must include a value of 0.5 ppm (0.5 mg/L), as this value is the lower limit considered healthy for drinking water by sanitary institutions in Peru.

4 Experiment

4.1 Materials and Set up

The images were taken using a commercial webcam (Microsoft Lifecam HD-3000) inside a dark room (see Fig. 2), which was made using a simple cardboard box, a white background using cardboard paper and aluminium foil on the sides. The illumination was generated using led strips (a total of 3 strips of 12 LEDs each).

The water samples were put in glass of 10 ml size, and DPD reagent was added (using commercial dose for samples of 10 ml). For reference purposes, the actual

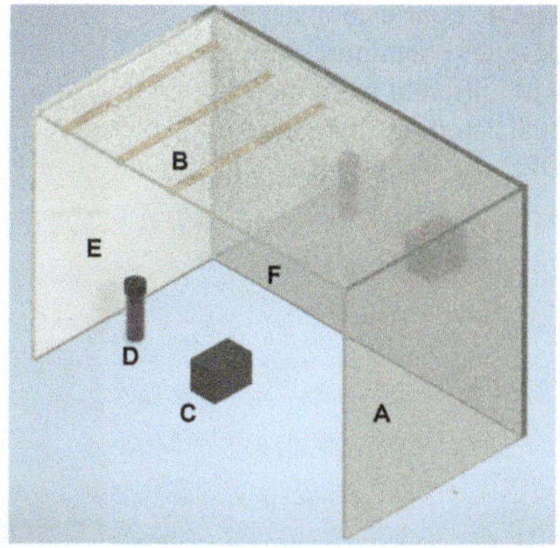

Fig. 2 Simple schematic cross-section of the experimental setup. Inside a cardboard box (A) there are LEDs at the top (B), a non-moving webcam (C), a cuvette with sample (D) in front of a white background (E) and reflective aluminium foil on the sides (F). Note that in this schematic, the box is shown as transparent to illustrate the interior setup

Fig. 3 Glass cuvette, DPD dose and Free Chlorine Checker from Hanna Instruments

value of free chlorine concentration was determined using a Free Chlorine Checker (see Fig. 3). All these items are produced by Hanna Instruments manufacturer.

To keep the distance between elements consistent, the positions of the cardboard box, the cuvette containing the water sample, and the webcam were marked in pencil on a white cardboard paper.

The water samples were made using diluted commercial bleach and tap water. The bleach had 5% of chlorine concentration (50000 ppm). The tap water free chlorine concentration was also measured using the Free Chlorine Checker, which in most cases resulted in a measurement of 0 ppm concentration (maximum result was 0.05 ppm). The bleach was diluted three times to produce a solution with a chlorine concentration of 2 ppm. This dilution was then mixed with tap water in various proportions, resulting in 10 ml samples with chlorine concentrations 2 ppm or less.

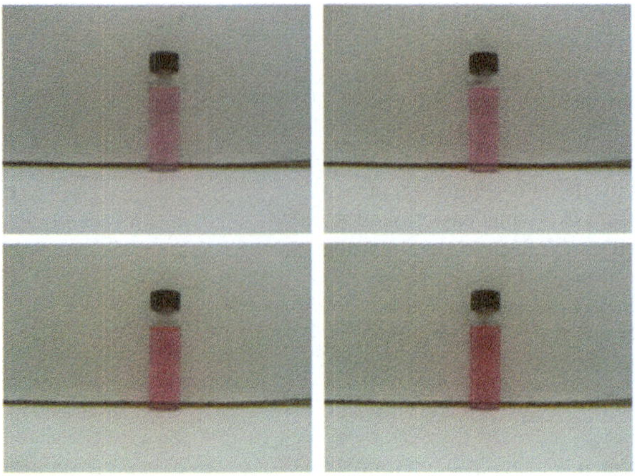

Fig. 4 Four examples of cuvettes containing samples with different free chlorine concentrations reacting to DPD

4.2 Execution

The DPD reagent was added to samples with chlorine concentrations of 2 ppm and less. These samples were evaluated using the Free Chlorine Checker and, immediately afterward, a photograph of the cuvette was taken inside the box with LEDs turned on (see Fig. 4). The images were taken using a Python-based software using the OpenCV library. which was running on the laptop PC where the webcam was connected. The pictures were saved using the free chlorine concentration in ppm as filename (for example, 1.80.jpg), for ease of use in later analysis.

4.3 Analysis

The quantification of free chlorine concentration was made using image processing based on the methods described in [19] and[20].

Region of Interest. In order to extract information from the pictures, the region of interest (ROI) consisted of three rectangular areas: one inside the cuvette and the other two on the sides of the image (i.e., from the white background) (Fig. 5).

Color Quantification. A method is required to quantify the change of color. By using a numerical representation of the image, a relation between this quantification number and the free chlorine concentration can be sought. Images in RGB color space (red (R), green (G), and blue (B)) are used; the mean RGB values were then calculated for each part of the ROI, as described below.

Quantification Method in the Literature. Quantification of the color change was performed using a method based on the approaches described in [19] and [20]. The method described in these publications calculates the mean value for each layer of

Fig. 5 ROI consisted of two areas over the white background (blue rectangles) and one over the cuvette (green rectangle)

the central area (Rs, Gs and Bs) and the areas on the sides (Rr, Gr and Br). The quantification is then calculated as follows:

$$CR = \frac{Rs/Rr + Gs/Gr + Bs/Br}{3} \quad (1)$$

Proposed Quantification Method Variation. The number that quantifies the color change is not a simple average (Color Ratio, CR), but rather a weighted average (Weighted Color Ratio, WCR). This publication proposes a weighted average, due to the observation that some ratios are more significant than others, having larger value changes, and therefore their values are more helpful in distinguishing between different tonalities of pink.

Layer Ratios Relevance Analysis. After inspection of the values of the ratios for all images, we concluded that the change in the ratio of the green layer has a larger range than the corresponding ratio change in the blue layer, making the former ratio more representative than the latter. Both of these ratios show a clear tendency: the higher the chlorine concentration, the lower the ratio. However, ratio changes in the red layer show a smaller value range. This ratio does not show as clear a tendency as the other two ratios (Fig. 6); this ratio was therefore discarded as a representative value for estimation calculations, as seen in Eq. (2).

Layer Ratios Usefulness Among the Whole Range. Another detail that can be seen in Fig. 6 is the observation that ratio values become more disperse for concentrations greater than 1 ppm. The value of 1 ppm was therefore assigned as the upper sensitivity limit of the working range. As a comparatively small absolute error value (e.g., 0.01 ppm) can represent a large magnitude of relative error when the target value is small, we selected a value of 0.25 ppm as our lower limit in the working range.

Based on filtering of the data shown in Fig. 6 and using the points that lie within the range of 0.25–1 ppm, the value ranges of the red, blue and green layer ratios are shown in Table 2.

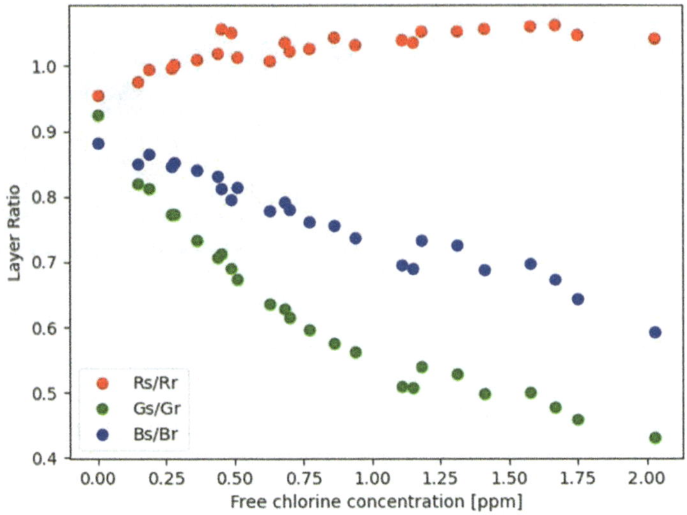

Fig. 6 Plot of the ratio for each color channel in a image vs the free chlorine concentration

Table 2 Value variation of ratios

Layer of the ratio	Min. value	Max. value
Red	1.00	1.06
Green	0.56	0.77
Blue	0.74	0.85

The ratio value range variation indicates how representative each one for estimating the concentration. Given that, the weighted color ratio (WCR) has the weights shown in (2).

$$WCR = \frac{2 * Gs/Gr + Bs/Br}{3} \qquad (2)$$

Once the algorithm was applied to all the images within the described concentration range, values were extracted from each. These values quantify the change in the color of the water sample in each cuvette, caused by the reaction with DPD. In other words, given a free chlorine concentration (coded in an image) there is some value (WCR) that quantifies this concentration, which can be plotted as points in 2D space (Fig. 7).

A similar process can be performed by directly applying the quantification method described in the literature (i.e., Color Ratio) with no modification to working ranges, and plotting the free chlorine concentration versus the color ratio (see Fig. 8) rather than the weighted color ratio. This approach allows a comparison between both quantification methods.

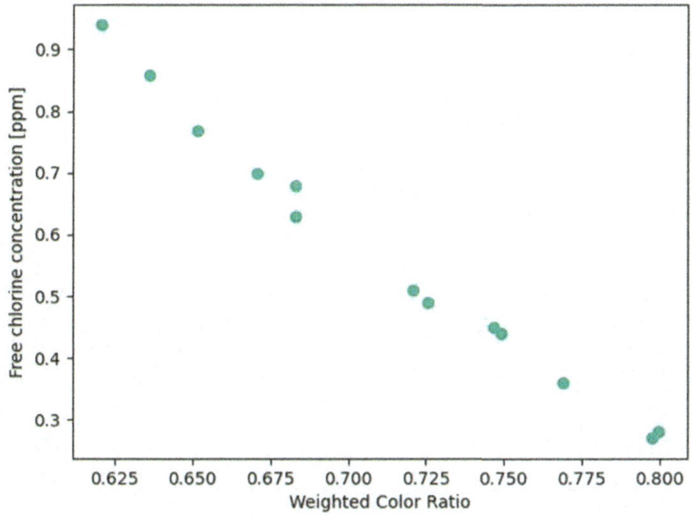

Fig. 7 Plot of the free chlorine concentration vs WCR

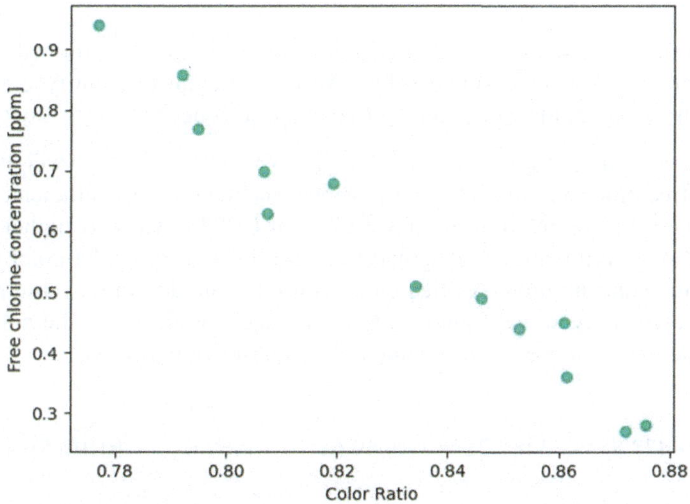

Fig. 8 Plot of the free chlorine concentration vs CR

Visual Comparison. One of the key differences that can be observed when comparing Figs. 7 and 8 is the distance between the points. Specifically, when using the weighted color ratio instead of the regular color ratio, the points are less dispersed. In both cases, there is a clear tendency: the higher the free chlorine concentration, the lower the WCR or CR value. However, this trend is more clearly exhibited using WCR, and this quantification approach was therefore selected for this study.

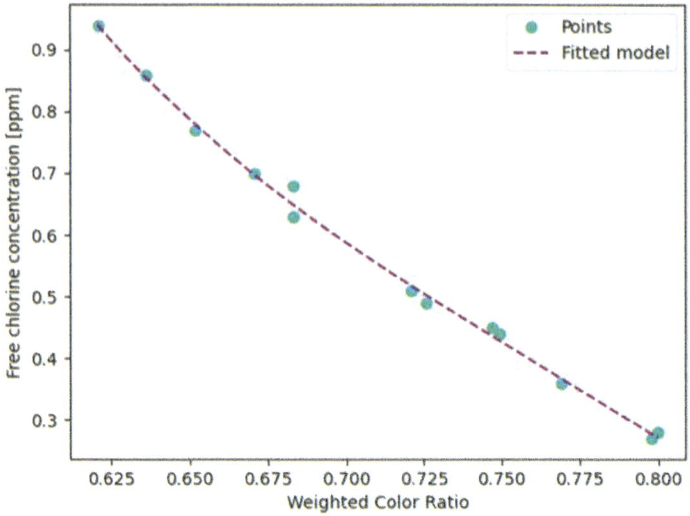

Fig. 9 Plot of the points and the fitted model

Calibration. As noted above, after a quantification method is determined, an image can be represented using a single value. We now attempt to identify a numerical relationship between this value and the free chlorine concentration.

Model Determination Approach. The relation between the calculated weighted color ratio and free chlorine concentration is shown in Fig. 9. The coefficient of determination (R^2) of this relationship has a value of 0.9873, which is higher than the coefficient of determination (calculated as 0.9609) between regular color ratio and free chlorine concentration. A fitted curve is used as model; in this case the curve is a fifth degree polynomial found using least-squares regression. The polynomial, with coefficients rounded to one decimal place, is shown in Eq. (3):

$$P(x) = -478.8x^5 + 3489.2x^4 - 10049.6x^3 + 14359.8x^2 - 10197.5x + 2881.2 \tag{3}$$

where x is the weighted color ratio.

The evaluation of this model and its performance will be presented in the results section.

4.4 Cloud Computing Implementation

Architecture Description. The free chlorine concentration estimation algorithm was implemented using serverless cloud computing resources from AWS (Fig. 10). The

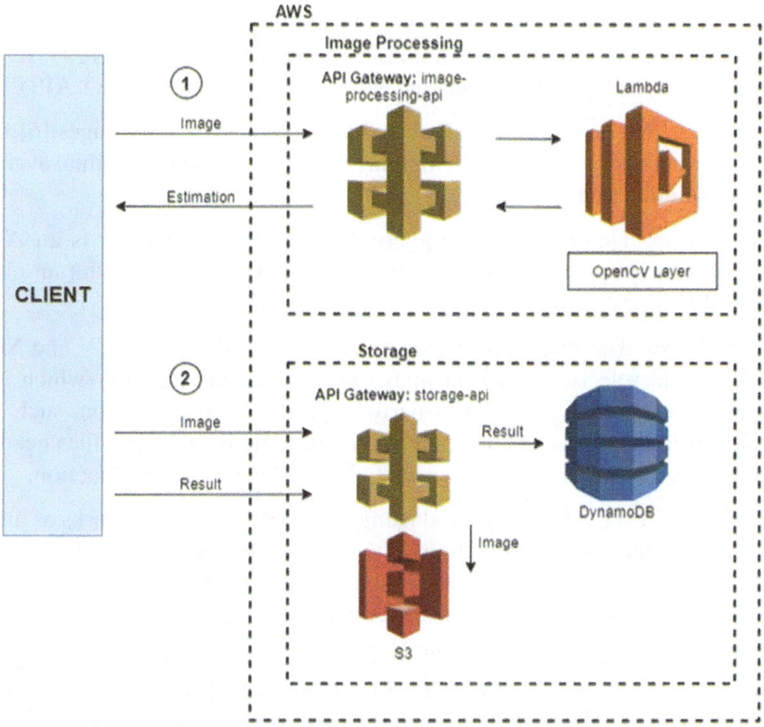

Fig. 10 Architecture used to implement the algorithm on the cloud and store results

system's functions have two stages; firstly, the client invokes the Lambda Function via the "image-processing-api" and secondly, as a final stage, the client takes the response (i.e., concentration estimation) and the image and sends them to Dynamo DB and S3 respectively through the "storage-api".

Computing Architecture. A REST API is created using API Gateway [1], which receives an HTTP Request containing an image in the body and transfers it to a computing service (Lambda) [4], before waiting for a result to return to the client. Lambda is used to host the algorithm code in a Python 3.8 runtime. This "Lambda Function" runs using a layer that contains the required libraries (Python packages), specifically OpenCV and NumPy.

Storage Architecture. The client can also save the results using a serverless implementation. The numeric result (free chlorine concentration estimation) is stored in DynamoDB [2] and the image that was processed is stored in S3 [3], which is a file/object storage service.

Lambda Function. A Lambda function was deployed to execute the algorithm (using a Python 3.8 runtime). The function takes an image as an input and returns the result from calculations. The image processing takes in the cloud to calculate a WCR from the input image; this result is then passed to a polynomial evaluation function to estimate the free chlorine concentration. The functions return an HTTP response containing the calculated WCR and concentration to its trigger (REST API).

Lambda Layer. A Lambda Layer is created to contain Python packages [6]. These packages are necessary to proceed with image processing and polynomial evaluation in the Lambda function.

Lambda Trigger. The action that triggers the function's execution is an API [8] (image-processing-api) (Fig. 10). The input is coded as an event with an element called "body".

Python Packages. The Python packages that are included are OpenCV and Numpy. OpenCV's main role is to perform image processing and NumPy (which is also an OpenCV dependency) is used to perform numerical computation, such as the aforementioned polynomial fitting and evaluation. These packages' files need to be installed in a Linux environment in order to be used in a Lambda Function.

REST API. A REST API is deployed using API Gateway, which works as an entry point to access the Lambda function. The API receives an image as binary data (encoded in base64) from the client and returns a JSON response that contains the calculated WCR and the free chlorine concentration in ppm. The API correct deployment was tested in software (Postman v7.36.5).

Method. The API can be called via a POST method, which allows the inclusion of a body in the request. The ability to do so is important because the request's body is where the binary image will be sent.

Input. The API settings are such that image formats (jpeg and png) are set as the default interpretation of incoming binary data [5].

Lambda Function Connection. The API connects to an specific AWS resource: the aforementioned Lambda function. The connection to this resource is via a method referred to in AWS as a Lambda-proxy [7].

5 Results

The model proposed for chlorine concentration estimation is optimized using least-squares, meaning that the polynomial P(x) has coefficients that minimize the misfit as shown in Eq. (4), where "n" is the number of points used to calculate the polynomial fit, and the points are (x_k, y_k):

$$\sum_{k=1}^{n}(P(x_k) - y_k)^2 = L \tag{4}$$

In this case n=13 and P(x) coefficients after least-squares optimization are shown in (3). This process was performed using NumPy library in Python.

Table 3 Model's evaluation

Concentration [ppm] (C)	Estimation [ppm] (E)	Squared error $\sqrt{(C-E)^2}$
0.27	0.2782	0.0082
0.28	0.2717	0.0083
0.36	0.3685	0.0085
0.44	0.4301	0.0099
0.45	0.4374	0.0126
0.49	0.5029	0.0129
0.51	0.5185	0.0085
0.63	0.6493	0.0193
0.68	0.6491	0.0309
0.70	0.6982	0.0017
0.77	0.7806	0.0106
0.86	0.8552	0.0048
0.94	0.9404	0.0004

Model Performance. The model is evaluated in two main ways: at each point and a general evaluation. The first approach is useful for evaluating the model and allows detection abnormalities at a specific point; the latter is useful to evaluate the model's performance as a whole.

Point-wise Evaluation. The model is evaluated along the points shown in Fig. 7. The actual value and estimation are compared and the difference (error) is calculated. These results are exposed in Table 3.

General Evaluation. The estimation model's error is shown to have a standard deviation lower than 0.0129 ppm in the working range from 0.25 ppm to 1.00 ppm. The value of the loss function (least-squares) after the optimization is 0.00016 ppm^2.

In order to visualize the method's precision, a graph is plotted. This graph includes both the fitted model but real value points with error-bars. Each error-bar is centered in the point and has size of four times 0.0129 ppm, which is the value of the error's standard deviation. This graph is depicted in Fig. 11, where it can be noticed that almost all the points lie in a range of four sigma error. This magnitude of the error-bars is chosen to represent what we consider an acceptable absolute error. Meanwhile, the relative error has mostly a value under 3.1%, surpassing this threshold only once with a value of 4.54%.

These results shows the effectiveness of the methodology. The technique applied in the experiment shows itself as promising to improve the calculation of the color ratio for color quantification. Thus, the enhancement allows reaching a better relationship regards to chlorine concentration than one found with traditional color ratio calculation.

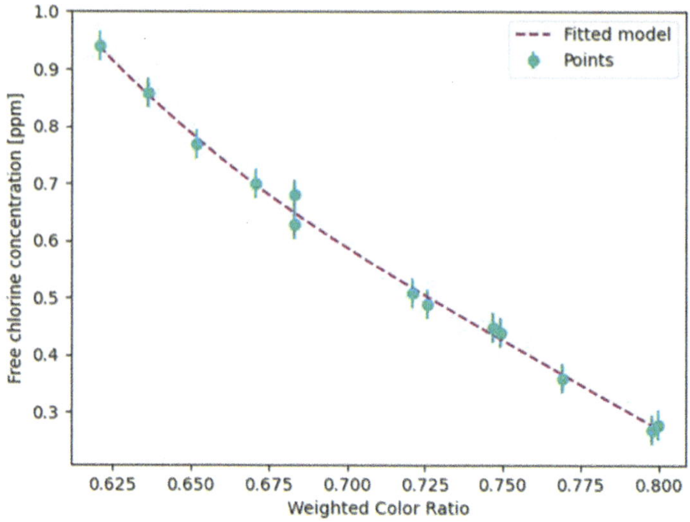

Fig. 11 Fitted model, points and error-bars

6 Conclusions

A literature review was undertaken to identify contributions such as the mentioned CR are shown with promising results. We propose a simple, new physical set-up for chlorine concentration using DPD, and a polynomial model was derived for this physical set-up. This model describes a relationship between the quantification value for an image and the measured concentration and shows a standard deviation of 0.0129 ppm and a maximum relative error of 4.54%.

Comparing the results with the literature, specially with the work of Sumriddetchkajorn et al. (whose method was modified to develop ours) these results in a lower standard deviation ($0.0129 < 0.02$) of the error and lower relative error ($4.54 < 7\%$). The values used in the quantification method described in the literature (CR) were analysed, based on our experimental results and after the comparison done before we devised an improved version of this method which applies a per-channel weighting (WCR), being this analysis process a promising way to improve the first described quantification method.

AWS cloud architecture was deployed and tested as functional, being able to process an image and responding with an estimation result. Future works may address the possibility of an improved physical set-up to reduce uncertainties due to cuvette position, and also to achieve a better lighting.

References

1. Amazon Web Services: "Amazon API Gateway". https://aws.amazon.com/es/api-gateway/
2. Amazon Web Services: "Amazon DynamoDB". https://aws.amazon.com/es/dynamodb/
3. Amazon Web Services: "Amazon S3". https://aws.amazon.com/es/s3/
4. Amazon Web Services: "AWS Lambda". https://aws.amazon.com/es/lambda/
5. Amazon Web Services: "Binary Data Now Supported by API Gateway". https://docs.aws.amazon.com/apigateway/latest/developerguide/api-gateway-payload-encodings-configure-with-console.html
6. Amazon Web Services: "Creating and sharing Lambda layers". https://docs.aws.amazon.com/lambda/latest/dg/configuration-layers.html
7. Amazon Web Services: "Set up Lambda proxy integrations in API Gateway". https://docs.aws.amazon.com/apigateway/latest/developerguide/set-up-lambda-proxy-integrations.html
8. Amazon Web Services: "Using AWS Lambda with Amazon API Gateway". https://docs.aws.amazon.com/lambda/latest/dg/services-apigateway.html
9. Arca RMM, López FMS, Delgado JMP (2020) Sistema ciber-físico en la nube para monitoreo de contaminación acústica urbana bajo plataforma iot. Revista Ibérica de Sistemas e Tecnologias de Informação E37:91–103
10. Bürkert: "Type 8232 - Chlorine sensor". https://www.burkert.com/en/type/8232"
11. Dou J, Shang J, Kang Q, Shen D (2019) Field analysis free chlorine in water samples by a smartphone-based colorimetric device with improved sensitivity and accuracy. Microchem J 150:104200
12. Endress+Hauser: "Memosens CCS51D: Digital free chlorine sensor". https://www.endress.com/en/field-instruments-overview/liquid-analysis-product-overview/free-chlorine-digital-sensor-ccs51d
13. Guzman D, Brito G, Naranjo JE, Garcia CA, Saltos LF, Garcia MV (2018) Virtual assistance environment for deaf people based on an electronic gauntlet. In: 2018 IEEE third Ecuador technical chapters meeting (ETCM), pp 1–6. https://doi.org/10.1109/ETCM.2018.8580281
14. Hsu LH, Hoque E, Kruse P, Ravi Selvaganapathy P (2015) A carbon nanotube based resettable sensor for measuring free chlorine in drinking water. Appl Phys Lett 106(6):063102
15. Mariño C, Vargas J (2020) Ergonomic postural evaluation system through non-invasive sensors. In: Advances and applications in computer science, electronics and industrial engineering. Springer, Cham, pp 274–286 (2020)
16. Quezada J, Siguenza-Guzman L, Llivisaca J (2020) Optimization of motorcycle assembly processes based on lean manufacturing tools. In: Advances and applications in computer science, electronics and industrial engineering. Springer, Cham, pp 247–259 (2020)
17. Sensorex: "FCL Amperometric Free Chlorine Sensor". https://sensorex.com/product/fcl-free-chlorine-sensor/
18. Shang J, Yu L, Sun Y, Chen X, Kang Q, Shen D (2019) On site determination of free chlorine in water samples by a smartphone-based colorimetric device with improved sensitivity and reliability. New J Chem 43(36):14409–14416
19. Sumriddetchkajorn S, Chaitavon K, Intaravanne Y (2013) Mobile device-based self-referencing colorimeter for monitoring chlorine concentration in water. Sensors Actu B Chem 182:592–597
20. Sumriddetchkajorn S, Chaitavon K, Intaravanne Y (2014) Mobile-platform based colorimeter for monitoring chlorine concentration in water. Sensors Actu B Chem 191:561–566
21. Suzuki Y, Endo M, Jin J, Iwase K, Iwatsuki M (2006) Tristimulus colorimetry using a digital still camera and its application to determination of iron and residual chlorine in water samples. Anal Sci 22(3):411–414
22. Yesmin S, Chetia D, Basumatary T, Singh H K (2018) Fiber optic sensor for detection of chlorine level in water. In: Advances in communication, devices and networking. Springer, Heidelberg, pp 429–437 (2018)

Formative Assessment Model Using an Analytical Rubric for Written Tasks

Johana Morales, Vicente Morales, Sussy Bayona, and Marjorie Morales

Abstract A formative assessment process provides students the opportunity to be involved in their learning process. Many teachers and students of English have problems in this type of process because they do not have a homogeneous evaluation procedure. The purpose of this study is to propose a formative evaluation model using an analytical rubric that standardizes the process of evaluating writing ability in the field of EFL. This study followed a quasi-experimental design that included a pre and post-test. Two groups of A2 level students were asked to write book reports where an intervention was applied to the experimental group, which included a self and peer-assessment process based on a rubric that was designed among A2 level teachers. The results showed that the experimental group outperformed the comparison group in the post test. The study reveals that this process helped students get closer to the rubric criteria due to recursive feedback that was shared in a collaborative setting. The research has practical implications for teaching book report writing among English learners of the same level, but with different teachers.

Keywords Formative assessment · Peer assessment · Self-assessment · Rubric · Recursive feedback

J. Morales
Pontificia Universidad Católica del Ecuador, Av. Manuelita Sáenz, 180207 Ambato, Ecuador
e-mail: johana.e.morales.1@pucesa.edu.ec

V. Morales (✉)
Universidad Técnica de Ambato, Los Chasquis y Río Payamino, 180103 Ambato, Ecuador
e-mail: jvmorales99@uta.edu.ec

S. Bayona
Universidad Nacional Mayor de San Marcos, Av. Germán Amézaga s/n, 15081 Lima, Peru

M. Morales
Ministerio de Salud Pública, Av. República del Salvador 36-64, 170515 Quito, Ecuador
e-mail: mmorales0011@uta.edu.ec

© The Author(s), under exclusive license to Springer Nature Switzerland AG 2022
M. V. Garcia et al. (eds.), *Advances and Applications in Computer Science, Electronics, and Industrial Engineering*, Lecture Notes in Networks and Systems 433,
https://doi.org/10.1007/978-3-030-97719-1_8

1 Introduction

The formative assessment in EFL (English as a Foreign Language) writing provides students with the opportunity to control their learning by self-regulating this process through feedback [1]. Therefore, student-centered learning emerges as a primary focus. Studies have shown that learning can become meaningful by actively involving students in the classroom [2–4].

Assessment for learning should be the goal rather than assessment of learning. According to [3] this can be achieved with a formative assessment that can improve learning behavior in a constructive environment. [5, 6] state that knowledge has to be a social interaction that has cycles where collaborative exchange improves learning with a process of self and peer-assessment during a formative evaluation, instead of obtaining grades without adequate feedback.

Peer review leads to teamwork where students can help each other to improve their written work. [7] state that students will be more involved in this type of feed-back rather than the one given from a teacher. Rubrics can lead to effective implementation of this process. Similarly, rubrics can promote self-assessment as students gain a clear understanding of the task criteria to enrich their written work [8–10].

Standards are necessary to achieve a specific skill, thus if writing tasks could be measured using a standardized assessment model, it could maintain a consistent process [11]. A significant way to evaluate writing activities is the use of rubrics which provide more effective feedback. A rubric is essential because it assesses student performance, improves instruction, and provides relevant information to program progress [12–14].

The objective of this study is to propose a formative evaluation model using a specific rubric that standardizes the process of evaluating the writing skill in the field of EFL. A quasi-experimental research design will be used with a comparison group and an experimental one. This approach allows testing whether an intervention produces any change in the performance [15]. Therefore, a pre-test, an intervention to the experimental group using the formative evaluation model and a post-test are included to measure the results of the intervention.

This research developed a proposal for a standard formative evaluation model that has five phases: planning, drafting, peer assessment, self-assessment and presentation of the final product through an analytical rubric for book reports that includes four components [16]. In conclusion, this model generated a standardized process that helps students understand what is requested in the activity to improve writing performance in a collaborative environment.

The application of the formative assessment model determined that students can have a considerable reach to the standards requested in the rubric criteria by becoming familiar with them while the phases of the process were developed. Thus, they were able to internalize their peers' comments and decide on ways to improve their written work. This process was viewed as positive by the students because they were able to achieve higher levels of mastery that support their learning process.

The rest of the article is organized as follows: Sect. 2 presents the background and motivation for the study. In Sect. 3 the research methodology is presented. The results are presented in Sect. 4. Finally, the conclusions are presented in Sect. 5.

2 Background and Motivation

2.1 *Writing Process*

Reading and writing are part of written communication, although the last one has a complex procedure. The writing process is crucial because it improves writing ability by rewriting, proofreading, and editing [17]. The stages in the writing process include: planning, writing, and proofreading; same that occur simultaneously with each other [18]. Teachers must frequently manage this process with formative writing assessments to turn it into an everyday teaching and learning process that provides direct feedback [19].

2.2 *Assessment*

Assessment is the collection of data on student performance in the classroom that includes the student's reaction to the feedback provided to their learning [20]. Assessment does not mean providing a test and scoring it; it aims is to benefit teachers and students through a complete view of the achievement of the course goals during the teaching and learning process by getting a constructive feedback [21].

The concept of formative assessment in the field of education focuses on the development of a certain educational process. This assessment helps students understand their learning objectives by reviewing their own learning progress [22, 23]. Formative assessment will be an assessment for learning in which evidence will be collected to provide feedback and analyze educational strategies [24]. This process increases motivation in students and promotes the importance of reviewing their own work and from others.

Self-assessment and peer-assessment are helpful in improving students' writing skills, helping them internalize the assessment criteria to understand what is expected of them. These types of assessments are more common during formative assessment [25, 26]. Self and peer-assessment are related, since they provide feedback on an activity that has a supported criterion [27].

During peer-assessment, students gain the benefit of receiving and providing feedback by making judgments based on criteria. By practicing this process, students will be able to provide more specific and relevant comments to their peers [28, 29]. Likewise, "self-assessment is an effective strategy to improve students' writing performance, since they have a good perception of it" [30].

2.3 Rubrics

A rubric is a powerful tool that helps during the teaching and assessment process by showing students the requirements they need to meet. During this process, students will have the opportunity to analyze their work in depth by comparing it with the criteria of the rubric provided [31, 32]. The analytical rubrics are seen as the best option since they will help during a formative process by providing details of what needs to be improved in each of the criteria since they are analyzed separately [33].

2.4 Rubric-Based Peer and Autonomous Formative Assessment

In the context of EFL, rubrics are considered effective tools for students to monitor their own work. Students can participate in the self-assessment and peer review process to analyze their own assignments and provide feedback to their peers through the rubric [34].

According to [5], knowledge is social due to interaction during reflections. Learning becomes more visible, since the evaluation is carried out from the collaborative exchange process, achieving recursive feedback instead of a classical linear evaluation. Students used to do their work and get a grade. Now, the intention is to get information in small cycles by leaving behind the difference between evaluation and learning.

Students will be able to view the rubric and use it during the writing process to get a clear idea of what to expect. Evaluation as recursive feedback includes five steps: planning, writing, examining, reviewing, and publishing. Students write and review other peers' work based on the same rubric, gaining other perspectives by receiving feedback and reflecting on their work [16].

3 Methodology

This section shows the general description of the type of investigation, as well as the design, instruments, procedure and data analysis.

3.1 Characterization of the Institution and Participants

This research was carried out in a private school in Ambato city. This institution owns an English program based on levels of language proficiency. In order to in-crease the acquisition of English skills, a literature project has been generated. This project includes a book report plan that can focus on any English skill, but writing is one of the priorities, as it has been experiencing drawbacks. Therefore, this new school project seeks to improve this skill, but no evaluation process has been previously defined. Consequently, teachers of the same level evaluate the writing work based on their personal criteria with different evaluation methods.

The research was developed with A2 level students due to the benefits of having direct access to them. Therefore, one group are sophomore while the other group is in their junior year of high school, but both are in the same English level. Unfortunately, their work is not being homogeneously evaluated, so far, any standard rubric has been used.

3.2 Research Approach

This study considered a quasi-experimental research that includes two groups where is not possible to randomize them, groups that already exist and are natural. These groups were called the experimental and comparison group [35]. Quasi-experimental research was applied to evaluate the efficacy of a treatment, especially in educational interventions, since it is an appropriate design to test whether an intervention causes changes in the performance of experimental groups [15, 36].

A quasi-experimental research has different types of designs, the non-equivalent group design was chosen in this study. This is like the pretest-posttest control group in an experimental investigation, although the difference is that a non-random assignment is part of the design [37]. Furthermore, in this design, subjects were considered in their natural environments rather than under controlled conditions in a laboratory [38].

3.3 Instruments

The measurement instrument used for the data analysis was an interview applied to the teacher of the sophomore class to find out the perspective of the formative assessment, previous knowledge about the use of rubrics during this type of assessment, and the current situation over assessing the written literature project in the A2 level class.

Likewise, an analytical rubric was designed that was verified by three English teachers to certify its validity. Four evaluation criteria were included: content (knowledge of the book), organization, language, and project requirements. Each of the categories uses a four-point scale scoring as follows: 1—Unacceptable, 2—Below Average, 3—Acceptable and 4—Excellent.

To statistically measure the differences between the comparison group and the experimental group, the frequency of errors and the differences in scores in the pre and post-test were used.

The pre-test was designed to assess the writing skills of students in their literature project. Therefore, this diagnostic test was performed with two groups of ten students each. During the third term of the 2020–2021 school year, the experimental group wrote book reports and the comparison group worked on emails regarding the books. The students were already familiar with both text types and the intention was to work on a written assignment where the students talked about the book, they have just read by paying attention to the project requirements.

Assessment for learning involves a diagnostic evaluation in order to know the strengths and weaknesses of students to provide solutions that facilitate learning [39]. Thus, the pre-test showed the students' problems based on the criteria included in the rubric.

During the fourth quarter of the same school year, the post-test was carried out where the experimental group (EG) used the formative assessment model to evaluate their work, which included five phases: planning, drafting, peer assessment, self-assessment and presentation of the final product. On the other hand, the comparison group (CG) did not have any intervention related to this training process.

As soon as the students in both classes finished reading the literature books, they wrote their book reports. The task included specific requirements to ensure similar conditions. This writing allowed calculating the performance of the students on predetermined criteria that were included in the rubric of this formative assessment model. Therefore, the application of this test allowed to evaluate the progress of the students [40] and to study the impact of the formative assessment with rubrics in the improvement of the writing skills of both groups.

3.4 Procedure and Data Analysis

This investigation had three phases: pre-test, intervention and post-test. The results of the students were compared to determine if the formative assessment process generated relevant differences in the scores between the comparison and experimental groups. The rubric was scored over 16 points. Since there were ten students per group, the total was over 160 points.

Phase 1. Pre-test Results. Only In this phase, a quantitative analysis was necessary for the pre-test results of the comparison group (CG) and the experimental group (EG). Therefore, the mean and standard deviation were calculated. The mean ($\bar{x}$) which is "another average of all the scores in a distribution was determined by adding up all the scores and then dividing this sum by the total number of scores" [41].

$$\bar{x} = \frac{\Sigma X}{n} \quad (1)$$

The standard deviation (SD) which is "the most useful index of variability is a unique number that represents the spread of a distribution, and as with the mean, all the scores of the distribution were used to calculate it." Furthermore, "The more spread-out scores are, the greater the deviation scores will be and hence the larger the standard deviation, and the closer the scores are to the mean, the less spread out they will be and hence the smaller the standard deviation" [41].

$$SD = \sqrt{\frac{\Sigma(x - \bar{x})^2}{n}} \tag{2}$$

SD: standard deviation
Σ: sum of
X: raw score
($\bar{x}$): mean
n: number of scores in the distribution

The writing tasks showed some difficulty for both groups (CG/EG) because the CG got 114 (71.25%) and the EG scored 104 (65%) out of the total score 160 ($X_n = 16 \times 10$). The CG total scores in every criterion go from 25 to 36 and the EG scores between 23 and 30. As soon as the statistical calculations of the CG and EG results were done, it showed a mean of 11.4 out of 16 for the CG with a SD equal to 3.23. Also, the mean EG was 10.4 with a standard deviation of 3.10. In the Table 1, it is shown the CG and EG standard deviation are almost similar and they are closed to the mean.

The means of errors frequency are presented for each criterion. The CG committed 46 (28.75%) errors out of 160 (100%) where the criterion with the most issues is the organization (1.5 frequency of errors), then content (1.4), language (1.3) and project requirements (0.4). On the other hand, the EG had 56 (35%) errors where organization included (1.7 frequency of errors), project requirements (1.5), language (1.4) and content (1) as shown in Fig. 1.

Table 1 CG/EG Pre-test results per criterion

Criteria	Pre-test					
	CG			EG		
	Score X_n	Mean ($\bar{x}$)	SD	Score X_n	Mean ($\bar{x}$)	SD
Content (Knowledge of story)	26	2.6	1.07	30	3.0	1.05
Organization	25	2.5	0.97	23	2.3	0.82
Language	27	2.7	0.67	26	2.6	0.70
Project requirements	36	3.6	0.52	25	2.5	0.53
Total	114	11.4	3.23	104	10.4	3.1

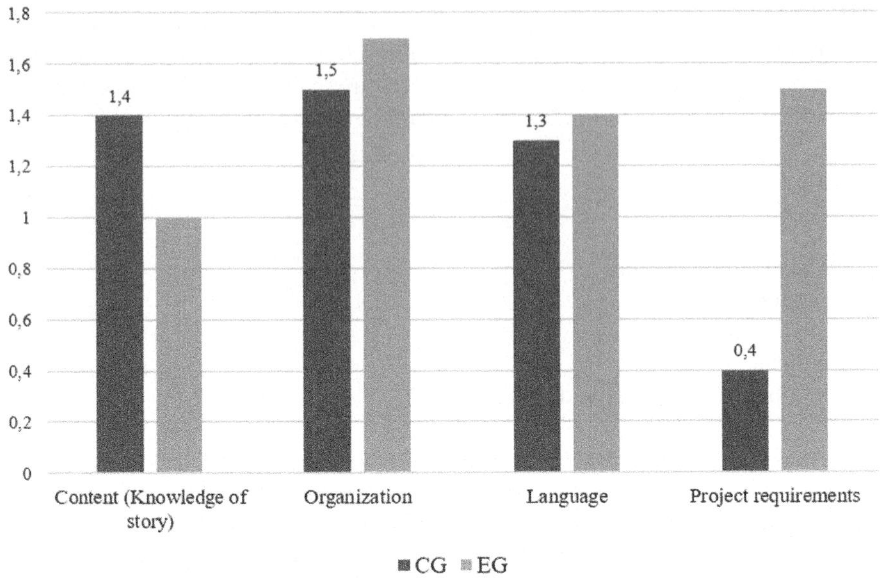

Fig. 1 CG/EG pre-test error frequency

Phase 2. Intervention. The treatment period was for two weeks. The experimental group had a regular practice with the formative assessment model by becoming familiar with the rubric. On the other hand, the comparison group was not included in this phase.

The first session helped students understand the rubric for book reports and its importance. Students wrote their drafts of the book report according to the rubric criteria. In the second session, students worked on the peer assessment process anonymously. An example was presented for a better understanding of the process.

In the last session the objective and the process of formative assessment were explained. Students reviewed their peers' comments, internalized the comments, worked on the final product and checked it against the rubric again.

Phase 3. Post-test Results. For the CG/EG post-test results, the same statistical process was used. The CG scored 120 (75%) out of the total scores 160 (100%). On the other hand, the EG got 142 (88.75%) and missed 18 (11.25%). In Table 2, it is shown the meaningful progress of EG. The CG scores range from 26 to 36 while the EG the scores vary from 30 to 40. The EG mean is 14.2 including a standard deviation of 1.95. The CG/EG standard deviation are low which means the scores are close to the total group mean.

Table 2 CG/EG post-test results by criteria

Criteria	Pos-test					
	CG			EG		
	Score X_n	Mean ($\bar{x}$)	SD	Score X_n	Mean ($\bar{x}$)	SD
Content (Knowledge of story)	31	3.1	0.74	40	4.0	0.00
Organization	26	2.6	0.84	30	3.0	0.94
Language	27	2.7	0.82	35	3.5	0.53
Project requirements	36	3.6	0.52	37	3.7	0.48
Total	120	12	2.92	142	14.2	1.95

The CG committed 40 (25%) errors out of 160 (100%). The most problematic criterion is the organization with 1.4 frequency of errors, then language (1.3), the content (0.9) and project requirements (0.4). On the contrary, EG had 1 frequency of errors in organization, language (0.5), project requirements (0.3) and no errors in content as expressed in Fig. 2.

The post-test results show a significant decrease in the EG students' performance errors while CG error frequency presents almost the same results of the pre-test (see Table 2). For CG, the most problematic criteria are still organization and language. In contrast, EG students seemed to have overcome their issues as presented in Fig. 2.

4 Results

This section includes the analysis of the results based on the researcher's proposal considering the objectives of the investigation. In addition, it includes the validation of the results.

4.1 Analysis, Interpretation and Validity of the Results

This section presents the differences between the results collected before and after the intervention. These data come from the pre-test results of the rubric that included the criterion: content, language, organization and requirements of the project; as well as the post-test results of the task carried out after the intervention that included the use of the same rubric through a formative assessment process with the experimental group.

Mean scores and the standard deviation of the pre and post-tests from the comparison and experimental groups are achieved using the t test. These are made up of the pre and post-test rubric scores of each group. CG and EG performance are

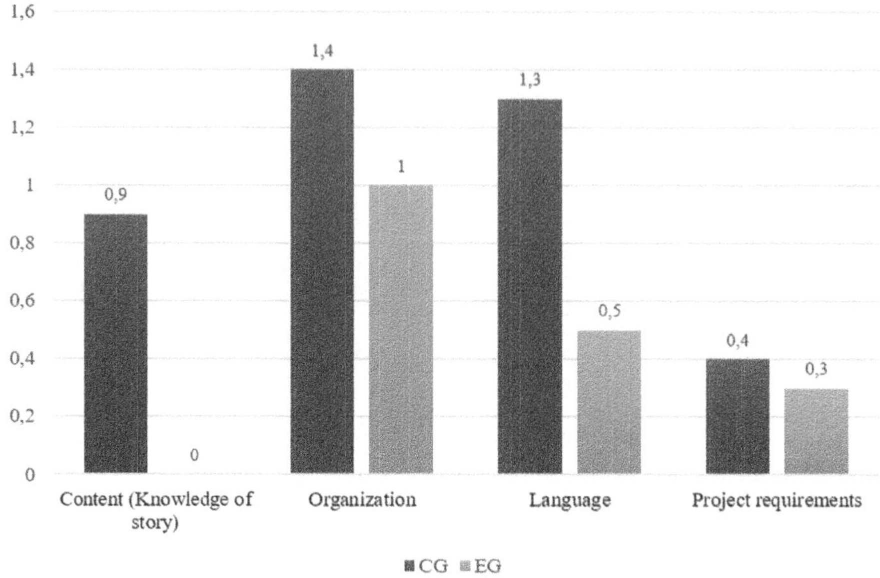

Fig. 2 CG/EG post-test frequency of errors

illustrated by bar graphs. In addition, the students' perception over the intervention is presented in a pie graph based on the results of a survey conducted at the end of the process. Likewise, descriptive statistics are used to analyze information regarding the intervention.

4.2 The Performance of the Comparison and Experimental Group

This study explores the difference between CG and EG performance before and after the intervention proposal and consists of working on a written task while applying a formative evaluation model that includes a standard rubric. To show the effectiveness of the intervention, the pre and post scores from the students during the written task are presented below to demonstrate the performance's difference among both groups as indicated by the mean difference of the groups that were tested (see Table 3).

The results show an improvement in the scores in the two groups, although "the considerable increase in scores" of the EG students is notable. As presented in Table 3, a great growth in the EG scores on the written task is evident. The EG increased its mean by 3.8, going from 10.4 to 14.2 in the post-test. Figure 3 shows the almost identical performance of the two groups in the pre-test, although a significantly different result in the post-test.

Table 3 Mean of the comparison and experimental group

		CG	EG	Mean difference
		Mean	Mean	
Written task	Pre-test	11.4	10.4	−1
	Post-test	12.0	14.2	2.2
	Mean	0.6	3.8	

Table 4 Test of difference between pre/post-test mean of the EG

	Mean	SD	Computed t-value	Tabular t-value	Decision	Interpretation
Pre-test	10.4	3.10	5.019	1.833	Reject Ho	Significant
Post-test	14.2	1.95				

4.3 *Differences in the Performance of the Written Activity Before and After the Intervention in the EG*

There is a significant difference in the scores of the participants during the performance of the written task before and after the implementation of the formative assessment process as shown in Table 4.

The group has a mean score of 10.4 in the pre-test and a 14.2 mean score in the post-test. By focusing on a 0.05 level of significance the computed t-value was 5.019 which was higher than the tabular t-value of 1.833. This implies that there was significant gain in the pre-test and post-test mean scores of the EG after the application of the formative assessment model through a rubric. Therefore, the alternative hypothesis is accepted since a standard rubric for the book reports is helping teachers of the same level to maintain a homogeneous assessment due to the high scores that have been achieved.

Furthermore, it can be stated that A2 level students in the experimental group can continue to improve their writing performance by receiving more training sessions on how to handle the formative assessment process with an analytical rubric. This research provides more evidence to support the effectiveness of the formative assessment proposal since the rubric criteria achieved a significant increase in the post-test scores.

EG students obtain higher scores which seems to be the result of the application of the formative assessment process. For example, in the post-test, both groups are asked to write a book report with the same requirements and the same rubric for the assessment; but the experimental group uses the five phases of formative evaluation.

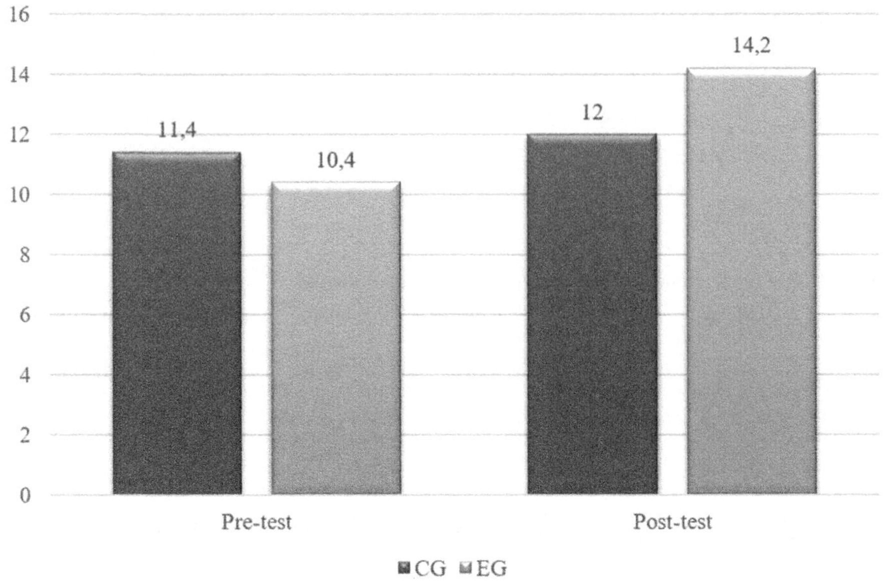

Fig. 3 Comparison and experimental groups mean

The findings of this research confirm some previous studies such as the one carried out by [16] who states that the use of recursive feedback can help students to be aware of what is expected of their performance, as well as how students can achieve the criteria required thanks to the existence of collaborative work. Furthermore, the importance of rubrics is confirmed since [8] he points out that rubrics are useful tools to support the self-monitoring process.

4.4 Perceptions of EG Students on the Application of the Formative Assessment Process Through a Standard Rubric

The measurement instrument used for data analysis was a survey in order to know the perception of EG students about the application of the formative assessment process through a standard rubric. The results of the survey carried out on the experimental group reflect a positive response from the students towards the formative evaluation proposal. As illustrated in Fig. 4, 70% of EG students strongly agree that a standardized assessment proposal helped them meet the criteria for the book report task. In addition, the remaining 30% of respondents agree that the formative assessment model is useful. These results mean that no one perceives the model as a worthless process.

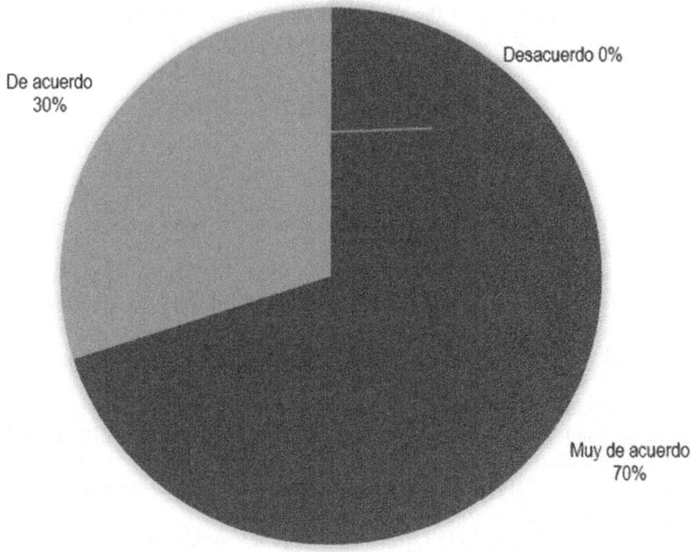

Fig. 4 EG students' perception of the application of the intervention

5 Conclusions

The analysis of the current assessment model for writing activities in this private school let to identify the deficiencies that prevent keeping a homogeneous assessment between the same levels of English with different teachers. For this phase, an interview was conducted that allowed diagnosing problems such as: the lack of use of a standard evaluation tool and an evaluation process that only focused on providing ratings instead of recursive feedback.

The design of a standard formative assessment model proposal was achieved through an analytical rubric for book reports that has four components, which was a key element during the development of the five phases that this model owns. This model achieved a standardized process that help students understood the parameters to be met to increase their performance in writing skills through recursive feedback.

The application of the formative assessment model allowed students to achieve the rubric criteria more easily by having a better understanding of how to achieve it. This model had a more viable execution during the peer assessment, as the self-assessment cost them more work since they are not aware of the importance of developing their reflective and critical skills when giving feedback to themselves.

The research was carried out in an online classroom environment, but in the future, it would be very interesting to replicate the study in a face-to-face environment to compare the results and analyze its impact. Likewise, self-assessment work during formative assessment would need further investigation since students can more easily comment their peer works rather than their own.

Acknowledgments The gratitude for the support provided in the research to the Technical University of Ambato, and to the Research and Development Directorate (DIDE).

References

1. Mohamadi Z (2018) Comparative effect of online summative and formative assessment on EFL student writing ability. Stud Educ Eval 59:29–40
2. Vasu KA, Mei Fung Y, Nimehchisalem V, Md RS (2020) Self-regulated learning development in undergraduate ESL writing classrooms: teacher feedback versus self-assessment. RELC J 1:1–15
3. Wiliam D, Black P (1996) Meanings and consequences: a basis for distinguishing formative and summative functions of assessment? Br Edu Res J 22(5):537–548
4. García CA, Castellanos EX, Rosero C, Sánchez C, García MV, Pérez F, et al (2017) CPPS on low cost devices for batch process under IEC-61499 and ISA-88. In: 2017 IEEE 15th international conference on industrial informatics (INDIN), pp 855–860
5. Kalantzis M, Cope B (2012) New learning: elements of a science of education. Cambridge University Press
6. Huapaya D, Marin D, Mauricio D (2021) TCO app: telemonitoring and control of pediatric overweight and obesity. In: Advances and applications in computer science, electronics and industrial engineering. Springer, Singapore, pp 79–97
7. Adachi C, Tai JHM, Dawson P (2018) Academics' perceptions of the benefits and challenges of self and peer assessment in higher education. Assess Eval High Educ 43(2):294–306
8. Wang W (2017) Using rubrics in student self-assessment: student perceptions in the English as a foreign language writing context. Assess Eval High Educ 42(8):1280–1292
9. Cockett A, Jackson C (2018) The use of assessment rubrics to enhance feedback in higher education: an integrative literature review. Nurse Educ Today 69:8–13
10. García MV, Irisarri E, Pérez F, Marcos M, Estévez E (2017) Engineering tool to develop CPPS based on IEC-61499 and OPC UA for oil & gas process. In: 2017 IEEE 13th international workshop on factory communication systems (WFCS), pp 1–9
11. Schmoker M (2002) Up and away. J Staff Dev 23(2):10–13
12. Burke K (2010) From standards to rubrics in six steps: tools for assessing student learning. Corwin Press
13. Chowdhury F (2019) Application of rubrics in the classroom: a vital tool for improvement in assessment, feedback and learning. Int Educ Stud 12(1):61–68
14. Wolf K, Stevens E (2007) The role of rubrics in advancing and assessing student learning. J Effect Teach 7(1):3–14
15. Gatcho ARG, Hajan BH (2019) Augmenting senior secondary ESL learners' reading skills through explicit instruction of metacognitive strategies. Online Submission 6(1):1–23
16. Kalantzis M, Cope B: Rubric-based peer and formative assessment. https://bit.ly/3ubZwJk
17. Ariyanti A (2016) The teaching of EFL writing in Indonesia. Dinamika Ilmu 16(2):263–277
18. Abas IH, Abd Aziz NH (2017) Classification of L2 writing process and writing strategies. Proc ICECRS 1(1):367–380
19. Graham S, Hebert M, Harris KR (2015) Formative assessment and writing: a meta-analysis. Elem Sch J 115(4):523–547
20. Ridhwan M (2017) Understanding formative and summative assessment for EFL teachers: theoretical reflections on assessment for learning. J-SHMIC J Engl Acad 4(1):40–50
21. Torres JO (2019) Positive impact of utilizing more formative assessment over summative assessment in the EFL/ESL classroom. Open J Mod Linguist 9(1):1–11
22. Scriven M (1967) The methodology of evaluation, vol 1. Social science education consortium
23. Black P (2015) Formative assessment-an optimistic but incomplete vision. Assess Educ Princ Policy Pract 22(1):161–177

24. Yüksel HS, Gündüz N (2017) Formative and summative assessment in higher education: opinions and practices of instructors. Eur J Educ Stud
25. Fathi J, Khodabakhsh MR (2019) The role of self-assessment and peer-assessment in improving writing performance of Iranian EFL students. Work 7(03)
26. Nogales R, Benalcazar F, Guilcapi J, Vargas J (2021) Selection of classifiers for hand gesture recognition using analytic hierarchy process: a systematic literature review. In: Advances and applications in computer science, electronics and industrial engineering. Springer, Singapore, pp 265–278
27. Wanner T, Palmer E (2018) Formative self-and peer assessment for improved student learning: the crucial factors of design, teacher participation and feedback. Assess Eval High Educ 43(7):1032–1047
28. Hislop J, Stracke E (2017) ESL students in peer review: an action research study in a university English for Academic Purposes course. University of Sydney Papers in TESOL, 12
29. Morales L, Silva D, Moreno V, Collantes S (2020) Symptomatology of musculoskeletal pain related to repetitive movements. preliminary study "post-harvest in floriculture companies". In: Advances and applications in computer science, electronics and industrial engineering. Springer International Publishing, pp 329–340
30. Ratminingsih NM, Marhaeni A, Vigayanti L (2018) Self-assessment: the effect on students' independence and writing competence. Int J Instr 11(3):277–290
31. Andrade HG (1997) Understanding rubrics. Educ Leadersh 54(4):14–17
32. Li C (2018) Constructing and applying rubrics in college-level EFL writing assessment in China. Syracuse University
33. Brookhart SM (2013) How to create and use rubrics for formative assessment and grading. Ascd
34. Ragupathi K, Lee A (2020) Beyond fairness and consistency in grading: the role of rubrics in higher education. In: Diversity and inclusion in global higher education. Palgrave Macmillan, Singapore, pp 73–95
35. White H, Sabarwal S, et al.: Quasi-experimental design and methods: methodological briefs-impact evaluation, no 8. https://bit.ly/3wGOk9g
36. Cook T, Campbell D (1979) Quasi-experimentation. Rand Mc-Nally, Chicago. CookQuasi-Experimentation
37. Dawson TE: A primer on experimental and quasi-experimental design. ERIC
38. Braddock K (2019) A brief primer on experimental and quasi-experimental methods in the study of terrorism. JSTOR
39. Nikolov M (2017) Students' and teachers' feedback on diagnostic tests for young EFL learners: implications for classrooms. In: García Mayo MP (ed) Learning foreign languages in primary school: research insights. Multilingual Matters, Bristol, pp 249–266. 10:9781783098118–014
40. Luce C, Kirnan JP (2016) Using indirect vs. direct measures in the summative assessment of student learning in higher education. J Scholarship Teach Learn 16(4):75–91
41. Fraenkel JR, Wallen NE, Hyun HH (2012) How to design and evaluate research in education. McGraw-Hill, New York. https://shortest.link/Ytw

Backtesting Recurrent Neural Networks with Gated Recurrent Unit: Probing with Chilean Mortality Data

Jorge M. Bravo and Vitor Santos

Abstract Understanding the survival prospects of a given population is essential in multiple research and policy areas, including public and private health care and social care, demographic analysis, pension systems evaluation, the valuation of life insurance and retirement income contracts, and the pricing and risk management of novel longevity-linked capital market instruments. This paper conducts a backtesting analysis to assess the predictive performance of Recurrent Neural Networks (RNN) with Gated Recurrent Unit (GRU) architecture in modelling and multivariate time series forecasting of age-specific mortality rates on Chilean mortality data. We investigate the best specification for one, two, and three hidden layers GRU networks and compare the RNN's forecasting accuracy with that produced by principal component methods, namely a Regularized Singular Value Decomposition (RSVD) model. The empirical results suggest that the forecasting accuracy of RNN models critically depends on hyperparameter calibration and that the two hidden layer RNN-GRU networks outperform the RSVD model. RNNs can generate mortality schedules that are biologically plausible and fit well the mortality schedules across age and time. However, further investigation is necessary to confirm the superiority of deep learning methods in forecasting human survival across different populations and periods.

Keywords Recurrent Neural Networks (RNN) · Gated Recurrent Unit (GRU) · Mortality modelling and forecasting · Pensions · Life insurance · Backtesting

1 Introduction

Understanding the dynamics of the survival prospects of a given population is vital in multiple research and policy areas, for instance, in public and private health care

J. M. Bravo (✉)
Universidade Nova de Lisboa - NOVA IMS & Université Paris-Dauphine PSL & MagIC & CEFAGE-UE & BRU-ISCTE, Lisbon, Portugal
e-mail: jbravo@novaims.unl.pt

V. Santos
Universidade Nova de Lisboa - NOVA IMS & MagIC, Lisbon, Portugal
e-mail: vsantos@novaims.unl.pt

© The Author(s), under exclusive license to Springer Nature Switzerland AG 2022
M. V. Garcia et al. (eds.), *Advances and Applications in Computer Science, Electronics, and Industrial Engineering*, Lecture Notes in Networks and Systems 433,
https://doi.org/10.1007/978-3-030-97719-1_9

planning (e.g., of preventive actions, of long-term care needs, of epidemiological episodes), in demographic analysis (e.g., population projections, ageing assessment), in pension systems design, reform and solvency analysis, in the pricing and risk management of novel longevity-linked capital market instruments (e.g., q-forwards, longevity bonds, longevity swaps), or the valuation of life insurance and private (individual, occupational) retirement income schemes [2, 6–10, 19, 24, 35].

To reduce or eliminate the short- and long-term solvency concerns in retirement income schemes created by continuous life expectancy increases, an upward trend in old-age dependency ratios, and insufficient economic growth, in recent decades most countries have responded with parametric (e.g., increasing the retirement age) or structural pension reforms, including the switch from pay-as-you-go (PAYG) defined benefit (DB) plans towards mandatory fully-funded defined-contribution plans (e.g., the 1981 reform in Chile), the introduction of individual complementary funded accounts (e.g., Romania, Hungary, Poland, China) and the transition from classic DB PAYG plans towards Non-Financial Defined Contribution (NDC) schemes (e.g., Sweden, Italy, Latvia) (OECD 2019). Another major pension reform trend has been to link earnings-related pension benefits to life expectancy developments. For instance, several countries (e.g., The Netherlands, Slovakia, Denmark, Portugal) automatically indexed their normal and early retirement ages to period life expectancy observed at retirement [4, 5, 17]. Others have opted to link the first pension benefit to demographic or sustainability factors (e.g., Finland, Portugal) or to transformation (annuity) factors (e.g., Italy, Norway). In France and Italy, the eligibility requirements for a full pension now depend on the number of contribution years linked to longevity trends. Private (and public) retirement income schemes introduced longevity-linked life annuities which differ from the traditional level or inflation-linked annuities in that benefits depend on the dynamics of actual against forecasted survival probabilities [1, 9, 16]. Retirement income providers are substantially exposed to non-diversifiable longevity (and interest rate) risk.

The increasing use of longevity markers in public policy and private practice led to a growing interest in the development of mortality and longevity forecasting methods. In the actuarial, financial, and demographic literature, the traditional approach to age-specific mortality forecasting is to pursue an empirical identification strategy by which, given some criteria (e.g., BIC information criteria) a unique discrete-time or continuous-time parametric or non-parametric stochastic mortality model is selected from a limited number of methods (see, e.g., [18, 26, 29, 31, 33, 37] and references therein). Empirical studies show, however, that there is no single universal mortality forecasting method that performs consistently better across populations. Because of that, and to account for model uncertainty, a recent competing research line recommends the use of model combinations (e.g., Bayesian Model Ensembles) of heterogeneous models [3, 13–15, 20, 32]. An emerging modelling approach is to use machine learning and deep learning methods to predict age-specific mortality rates [11, 12, 25, 28, 34].

This paper conducts a backtesting analysis to assess the predictive performance of Recurrent Neural Networks (RNN) with Gated Recurrent Unit (GRU) architecture in modelling and multivariate time series forecasting of age-specific mortality rates on

Chilean mortality data. RNNs are dynamic neural networks extending Feedforward Neural Network (FNN) to tackle network problems when handling time series. The algorithm incorporates an internal working memory (a loop) to step through learning problems that involve sequential input data. This overcomes the limitations of plain vanilla RNN which tend to exhibit reduced capacity to seize long-term trends dependencies in mortality data, generating poor forecasts. RNNs with GRU architecture is one of the most popular RNN structures that aim to solve the vanishing gradient problem when training networks using back-propagation.

We adopt a fixed horizon backtesting approach considering a common medium-term (10-year) lookforward window to train the networks and to produce forecasts of age-specific mortality rates by sex. It is well-known that neural networks are critically dependent on the choice of hyperparameters. To identify the optimum hyperparameters combination for the RNN-GRU neural network, we carried a preliminary fine-tuning round and empirically investigated the sensitivity of the forecasting results against alternative choices in one, two, and three hidden layers RNN-GRU models (number of hidden neurons, number of epochs, optimizer, batch size). To evaluate the predictive precision of RNN-GRU models, we selected a traditional principal component method, namely the Regularized Singular Value Decomposition (RSVD) model proposed by Huang et al. [29] and Zhang et al. [37]. The in-sample and out-of-sample forecasting error is measured by the Mean Squared Error (MSE) metric. The study deepens the preliminary investigations in Bravo [11] by examining the importance of the number of hidden neurons in the overall network performance. The empirical strategy is based on mortality (deaths classified by sex, age, calendar year, and birth cohort) and population (exposure-to-risk) data for Chile from 1992 to the latest available year (2017). The data source is the Human Mortality Database [30]. A common difficulty when modelling Latin American countries' longevity is the lack of sufficient past information on mortality trends disaggregated by individual age. Chile is the sole Latin American country available in HMD and this exercise could serve as a point of reference for life table preparation in this region.

The empirical results suggest that the performance of the RNN-GRU network depends on the number of hidden layers and the choice of the hyperparameters. The addition of hidden layers contributes to improving the model performance (minimizes the forecasting error) up to a certain point, after which further addition of layers reduces the model's accuracy because of overfitting. Two and three hidden layer networks outperformed the RSVD model in the validation dataset. The best RNN-GRU networks can produce consistent and biologically plausible mortality schedules across most ages of the human lifespan. We believe further investigation considering alternative RNN networks, different backtesting approaches (e.g., rolling fixed-length horizon backtests, jumping fixed-length horizon backtests), and alternative populations are however needed to confirm or reject the pilot results obtained in this study.

The remainder of the paper is organized as follows. In Sect. 2, we describe the data and model specification used in this study, namely the Recurrent Neural Networks with Gated Recurrent Unit structure, the RSVD model, the methods used to compute period life expectancy, and the learning data. Section 3 presents and briefly discusses the empirical results. Section 4 concludes and sets up the key areas for further research.

2 Data and Model Specification

2.1 *Recurrent Neural Networks with Gated Recurrent Unit Architecture*

Cho et al. [23] introduced Gated Recurrent Unit (GRU) to, like LSTM, solve the short-term memory problem of plain vanilla RNN. GRU is slightly less complex but is approximately as good as an LSTM performance-wise. GRU share many characteristics with the more complex structure of LSTM networks, namely the basic idea of using a gating mechanism to learn from long-term dependencies in the data and to decide which and how much past information on the time series should be forwarded to the output, but there are some important differences (Fig. 1). For instance, contrary to LSTM that have three gates (an input gate, a forget gate, and an output gate), a GRU has only two gates (a reset gate and an update gate) and does not have the output gate that determines how much to reveal of a cell. Second, a GRU does not include an internal memory differing from the exposed hidden state. The input and forget gates are connected by an update gate z_t and the reset gate r_t is applied directly to the previous hidden state.

Following Richman and Wüthrich [34], Zhang et al. [36] and Bravo [11], let $(x_1, \ldots, x_T)$ denote a time series of data (age-specific mortality rates) with components $x_t \in \mathbb{R}^{\tau_0}$ observed at times $t = 1, \ldots, T$. Our goal is to use this data as

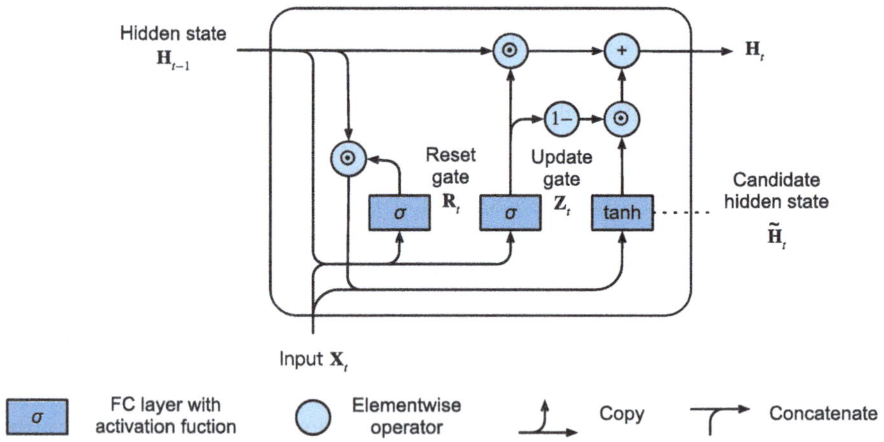

Fig. 1 Schematic representation of a Gated Recurrent Unit (GRU) block structure. *Source* Zhang et al. [36]

explanatory features to forecast a given output data $y \in \mathcal{Y} \subset \mathbb{R}$. Let $W \in \mathbb{R}^{\tau_0 \times h}$ and $U \in \mathbb{R}^{h \times h}$ denote the weight matrices for the input and the previous hidden-state result gates, respectively, with $h \in \mathbb{N}$ denoting the number of GRU blocks in a hidden layer. The unit receives as initial information flow the output from the previous GRU unit $h_{t-1} \in \mathbb{R}^h$ and the current input $x_t \in \mathbb{R}^{\tau_0}$. The reset gate controls how much of the previous hidden state we want to remember, capturing short-term dependencies in the time series. The update gate controls how much of the new state resembles the old one, i.e., it captures the long-term dependencies in the data sequences. We use the sigmoid activation function to engineer input values to be in the interval (0, 1). The RNN with GRU architecture can be formally described by the following set of equations:

$$r_t = \sigma(W_r x_t + U_r h_{t-1} + b_r), \tag{1}$$

$$z_t = \sigma(W_z x_t + U_z h_{t-1} + b_z), \tag{2}$$

$$\tilde{h}_t = \phi(W_h x_t + (r_t \circ h_{t-1}) U_h + b_h), \tag{3}$$

$$h_t = z_t \circ h_{t-1} + (1 - z_t) \circ \tilde{h}_t, \tag{4}$$

$$\sigma(x) = \frac{1}{1 + e^{-x}} \in (0, 1), \tag{5}$$

$$\phi(x) = \frac{e^x - e^{-x}}{e^x + e^{-x}} \in (-1, 1) \tag{6}$$

where $\sigma(\cdot)$ is the sigmoid function deciding how much input data should be used to update the memory of the network, $\phi(\cdot)$ is the hyperbolic tangent function controlling for the importance of the values which are passed, and b_r, b_z and b_h are biases. Equations (1) and (2) define, respectively, the reset gate (short term memory) and the update gate (long-term memory) mechanisms. Equation (3) describes the dynamics of the candidate hidden state, integrating the reset gate with the regular hidden state updating mechanism. Equation (4) determines the hidden state updating mechanism.

We empirically investigate different choices of the hyperparameters of one, two, and three hidden layers GRU networks (e.g., the number hidden neurons) considering for a fixed 10-year look-forward window, alternative values for the number of epochs, alternative optimizers, and the Mean Squared Error (MSE) as loss function. To calibrate the models, all RNN and principal component approaches are trained on the training set years $\mathcal{D}_1^{train} = \{t \in \mathcal{D}, 1992 \leq t \leq 2007\}$. Figure 2 illustrates the decomposition of the training data into the test and validation data as part of the backtesting exercise. The predictive accuracy was assessed on $\mathcal{D}_1^{test} = \{t \in \mathcal{D}, 2008 \leq t \leq 2017\}$ using the MSE, computed as

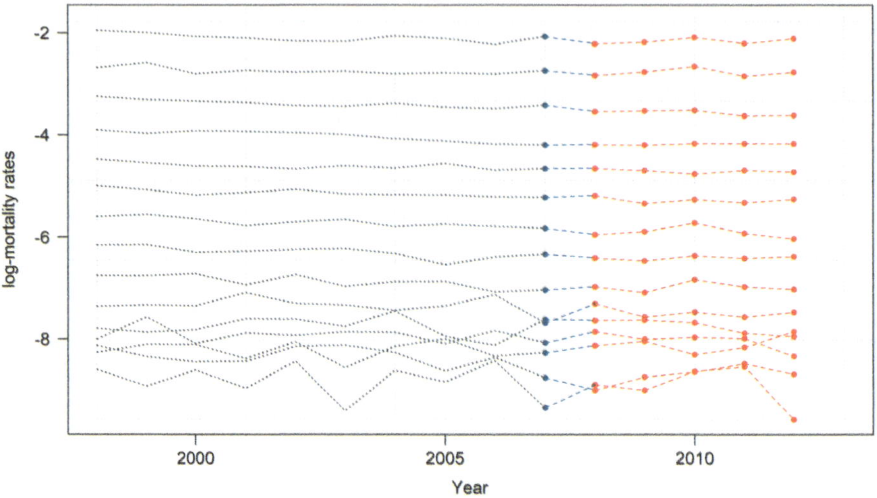

Fig. 2 Backtesting forecasting methods: training data illustration

$$MSE_g = \frac{1}{N}\sum\nolimits_{t=t_{min}}^{t_{max}}\sum\nolimits_{x=x_{min}}^{x_{max}}\left(\mu_{x,t,g} - \widehat{\mu}_{x,t,g}\right)^2, \tag{7}$$

with $N = (x_{max} - x_{min} + 1)(t_{max} - t_{min} + 1)$. Taking the multi-step 10-year forecasts of age-specific mortality rates, we then estimate period life expectancy as follows [4]:

$$e_{x,g}^{P}(t) := 0.5 + \sum\nolimits_{k=1}^{\omega-x} {}_{k}p_{x,t,g}, \tag{8}$$

where ${}_{k}p_{x,t,g}$ is the k-year survival probability for an individual aged x at time t, and ω is the highest attainable age in the life table, set at age 120 for all years and both sexes.

2.2 *Regularized Singular Value Decomposition Model*

This paper considers the Regularized Singular Value Decomposition (RSVD) model proposed by Huang et al. [29] and Zhang et al. [37] as a benchmark for the RNN-GRU architecture forecasting accuracy. The RSVD model forecasts age-specific mortality rates by extending the one-way functional principal component analysis (PCA) to two-way functional data. This is achieved by introducing regularization of both left and right singular vectors in the SVD of the data matrix. Formally, following Bravo et al. [20], let $D_{x,t,g}$ be the number of deaths observed at age x during calendar year t from the population (country, sex) g initially ($E_{x,t,g}^{0}$) or centrally ($E_{x,t,g}^{c}$) exposed-to-risk. Let $X = \left(m_{x,t}\right)_{n\times p}$ be a data matrix of mortality rates with n ages and p years.

The RSVD model assumes that the central mortality rate $m(x, t)$ can be explained in terms of both period t and age x effects as follows:

$$m(x, t) = \sum_{j=1}^{q} \lambda_j U_j(t) V_j(x) + \varepsilon(x, t), \tag{9}$$

where λ_q is the singular value, $U_i(\cdot)$ and $V_j(\cdot)$ are smooth functions of period and age, respectively, and $\varepsilon(x, t)$ is a mean zero random noise. The model is fitted iteratively. For instance, the first pair of singular vectors of X, $U_1(t)$ and $V_1(x)$, solves the following least-squares problem

$$(\widehat{\boldsymbol{u}}, \widehat{\boldsymbol{v}}) = \underset{(u,v)}{\operatorname{argmin}} \|\boldsymbol{X} - \boldsymbol{u}\boldsymbol{v}^T\|_F^2, \tag{10}$$

where $\|\cdot\|_F$ is the Euclidean norm of a matrix. The next pairs are extracted sequentially by removing the effect of the preceding pairs. For two-way functional data, the RSVD of Huang et al. [29] defines the regularized singular vectors as

$$(\widehat{\boldsymbol{u}}, \widehat{\boldsymbol{v}}) = \underset{(u,v)}{\operatorname{argmin}} \left\{ \|\boldsymbol{X} - \boldsymbol{u}\boldsymbol{v}^T\|_F^2 + \mathcal{P}_\pi(\boldsymbol{u}, \boldsymbol{v}) \right\}, \tag{11}$$

where $\mathcal{P}_\pi(\cdot)$ is a regularization penalty, defined as:

$$\mathcal{P}_\pi(u, v) = \pi_u u^T \Omega_u u \cdot \|v\|^2 + \pi_v v^T \Omega_v v \cdot \|u\|^2 + \pi_u u^T \Omega_u u \cdot \pi_v v^T \Omega_v v, \tag{12}$$

where Ω_u $(n \times n)$ and Ω_v $(p \times p)$ are symmetric and nonnegative definite domain-specific penalty matrices. Their objective is to balance the model goodness-of-fit against smoothness in mortality across age and time; π is a vector of regularization parameters optimally estimated based on generalized cross-validation criterion. To forecast mortality rates and derive confidence intervals, we use general univariate ARIMA processes to model the time functions $U_i(t)$.

2.3 *Mortality Data*

The Chilean mortality data used in this study are publicly available from the HMD. The datasets comprehend the number of recorded deaths together with the corresponding resident population counts (exposure-to-risk), classified by individual age x $\mathcal{X} = \{x \in \mathbb{N}, 0 \le x \le 110+\}$, calendar year $\mathcal{T} = \{t \in \mathbb{N}, 1992 \le t \le 2017\}$, year of birth $c = t - x$ and sex. Figure 3 plots the raw log-mortality rates $\widehat{m}_{x,t,g}$ by age in the range 0 to 100 years old and sex (left panel: Male; right panel: Female).

The longevity trends observed in Chile in the last 25 years are very similar to those observed in developed countries, with a clear downward trend in mortality rates at all ages for both sexes. Chile is one of the countries with the highest life expectancy at all ages on the American continent. The most significant longevity improvements

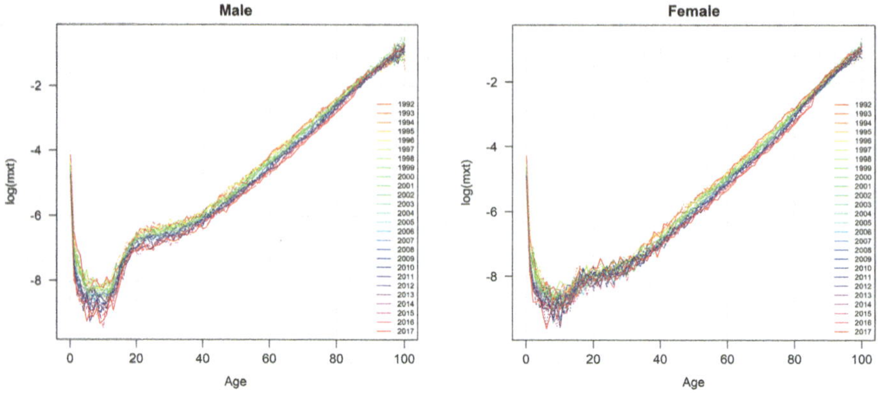

Fig. 3 Crude log-mortality rates by age and sex, Chile, 1992–2017

observed in Chile were recorded at younger ages, the exception being the accident hump in the male population in the age range 15–25 years old, and between women. Like in most countries of the world, Chilean women exhibit, on average, higher survival prospects than men of all ages. Increases in life expectancy at adult and old ages are also important. Figure 4 highlights the dynamics of longevity at all ages through a heatmap and a contour plot of the crude log-mortality rates by age, year, and sex. For both sexes, blue (orange) color represents low (high) mortality. We can observe that longevity gains have been shifting progressively and consistently from early ages to adult and old ages.

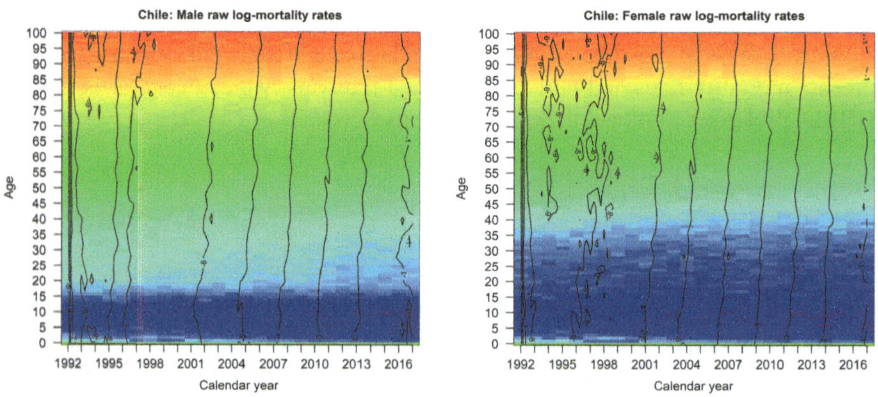

Fig. 4 Heatmap and contour plots of raw log-mortality rates by age and sex, Chile, 1992–2017
Notes for both sexes, blue (orange) color represents low (high) mortality

3 Results

3.1 Hyperparameter Calibration

We calibrate models for the male and female populations separately, i.e., we assume as usual that the mortality schedules of both sexes are independent. For very long-term forecasts, this may generate incoherence but given the lookforward window considered in this study, we believe it is a reasonable assumption.[1] For model training and learning, we randomly split the training data into a learning (test) dataset containing 80% of the data and a validation dataset comprehending the residual 20%. To calibrate the Stochastic Gradient Descent algorithm (SGD), we have experimented with different batch sizes and the number of epochs, controlling for the numbers of training samples processed before the model's internal parameters are updated and the number of complete passes through the training dataset. Based on the hyperparameter calibration, we finally opted to run 500 epochs (in batches of 100) of the SGD on the learning dataset. We have experimented with alternative optimizers (Nadam, Adamax, Adam,...) opting finally to use the Adaptive Moment Estimation (Adam) which revealed to be computationally efficient and more reliable, reaching a global minimum when minimising the cost function in training neural nets, and requiring reduced memory given the large nature of the problem in terms of data and number of parameters. For each model, the optimal calibration is identified by selecting the parameter combination with the lowest MSE loss in the test set.

For the one, two, and three hidden layers RNN-GRU architectures, we calibrated 6, 12, and 24 different networks $\mathcal{M}_j$, respectively, considering all possible combinations in the array $\mathcal{M}_j = \{\tau_0 = \{1, 3\}; \tau_1 = \{5, 10, 20\}; \tau_2 = \{10, 15\}; \tau_3 = \{5, 10\}\}$. The RNN-GRU networks with the lowest forecasting error were re-trained on $\mathcal{T}_1^{train}$, from which forecasts of age-specific mortality rates on $\mathcal{T}_1^{test}$ were produced. The model fitting and forecasting procedures have been implemented using a routine running on R software. Table 1 summarizes the average fitting and forecasting loss metrics for all the RNN-GRU hyperparameter combinations tested for the female population of Chile.[2]

We also report the run times for each model, measured in seconds on a personal laptop with Intel(R) Core(TM) i7-10510U CPU@2.30 GHz with 16 GB RAM. For the one hidden layer model (Panel A), the best performing network includes five hidden neurons, i.e., RNN-$GRU_1(\tau_0 = 1; \tau_1 = 5)$. For two hidden layer models (Panel B), the highest accuracy is obtained with the specification RNN-$GRU_2(\tau_0 = 1; \tau_1 = 20; \tau_2 = 15)$. Finally, for three hidden layer models (Panel C), the lowest forecasting error is obtained with the RNN-$GRU_3(\tau_0 = 1; \tau_1 = 5; \tau_2 = 15; \tau_3 = 5)$ model.

---

[1] For some examples on the joint modelling of both sexes' mortality schedules see, e.g., Hyndman et al. [31], Richman and Wüthrich [34] and Bravo [11].

[2] Due to space constraints, the results for the male population are not included in the main manuscript but are available from the authors upon request.

Table 1 RNN-GRU: fitting/forecasting loss metrics for different hyperparameter combinations

Parameters				MSE		CPU	Parameters				MSE		CPU
τ_0	τ_1	τ_2	τ_3	Fit	For.	time	τ_0	τ_1	τ_2	τ_3	Fit	For.	time
Panel A: $RNN - GRU_1(\tau_0;\ \tau_1)$													
1	5			0.54	2.74	35.5	3	10			0.69	4.41	61.2
3	5			1.31	9.29	67.8	1	20			0.53	3.02	39.9
1	10			1.02	5.75	46.2	3	20			0.46	2.84	36.7
Panel B: $RNN - GRU_2(\tau_0;\ \tau_1;\ \tau_2)$													
1	5	10		0.54	2.46	71.7	1	5	15		1.50	6.42	32.2
3	5	10		0.93	5.47	42.1	3	5	15		0.73	1.95	33.9
1	10	10		0.71	4.64	37.7	1	10	15		0.81	1.91	30.5
3	10	10		0.58	2.61	31.1	3	10	15		1.41	8.38	31.8
1	20	10		1.54	1.82	30.5	1	20	15		1.56	1.38	31.3
3	20	10		0.65	4.48	33.9	3	20	15		0.73	4.53	33.2
Panel C: $RNN - GRU_3(\tau_0;\ \tau_1;\ \tau_2;\ \tau_3)$													
1	5	10	5	0.55	3.31	38.3	1	5	10	10	1.65	7.37	42.3
3	5	10	5	0.73	2.52	42.4	3	5	10	10	0.52	3.68	42.7
1	10	10	5	0.58	3.69	41.8	1	10	10	10	0.69	4.47	43.7
3	10	10	5	0.57	2.63	42.7	3	10	10	10	0.79	4.56	45.5
1	20	10	5	1.03	1.95	43.8	1	20	10	10	0.51	3.33	44.4
3	20	10	5	0.55	2.81	46.1	3	20	10	10	0.51	3.79	45.7
1	5	15	5	0.99	1.74	41.9	1	5	15	10	0.68	4.54	43.0
3	5	15	5	1.34	6.82	43.9	3	5	15	10	1.02	6.63	43.9
1	10	15	5	2.66	9.70	42.7	1	10	15	10	0.54	3.69	44.4
3	10	15	5	0.54	2.94	45.5	3	10	15	10	1.12	1.92	48.3
1	20	15	5	0.59	2.83	45.4	1	20	15	10	0.57	3.23	50.2
3	20	15	5	1.93	2.54	47.7	3	20	15	10	0.55	3.15	53.8

Notes τ_0 . τ_1. τ_2 and τ_3 denote the number of hidden neurons in the hidden GRU layers; Average results for the female population considering for 10-year forecasting horizons. MSE values in 10^{-5}. CPU time in seconds

Figure 5 illustrates the learning strategy on the best RNN-GRU networks, plotting the early stopping in-sample and the out-of-sample loss on the test dataset. We can observe that the addition of hidden layers helps improve the model performance up to a certain point, after which further addition of layers reduces the model's performance, showing signs of clear overfitting.

Table 2 summarizes the predictive accuracy metrics of the best RNN-GRU models tested in this study and of the RSVD used as the benchmark.

The empirical results suggest that the performance of the RNN-GRU_2 network depends on the number of hidden layers and the choice of the hyperparameters, chiefly the number of hidden neurons per layer. The two and three hidden layer networks outperformed the RSVD model in the validation dataset. The extra computation time required by some RNN networks is not sufficient, at this stage, to discard this model as a valid option for longevity risk modelling.

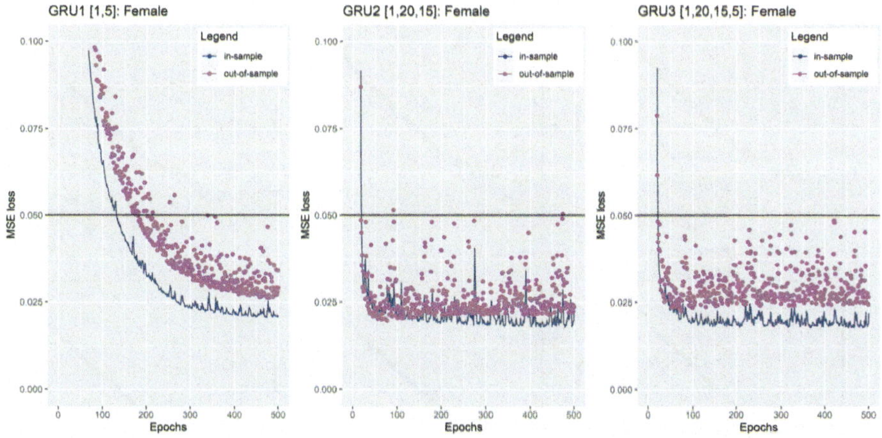

Fig. 5 Best RNN-GRU architectures with one, two, and three hidden layers, women

Table 2 Forecasting accuracy metrics of the alternative RNN and GAPC models tested

Model	GRU: #hidden neuron parameters				MSE		CPU
	τ_0	τ_1	τ_2	τ_3	Fit	Forecast	Time
RNN-GRU$_1$	1	5			0.54	2.74	35.5
RNN-GRU$_2$	1	20	15		1.56	1.38	31.3
RNN-GRU$_3$	1	5	15	5	0.99	1.74	41.9
RSVD					1.19	1.85	21.2

Notes Results obtained considering 10-year look-forward periods; MSE values in 10^{-5} units. CPU time in seconds

3.2 Forecasts of Period Life Expectancy

Figure 6 illustrates the observed (black dots) and the forecasted (orchid color line) log-mortality rates by individual year generated by the best RNN-GRU for the female population of Chile.

Figure 7 represents the corresponding (aggregate) forecasts using the RSVD model. The best RNN-GRU networks were able to produce consistent and biologically plausible mortality schedules across the entire lifespan spectrum, including at younger ages where the volatility is normally higher. Finally, to illustrate the application of forecasts of age-specific mortality in life table computation, we exhibit in Fig. 8 the estimates of the period life expectancy computed at birth ($x = 0$) and the benchmark retirement age of 65 ($x = 60$) for Chilean women. The vertical cyan line marks the split between the training and validation datasets. We can observe that the average remaining lifetime in the country has been increasing consistently over the last quarter of a century, from 77.28 (17.57) years at birth (age 65) in 1992 to 81.70 (20.57) years in 2017. Similar results (at a lower level) were obtained for the male

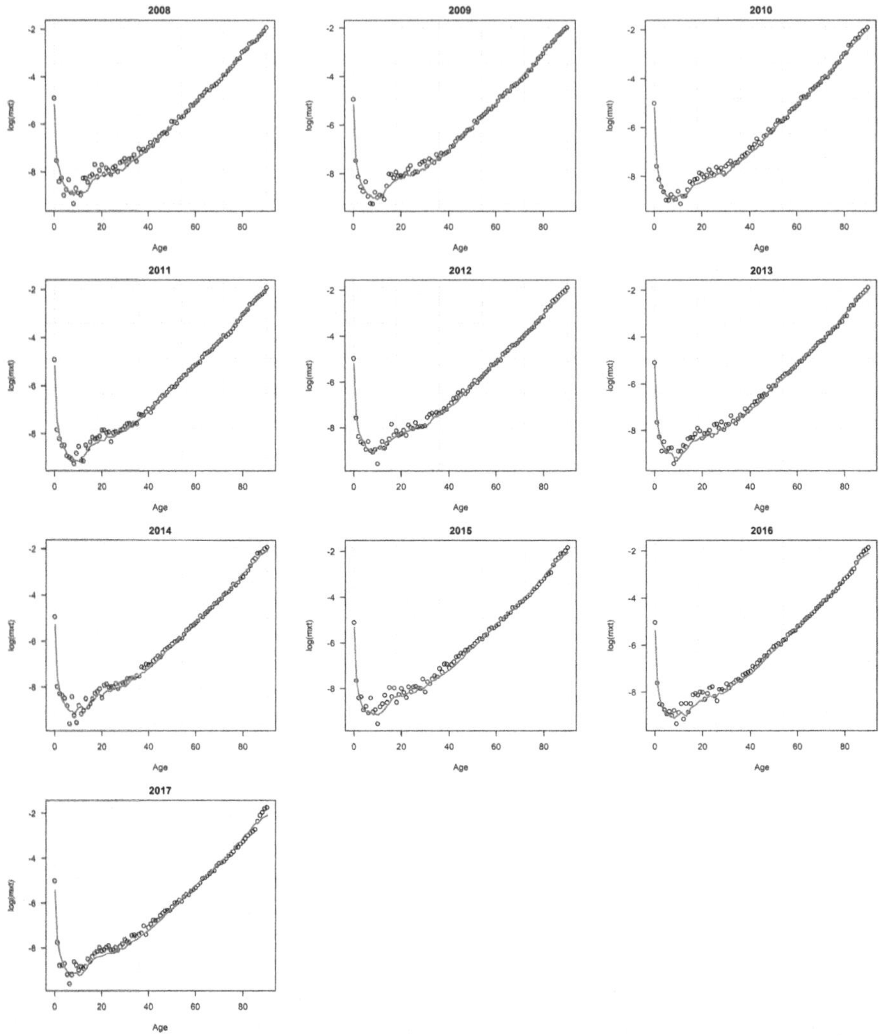

Fig. 6 Best RNN-GRU_2 model: forecasts of log-mortality rates by age and year. Women

population. The increasing life expectancy challenges the solvency of public and private pension schemes and has important implications in individual consumption, saving and, labour market decisions.

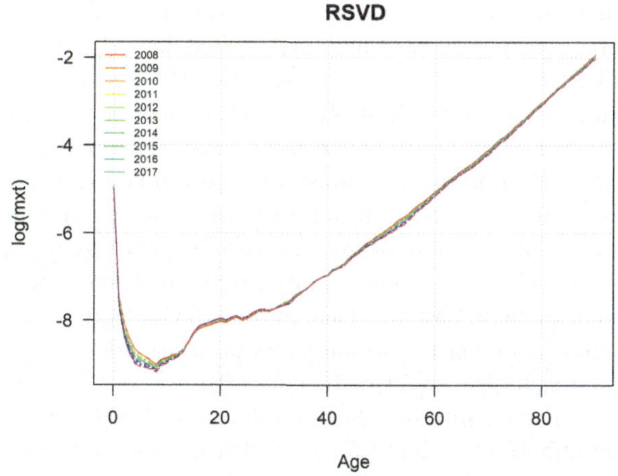

Fig. 7 RSVD model: forecasts of log-mortality rates by age, women

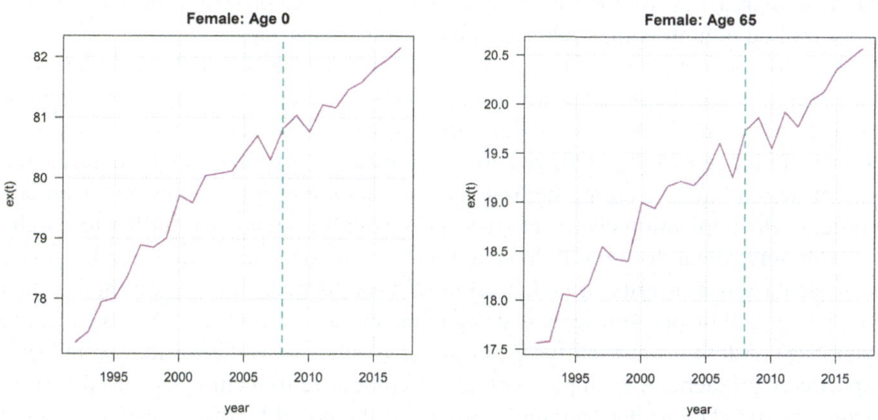

Fig. 8 RNN-GRU2: estimates of the life expectancy at birth and at the age of 65, women

4 Conclusion

Model selection and model ensembles are currently the two main competing approaches when modelling and forecasting age-specific mortality for actuarial, financial, and demographic applications. The pool of individual candidate models includes generalized age-period-cohort stochastic mortality models, principal component methods, and smoothing approaches. More recently, attempts have been made to use machine learning and deep learning methods for multivariate forecasting. This paper follows this latter research trend and conducts a backtesting analysis to assess the predictive performance of RNN with GRU architecture in multivariate time series forecasting of age-specific mortality rates on Chilean mortality data. We compare the RNN performance with that offered by traditional principal component methods (RSVD model), widely used in actuarial and demographic studies. The empirical results obtained on a limited dataset suggest that the forecasting accuracy of RNN-GRU networks outperforms the RSVD model. However, the results also suggest that the RNN-GRU predictive accuracy is critically dependent on hyperparameter calibration and that adding extra hidden layers may lead to model overfitting. This is important since most longevity-linked securities and insurance contracts are typically very long-term contracts with pricing fixed at contract initiation and without the possibility of revision if observed longevity trends deviate from assumed improvements. Conceptual uncertainty (model risk) must be incorporated into pricing, risk management, and inference purposes. One way of doing this is to combine heterogeneous stochastic mortality models using, for instance, a Bayesian model ensemble approach (see, e.g., Bravo et al. [20]).

The mortality schedules produced by RNNs are biologically plausible, which is an important criterion for a good stochastic mortality model, and consistent across all ages of the human lifespan. This is an advantage when compared to some of the classical approaches to mortality forecasting which have proved to perform poorly when applied to both young, adult, and oldest-old age groups.

However, to be able to confirm or reject the claim that RNN models can be added to the toolkit of researchers and professionals working in longevity risk management or public policy analysis, we believe that further investigation is required to investigate extensively the sensitivity of the results to, for instance, hyperparameter choices, the type of network architecture, the lookback and lookforward window, the accuracy metric, or the population characteristics. This is on the agenda for further research.

Extending research to multiple state mortality models accounting for longevity heterogeneity is a priority to tackle actuarial fairness considerations in both social policy (e.g., public pension scheme design) and private insurance contracts (e.g., life insurance). Adopting non-uniform policy approaches considering the ex-ante life expectancy gradient, e.g., implementing differential retirement ages, sustainability factors, or social contribution rates, are some of the possible reform avenues aiming at reducing the redistributive distortions created by longevity heterogeneity [15]. This can be done by using, e.g., multi-state models, which have been successfully applied to other problems such as long-term care and credit risk modelling [21, 22].

Despite the high predictive power of RNN-GRU against RSVD models found in this paper, the unsatisfactory interpretability of neural networks in mortality forecasting is still one of the key obstacles of deep learning techniques in its wide acceptance by the financial industry. For example, the European Union regulations provide customers impacted by tailored pricing algorithms the right to ask and receive an explanation for why a model makes a particular decision under specific circumstances, and the chance to benefit from fair algorithmic competition. Auditors and supervisors need to understand and approve internal risk models. This creates a big challenge when communicating deep learning algorithms and results.

Acknowledgements The authors express their gratitude to the editors and the anonymous referees for his or her careful review and insightful comments, which helped strengthen the quality of the paper. The authors were supported by Portuguese national funds through FCT under the project UIDB/04152/2020—Centro de Investigação em Gestão de Informação (MagIC) and grant UIDB/00315/2020 (BRU-ISCTE).

References

1. Alho J, Bravo JM, Palmer E (2013). Annuities and life expectancy in NDC. In: Holzmann R, Palmer E, Robalino D (eds) Nonfinancial defined contribution pension schemes in a changing pension world, vol 2. Gender, Politics, and Financial Stability. World Bank Publications, pp 395–436. https://doi.org/10.1596/9780821394786_CH22
2. Ashofteh A, Bravo JM (2020) A study on the quality of Novel Coronavirus (Covid-19) official datasets. Stat J IAOS 36(2):291–301. https://doi.org/10.3233/SJI-200674
3. Ashofteh A, Bravo JM (2021) Life table forecasting in COVID-19 times: an ensemble learning approach. In: CISTI 2021 - 16th Iberian conference on information systems and technologies, pp 1–6. https://doi.org/10.23919/CISTI52073.2021.9476583
4. Ayuso M, Bravo JM, Holzmann R (2021) Getting life expectancy estimates right for pension policy: period versus cohort approach. J Pension Econ Finance 20(2):212–231. https://doi.org/10.1017/S1474747220000050
5. Ayuso M, Bravo JM, Holzmann R, Palmer E (2021) Automatic indexation of pension age to life expectancy: when policy design matters. Risks 9(5):96. https://doi.org/10.3390/risks9050096
6. Blake D, Cairns AJG, Dowd K, Kessler AR (2019) Still living with mortality: the longevity risk transfer market after one decade. Br Actuar J 24:1–80
7. Bravo JM (2016) Taxation of pensions in Portugal: a semi-dual income tax system. CESifo DICE Rep J Inst Comp 14(1):14–23
8. Bravo JM (2019) Funding for longer lives: retirement wallet and risk-sharing annuities. Ekonomiaz 96(2):268–291
9. Bravo JM (2021) Pricing participating longevity-linked life annuities: a Bayesian Model Ensemble approach. Eur Actuar J. https://doi.org/10.1007/s13385-021-00279-w
10. Bravo JM (2021b) Pricing survivor bonds with affine-jump diffusion stochastic mortality models. In: 2021b the 5th international conference on E-commerce, E-business and E-government (ICEEG 2021). Association for Computing Machinery (ACM), New York, pp 91–96. https://doi.org/10.1145/3466029.3466037
11. Bravo JM (2021c) Forecasting longevity for financial applications: a first experiment with deep learning methods. In: Kamp M et al (eds) Machine Learning and Principles and Practice of Knowledge Discovery in Databases. ECML PKDD 2021. Communications in Computer and Information Science, vol 1525. Springer, Cham. pp. 232–249. https://doi.org/10.1007/978-3-030-93733-1_17
12. Bravo JM (2021d) Forecasting mortality rates with Recurrent Neural Networks: a preliminary investigation using Portuguese data. In: CAPSI 2021d proceedings (Atas da 21ª Conferência da Associação Portuguesa de Sistemas de Informação 2021), in press
13. Bravo JM, Ayuso M (2020) Mortality and life expectancy forecasts using Bayesian model combinations: an application to the Portuguese population. RISTI - Revista Ibérica de Sistemas e Tecnologias de Informação E40:128–144. https://doi.org/10.17013/risti.40.128-145
14. Bravo JM, Ayuso M (2021a) Forecasting the retirement age: a Bayesian model ensemble approach. In: Advances in intelligent systems and computing, AIST, vol 1365. Springer, Cham, pp 123–135 [2021a World Conference on Information Systems and Technologies, WorldCIST 2021]. https://doi.org/10.1007/978-3-030-72657-7_12
15. Bravo JM, Ayuso M (2021b) Linking pensions to life expectancy: tackling conceptual uncertainty through bayesian model averaging. Mathematics 9(24): 3307, p. 1-27. https://doi.org/10.3390/math9243307

16. Bravo JM, El Mekkaoui de Freitas N (2018) Valuation of longevity-linked life annuities. Insur Math Econ 78:212–229
17. Bravo JM, Herce JA (2022) Career breaks, broken pensions? Long-run effects of early and late-career unemployment spells on pension entitlements. J Pension Econ Financ 21(2): 191–217.https://doi.org/10.1017/S1474747220000189
18. Bravo JM, Nunes JPV (2021) Pricing longevity derivatives via Fourier transforms. Insur Math Econ 96:81–97
19. Bravo JM, Silva CM (2006) Immunization using a stochastic process independent multifactor model: the Portuguese experience. J Bank Finance 30(1):133–156
20. Bravo JM, Ayuso M, Holzmann R, Palmer E (2021) Addressing the life expectancy gap in pension policy. Insur Math Econ 99:200–221. https://doi.org/10.1016/j.insmatheco.2021.03.025
21. Chamboko R, Bravo JM (2016) On the modelling of prognosis from delinquency to normal performance on retail consumer loans. Risk Manage 18(4):264–287
22. Chamboko R, Bravo JM (2020) A multi-state approach to modelling intermediate events and multiple mortgage loan outcomes. Risks 8:64
23. Cho K, Van Merriënboer B, Bahdanau D, Bengio Y (2014) On the properties of neural machine translation: encoder-decoder approaches. arXiv preprint arXiv:1409.1259
24. Coughlan GD, Epstein D, Honig P (2007) Q-forwards: derivatives for transferring longevity and mortality risks. Working paper. J. P. Morgan Pension Advisory Group, London
25. Deprez P, Shevchenko P, Wüthrich M (2017) Machine learning techniques for mortality modeling. Eur Actuar J 7:337–352
26. Dowd K, Cairns A, Blake D, Coughlan G, Epstein D, Khalaf-Allah M (2010) Backtesting stochastic mortality models. North Am Act J 14(3):281–298
27. Hochreiter S, Schmidhuber J (1997) Long short-term memory. Neural Comput 9(8):1735–1780
28. Hong WH, Yap JH, Selvachandran G et al (2021) Forecasting mortality rates using hybrid Lee-Carter model, artificial neural network and random forest. Complex Intell Syst 7:163–189
29. Huang JZ, Shen H, Buja A (2009) The analysis of two-way functional data using two-way regularized singular value decompositions. J Am Stat Assoc 104(488):1609–1620
30. Human Mortality Database (2021) University of California, Berkeley (USA), and Max Planck Institute for Demographic Research (Germany)
31. Hyndman RJ, Booth H, Yasmeen F (2013) Coherent mortality forecasting: the product-ratio method with functional time series models. Demography 50(1):261–283
32. Kontis V, Bennett J, Mathers C, Li G, Foreman K, Ezzati M (2017) Future life expectancy in 35 industrialised countries: projections with a Bayesian model ensemble. Lancet 389(10076):1323–1335
33. Lee RD, Carter L (1992) Modeling and forecasting U.S. mortality. J Am Stat Assoc 87:659–671
34. Richman R, Wüthrich M (2019) Lee and Carter go machine learning: recurrent neural networks. SSRN https://ssrn.com/abstract=3441030. Accessed 10 Jan 2021
35. Simões C, Oliveira L, Bravo JM (2021) Immunization strategies for funding multiple inflation-linked retirement income benefits. Risks 9(4):60. https://doi.org/10.3390/risks9040060
36. Zhang A, Lipton Z, Li M, Smola A (2021) Dive into deep learning. arXiv:2106.11342.
37. Zhang L, Shen H, Huang JZ (2013) Robust regularized singular value decomposition with application to mortality data. Ann Appl Stat 7(3):1540–1561

Utility of Computer Hardware Recycling Technique for University Learning: A Systematic Review

Jefferson Osorio-Carlozama and Joe Llerena-Izquierdo

Abstract Technological recycling contributes to the reduction of environmental pollution. It is urgent, an adequate knowledge of techniques on the management of waste management of electronic devices and the environmental effect it can cause in society. The results obtained contributed to the knowledge in the following areas: Different ways that technological resources are used, access allowed to various technological resources to enhance availability and extension of the useful life of technological work tools. The continuous use of technological equipment entails responsibility in the adoption of techniques for its proper disposal or reuse. The objective of this study was to determine the usefulness of the technique of recycling computer equipment in university learning. The deductive analytical methodology was used with a quantitative approach through the technique of previous scientific literature review. In addition, a survey was applied to a group of first year engineering students at the Salesian Polytechnic University, to determine the knowledge about the reuse of technological resources and the impact of electronic waste in the environment where they live, being greater the benefits of having a learning that manages to channel correctly, a correct process of waste management in the environment of educational influence. It is concluded that having access to knowledge of various techniques for recycling technological media improves the selection of recycling techniques, resulting in commitment to proper disposal and management of waste.

Keywords E-recycling techniques · E-waste management · e-waste

J. Osorio-Carlozama · J. Llerena-Izquierdo (✉)
Universidad Politécnica Salesiana, Guayaquil, Ecuador
e-mail: jllerena@ups.edu.ec
URL: https://gieaci.blog.ups.edu.ec/

J. Osorio-Carlozama
e-mail: josorioc@est.ups.edu.ec

GIEACI Research Group and GieTICEA Educational Innovation Group, Guayaquil, Ecuador

© The Author(s), under exclusive license to Springer Nature Switzerland AG 2022

M. V. Garcia et al. (eds.), *Advances and Applications in Computer Science, Electronics, and Industrial Engineering*, Lecture Notes in Networks and Systems 433,
https://doi.org/10.1007/978-3-030-97719-1_10

1 Introduction

With the advent of technology, the growth of Electrical and electronic equipment (EEE) components has been increasing [1]. Different products that improve the quality of life of people such as smartphones, high-performance laptops, devices for health, entertainment, home or for the development of a company have been having a wide use and penetration in everyday life and in turn, the wear produced on them is evidenced in the increase and its measurement of Waste Electrical and Electronic Equipment (WEEE) generated by a population [2]. Not having the information on the amount of electronic waste that is collected, is to ignore the impact that a medium-term threat can lead to a global problem [3]. Faced with this situation, the recycling of computer elements, contributes to specify and manage more responsibly the choice of one device or another [4]. The technologies of previous generations show that new products with better characteristics related to processing capacity, with benefits for the user related to quality, as well as the generation of specialized knowledge with the improvement of the user experience, are remarkably accepted by the new generations of societies [5–7]. The treatment, management or recycling of electronic products in the world is considered as part of the environmental responsibility of every citizen [8, 9]. This intention to do something with the management of electronic waste has been increasing in different developed countries and is supported by non-governmental organizations (NGOs), governments and multinationals [10].

1.1 The Impact of E-Waste from the Point of View of Educational Institutions

Continuous improvement projects in educational institutions worldwide have opted to use technological resources in different areas to replace chalk and paper [11, 12]. The useful life of technological equipment in such establishments, due to continuous use and wear and tear during comprehensive digital learning through the use of the Internet, allows the production of technological support materials for learning [13, 14]. The obsolescence of technological waste and other types of elements in disuse, damage or defect, turn the study equipment, a set of parts and devices that generate mounds that occupy work spaces, similarly, occurs in companies and homes, with the use of reused components based on polychlorinated biphenyls (PCB), which were discarded at the end of their useful life (i.e., electronic waste) and after treatment, are sold as new on the black market [15]. The consumption of electronic products is vertiginous and companies respond to consumption with new products, useful for the educational area [16]. Technological equipment and tools contain toxic elements that should not be discarded like other waste, as is the case of single-use plastic waste or waste used in the industrial area, which is also included in programs to help minimize waste in cities and municipalities [17].

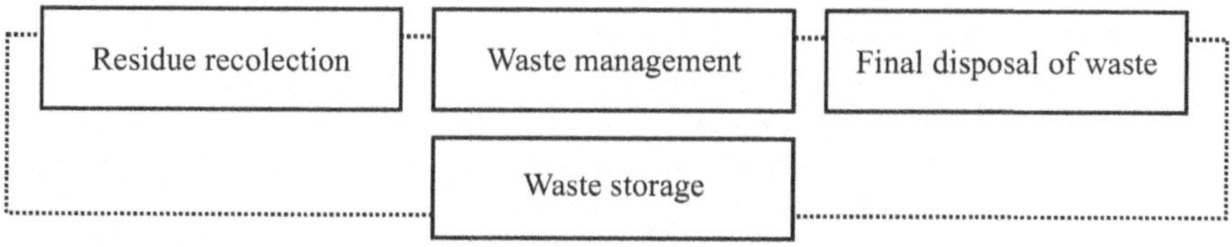

Fig. 1 Waste management cycle

Educational institutions, through plans for the use of resources for an environmentally friendly relationship with the environment, in the short and long term with the change of attitude and awareness of the importance of improving the quality of life with the proper use of technological resources [18, 19]. However, because sustainability is related to the responsibility of the profession to-wards society, this allows waste to be directed and converted into useful products [20].

This makes engineers individually responsible for considering sustainability in their work, framing sustainable engineering as a macro ethical issue [21, 22]. Reuse is a way to decrease the cost of the production cycle for environmental care, by reducing the volume of artificial toxic waste [23].

1.2 Recycling or E-Waste Management Techniques

It is evident that IT departments in educational institutions have taken advantage of the recurrence of technical maintenance with recycling and are responsible for the application of recycling or e-waste management techniques. For example, the percentage of each of the stages in the life phases of computer equipment in an educational institution goes through: Technical Department, Acquisition, Depreciation, Obsolescence, (see Fig. 1). Thus, new technological equipment replaces old equipment, which in turn enters a stage of reallocation of its use or is discarded [24].

The waste generated by electronic and electrical equipment is a local and international problem, the solid plastic elements, such as those chemicals frequently found in single-use disposable parts, pollute developing countries, which are the largest consumers of devices in the technological market [25].

The reduction of plastic waste, focusing its efforts on preventing waste from being sent to landfills, due to the high cost, high demand for space, generating rejection by society for the environmental damage that this entails. These initiatives have led to a significant reduction at the international level [26].

The implementation of technological recycling treatments would have a high impact on the economy. On the one hand, mechanical recycling would make it possible to use the discarded elements, which is why the first choice is to extend the useful life of technological equipment before resorting to these processes. It is important to consider that not all the technological equipment in a study center is out of ser-

Table 1 Types of technological recycling treatments

Type of treatment	Positive aspect	Negative aspect
Mechanical [26, 29–31]	Protects the environment, if it is based on eco-design, without molecular alteration	Discards should contain as much of the original plastic as possible, clean, and not derived from or contaminated with other substances. Collection in sufficient quantities to make it industrially and economically viable
Solovolysis [32–35]	Generates a source of plastic feedstock by producing a carbon fiber derivative	It is necessary to use chemicals for treatment. Requires prior inventory review (ecoinvent)
Hydrometallurgical [36–38]	It can be reused in various industries such as ceramics, textiles, fertilizers, among others	It lies in the use of acidic agents, if organic products are not used, it could cause significant environmental impacts
Eddy currents [25, 39–41]	Allows to separate and classify non-ferrous metals by using eddy current techniques text follows	Eddy current treatment, requiring equipment and conditioning to prevent unwanted energy losses
Crown effect [39, 42–44]	It allows separating and classifying non-ferrous metals by ionization of gaseous elements surrounding the charged conductor	Audible noise, deterioration of materials due to corrosion and erosion

vice due to damage; some only need to be reassigned to the function for which they supported the workload.

The techniques are aimed at extending the useful life of the equipment, so that processing is no longer necessary, but learning how to maximize the reinsertion of the elements to contribute to the educational work.

In recycling, the reuse operations allow that its usefulness does not lie in its transformation, but in its reuse, but in another way, this would be called technological evolution, changing the mission by which an expired device now fulfills new functions [27, 28]. At different stages of the use of technology, getting rid of electronic waste, becomes a matter of conscience, to reduce the environmental impact given by the good use of such waste (see Table 1).

1.3 Contribution of Educational Institutions to the E-waste Management Problem

The different areas of study, position the processes of integration of education with measures that satisfy the use of the means of work, as well as, the conditions of improvements through the low cost of the use of technological equipment that affect educators during the exercise of educational activities [45]. This produces reciprocity between the usefulness of technologies and the need for reflection on their proper use. On the other hand, the set of learning strategies and methodologies, allows digital literacy to have a space within the educational area [46].

Continuous learning is behind the use of technological equipment, this equipment needs to be well used before continuing with the paradigm shift. Both for recycling, as for technology and learning, it is necessary to know and focus the knowledge collected in that direction. Distributing the technologies, allows to maintain the digital communication products, as required in the assigned areas.

2 Materials and Methods

The deductive analytical methodology is used with a quantitative approach through the technique of previous scientific literature review. This research process is supported by stages such as: a) Research questions, b) Search for relevant information, c) Satisfaction relation of the process to find the data, d) Environment in which it relates, e) Discrimination of useful data, f) Sample of information in graph, g) Research planning, h) Results of data to be chosen.

The research questions serve as a guide to learn how technological waste has been treated and how it influences the educational process; scientific papers are reviewed, which provide relevant information regarding electronic recycling. Three questions are considered, Question 1 (Q1): What is the benefit of bibliographic and documentary information for the research process? Question 2 (Q2): How is computer recycling carried out? Question 3 (Q3) How does the treatment of technological waste contribute to the educational process?. In relation to each question, related topics are established and for this purpose it is considered to verify that the information found is relevant, then the way in which the recycling of obsolete computer equipment is carried out is described and in this way the contribution made by computer recycling within the educational field is determined.

The bibliographic review allows to obtain answers to these questions, since they are the means to achieve the proposed and systematically sustained objectives. The topics arise from the verification of the information, necessary to know the ways in which benefits can be obtained with technological recycling. The studies carried out reveal relevant aspects about digital transformation in all areas and, importantly, in the educational field, as well as an aspect of relevance related to technological recycling as an innovative measure for environmental care. To this end, it is necessary to maintain technological investment policies, improve technological knowledge, strategically plan the measures to be taken to execute these plans and avoid the unsuccessful recycling of non-useful components. In this way it is possible to walk as digital environments evolve, applying measures that benefit students and achieving better learning in the digital world and producing better educational quality. The analysis of the most relevant aspects is the result of the review of forty bibliographic samples, which have been compiled from thematic areas suitable for the usefulness of computer recycling and its collaboration in education; the references used are composed in detail of the thematic areas.

The articles studied provide information about technological advances and measures to be taken for the benefit of the environment and education, with the understanding that a process of continuous improvement must be followed to reduce the events produced by the extraction of contaminating elements and that they can also be replicated in other contexts. Learning is potentiated with the use of technology, updates require keeping within the limits that are facilitated by technological means, where the circumstances of changes in computer equipment are related to student development through effective methods to evoke them with experiential learning [47–51]. Environmental measures, together with continuity in the use of resources, suggests preventive maintenance that reduces damage and, in this area, good practices in the use of technological equipment, infrastructure and applications used would succeed in building environmentally friendly learning areas, with processes that modernize and generate greater opportunities for collaborative work, capable of improving people's quality of life [52].

Technological resources, within learning environments, are constantly updated, and the massification of the continuous use of technological tools makes this task a commitment that requires continuous support for their proper use, in addition to avoiding their rapid deterioration, as explained in all the research that has been reviewed and that has been reflected in a matrix, considering the relevance of these. Out of a total of 40 relevant articles, they consider membership and its percentage mean for this study. The congruence of their thematic areas is evident, such as: Sustainable technology 32.25, Culture aligned to digital change 24.5, Tics in education 24.5, Recycling and environment 30.75. On the other hand, sustainable technology is oriented to several sub-themes and their percentage equivalence, as well as their code, among them: digital transformation code 1 (C1) with 95%, technological recycling C2 with 57.5%, educational innovation C3 with 60%, technological knowledge C4 with 70%. Also, regarding the culture aligned to digital change, the subthemes found were technological investment policies C5 with 75%, recycled counterfeits C6 with 62.5%, recycled components C7 with 60%, strategic planning C8 with 72.5%. Another thematic mentioned is Tics in education, which shows studies with the subthemes: Evolution of digital environments C9 with 77.5%, resource allocation C10 with 57.5%, educational quality (C11) with 67.5%, digital learning C12 with 82.5%. Finally, the thematic on recycling and environment with subthemes: social responsibility C13 52.5%, recycled waste policy C14 with 50%, teacher training C15 with 57.5%, environmental educators C16 with 77.5%.

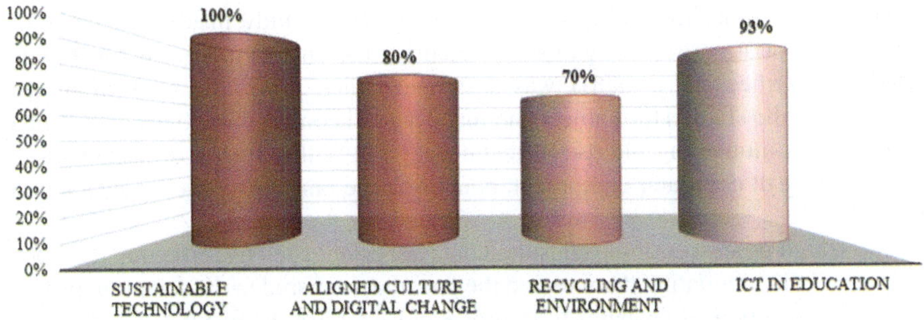

Fig. 2 Systemic matrix topics

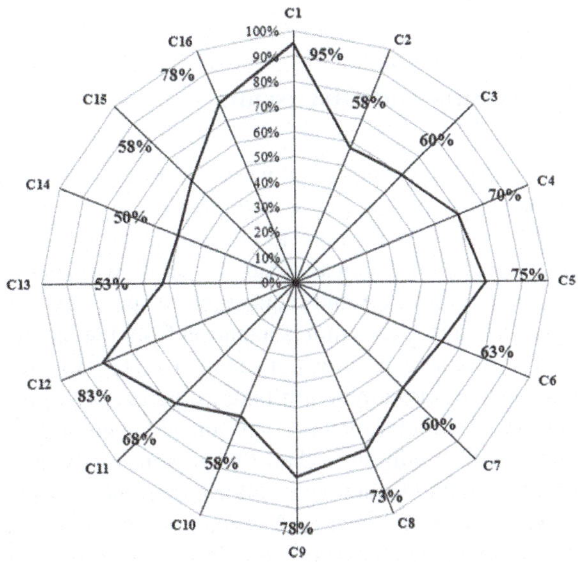

Fig. 3 Percentage of relevance and items according to the established code

Technology is vertiginous, growing and demands change, which also translates into costs, (see Fig. 2). It is necessary to consider the thematic based on the different areas of knowledge, and the matrix specifies technology, culture, environment, and benefit in education, which are part of the educational process in which it participates

The mentioned topics contribute to the investigation of the subject and allow to know that the reused technology becomes sustainable and benefits education. The reality of the educational environment allows to be in accordance with the world demands and responds to the need of the environment to maintain it with greater care. Of the topics reviewed, a total of 100% of the references argue about sustainable technology, 80% about aligned culture and digital change, 70% about recycling and the environment, and 92.50% about ICTs in education. The detail of the data that have been considered for the exposition of the information within the matrix evidences the relevance for the elaborated research (see Fig. 3).

The information chosen allows to know the subject of study, in addition to obtaining an appropriate criterion against the need for the environment and for this the data were verified and a matrix was designed, which shows the work of several researchers who studied the subject and made a good contribution to the educational environment, motivating the use of technology but with a vision of sustainability and recycling of computer equipment, benefiting the university level and society in general.

The search processes of the consulted data are established by retrieving the information and discriminating them to find the data that is shared in the scientific papers, so that this mapping contributes to improve and establish the result of the research, which allows the discussion of the topic. Each topic is oriented to consider technical and theoretical approaches, which allow the standards to be met in a meaningful way.

3 Results and Discussion

The excess of technological waste has made it possible to reach all educational levels and statuses. The use of technological resources is increasing, and technological changes are showing more accelerated. Data and knowledge growth contribute to the increase of technological waste, the debris within educational institutions is becoming larger and larger and requires more space and investment in new infrastructure to replace it, which requires attention from the institutional and personal side.

The research questions, question 1 (Q1): How does the bibliographic and documentary information contribute to the research process? The indexed documents consulted, are relevant and contribute to the good use of electronic components from their initial condition to their obsolescence. About question 2 (Q2) Is it necessary to know if the documentary information mentions the pros and cons of computer recycling and its effect on learning? Each document is supported by the indexed repository in which it has been uploaded. The authors have expressly shown the benefits, responsibility, and effects of the good or bad use of technologies within their stage of usefulness. Question 3 (Q3) Do the data collected contribute to the scientific study to elaborate a systemic matrix? The documents made it possible to elaborate the matrix and to know in more detail the effects it has on learning, as well as on the processes involved in an environment where technology is used.

The research questions made it possible to find documentation that supported the data on the technological problem, allowing the information to be entered to elaborate and identify the most relevant topics and their percentage of relevance, as well as the number of articles on each specific topic considered in the study.

In addition, a survey was applied to a random sample of 100 students from a population of 134, from the first year of engineering careers, corresponding to four courses, at the Salesian Polytechnic University of the city of Guayaquil in Ecuador, with a confidence level of 95% and a margin of error of 5%, to determine the knowledge about the reuse of technological resources and the impact of electronic waste in the environment where they live.

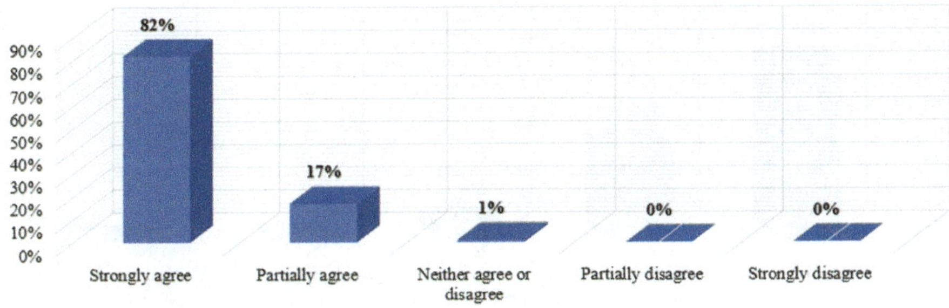

Fig. 4 Percentage of participants indicating that the use of recyclable elements creates a culture of care for the environment

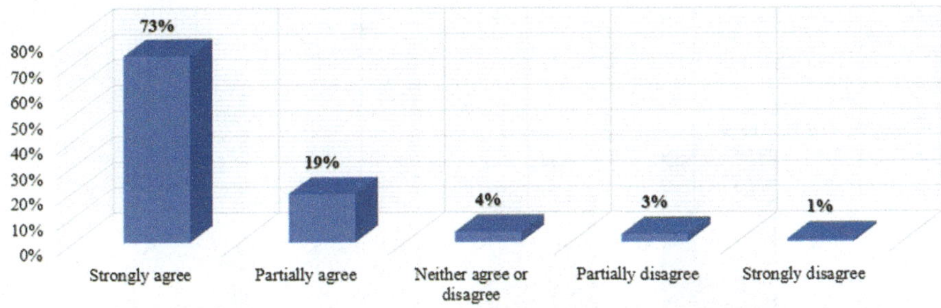

Fig. 5 Percentage of participants indicating that an adequate technological recycling education improves social coexistence

At the same time, data collection was considered through the collection of information from a set of structured questions with a Likert scale, which allowed us to understand the choice of the respondents. The results obtained contributed to the following areas: Different ways that technological resources are used, access allowed to various technological resources to enhance availability and extension of the useful life of technological work tools. For the question: Do you consider that the use of recyclable elements creates a culture of care for the environment? The results show that 82% totally agree that a culture of care for the environment is possible when using recyclable elements. In addition, each participant understands that recycling is a way to contribute to the improvement of the quality of life, which is reflected in an adequate education in the use of recyclable elements (see Fig. 4).

For the question: Do you believe that an adequate education on technological recycling can achieve a better social coexistence that avoids the accumulation of electronic waste? The results were 73% agree, 19% partially agree, 4% neither agree nor disagree, 3% partially disagree and 1% totally disagree (see Fig. 5). In other words, a total of 92% of the participants considered the possibility of improving social coexistence, not only perceived as improving the quality of life as in the previous question, but also agreed with the recycling process and the way in which respect for space and work tools can become an improvement in the study environment.

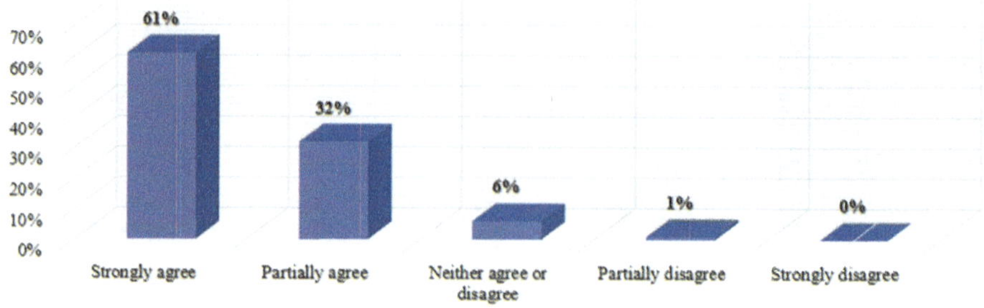

Fig. 6 Percentage of participants indicating that technological recycling improves understanding of the proper use of resources

Fig. 7 Percentage of relevant works that present proposals in the different dimensions

For the question, Do you think that technological recycling would improve the understanding of the good use of existing resources? The result was 61% agree completely, 32% partially agree, 6% neither agree nor disagree, and 1% partially disagree. A total of 93% agreed that technological recycling improves understanding in the good use of resources (see Fig. 6).

Finally, it is discussed that 65% of the works found present proposals in the area of social responsibility, which translates into wellbeing, improving the use of resources, both by the institutions and their members (see Fig. 7). 53% of the papers address solutions in the area of e-recycling policies that encourage the social responsibility of large countries to reduce e-waste and reuse it to extend its useful life. While 93% of the works address solutions in the area of training environmental educators to accompany students during the training process in schools, colleges and universities.

Finally, 98% define solutions in the training of teachers to address this issue and to incorporate an added value to the activities offered in learning and a culture that allows considering recycling as an important habit for the good use of technological resources by students.

Table 2 Recycling technology costs per tonne

Waste Kg/person	Waste 2020 per Mt	Cost
EU-28	11.430	$500
USA	10.050	$500
China	12.066	$500
Japan	3.200	$500
India	6.755	$500
Germany	1.974	$500

On the other hand, the management of electronic waste shows in the emergence of exports of technological equipment to derive responsibilities, where non-updated equipment is sent to less developed countries to finish their useful life there, this produces a series of costs to which the country where these technological products arrive must be responsible and must treat the different components as an acquired responsibility.

Since in Ecuador there is no specific company dedicated to the treatment of technological equipment such as cathode ray tubes or CRTs, which are presented in articles with several years of use, as well as those components such as cables and specific media such as circuits, it is there where service companies that process these components are responsible for proper handling which translates into costs to prevent them from being misused or reassigned to functions that are not convenient according to their capacity of use.

The data found for these countries (see Table 2) present values per ton in several countries considered developed, therefore, it is necessary to take into consideration the reception and storage, processing, wrapping, local sale, exports and international sale [25]. The approximate cost of e-waste recycling equipment is approximately $500 per ton and its handling of components such as copper, aluminum, gold, among others, awakened a new type of tenure to reduce e-waste, which is equivalent to 60% of the waste, knowing that polluting components are estimated at 2.7%, which allowed the emergence of urban mining.

4 Conclusion

The research data and sources consulted, support the conclusion that points to: It was determined the usefulness of introducing technological recycling through the adoption of sustainable technology guidelines, for learning with environmental educators to reduce the events that distort the lifespan of technological resources.

The different methods and techniques made it possible to reveal the positive and negative aspects of recycling, leaning towards the most environmentally friendly possible, as well as the bibliographic analysis supported by the systemic review. The methods and classification, as well as the actions used in the computer equipment, by means of the comparative table, could demonstrate that its use requires the use of high-cost technological equipment, as well as the manipulation of chemical and electrical elements for its use, as well as a very high volume of recycling that justifies the transformation of elements.

By evaluating the usefulness of recycling computer equipment for learning, it was proposed to promote it through improvement policies with the repowering of existing technological equipment and the reduction of the technological gap by means of ICT training to reduce environmental pollution and save resources.

This type of waste, if handled responsibly, becomes a source of employment; however, it requires a detailed economic feasibility study to optimally determine the break-even point until all the waste is processed or diverted to other functions. It should be noted that the waste management system must have responsibility in the polluter, being responsible for the cost of recycling in the use of technological equipment.

References

1. Boubellouta B, Kusch-Brandt S (2021) Cross-country evidence on environmental Kuznets curve in waste electrical and electronic equipment for 174 countries. Sustain Prod Consum 25:136–151. https://doi.org/10.1016/j.spc.2020.08.006
2. Ádám B, Göen T, Scheepers PTJ, Adliene D, Batinic B, Budnik LT, et al (2021) From inequitable to sustainable e-waste processing for reduction of impact on human health and the environment. Environ Res 194:110728. https://doi.org/10.1016/j.envres.2021.110728
3. Rautela R, Arya S, Vishwakarma S, Lee J, Kim KH, Kumar S (2021) E-waste management and its effects on the environment and human health. Elsevier B.V
4. Chen J, Huang S, BalaMurugan S, Tamizharasi GS (2021) Artificial intelligence based e-waste management for environmental planning. Environ Impact Assess Rev 87:106498. https://doi.org/10.1016/j.eiar.2020.106498
5. Islam MT, Dias P, Huda N (2021) Young consumers' e-waste awareness, consumption, disposal, and recycling behavior: a case study of university students in Sydney, Australia. J Clean Prod 282:124490. https://doi.org/10.1016/j.jclepro.2020.124490
6. Asibey MO, King RS, Lykke AM, Inkoom DKB (2021) Urban planning trends on e-waste management in Ghanaian cities. Cities 108:102943. https://doi.org/10.1016/j.cities.2020.102943
7. Gautam A, Shankar R, Vrat P (2021) End-of-life solar photovoltaic e-waste assessment in India: a step towards a circular economy. Sustain Prod Consump 26:65–77. https://doi.org/10.1016/j.spc.2020.09.011
8. Jaidev K, Biswal M, Mohanty S, Nayak SK (2021) Sustainable waste management of engineering plastics generated from E-waste: a critical evaluation of mechanical, thermal and morphological properties. J Polym Environ 29(6):1763–1776. https://doi.org/10.1007/s10924-020-01998-z
9. Ayala CR (2011) Pedagogical hope: a fresh and deep glance to educational experience from van Manen's approach. Revista Espanola de Pedagogia 69(248):119–143
10. Wang F, Yu L, Wu A (2021) Forecasting the electronic waste quantity with a decomposition-ensemble approach. Waste Manage 120:828–838. https://doi.org/10.1016/j.wasman.2020.11.006
11. Azeez NA, Van Der Vyver C (2019) Digital education: assessment of e-learning and m-learning adoption in tertiary institutions in South Africa. In: 2018 IEEE conference on e-learning, e-management and e-services, IC3e 2018, pp 23–28. https://doi.org/10.1109/IC3e.2018.8632654

12. Llerena-Izquierdo J, Ayala-Carabajo R (2021) University teacher training during the COVID-19 emergency: the role of online teaching-learning tools. In: International conference on information technology & systems. Springer, pp 90–99. https://doi.org/10.1007/978-3-030-68418-1_10
13. Maria IJ, Thirupathi D, Rajendran R, Velumani B (2019) Technologies, challenges and tools for digital learning. In: Proceedings - IEEE 10th international conference on technology for education, T4E 2019, pp 268–269. https://doi.org/10.1109/T4E.2019.00068
14. Llerena-Izquierdo J, Barcia-Ayala O, Ayala-Carabajo R (2020) Faculty training through crowdlearning for emerging online education. In: 2020 IEEE ANDESCON, pp 1–7. https://doi.org/10.1109/ANDESCON50619.2020.9272103
15. Kumari P, Talukder BMSB, Sakib S, Ray B, Rahman MT (2018) Independent detection of recycled flash memory: challenges and solutions. In: Proceedings of the 2018 IEEE international symposium on hardware oriented security and trust, HOST 2018, pp 89–95
16. Chang CT, Tsai CY, Yu PT, Tsai HH, Chang TW (2020) On the study of digital learning interactive method in situated learning. In: Proceedings - (2020) international symposium on educational technology, ISET 2020, pp 165–169. https://doi.org/10.1109/ISET49818.2020.00044
17. Claur WA, Cristie Montano J, Sarita E, Gumawa A, Guilleran J, Caceres K, et al (2019) Utilization of food waste as materials for composite plastic tile. In: 2019 IEEE 11th international conference on humanoid, nanotechnology, information technology, communication and control, environment, and management, HNICEM 2019, pp 3–6. https://doi.org/10.1109/HNICEM48295.2019.9072839
18. Soto C, Jiménez W, Ibarra M, Moreano L, Aquino M (2019) Digital educational resources to motivate environmental education in rural schools. In: Proceedings - 14th Latin American conference on learning technologies, LACLO 2019, pp 265–271. https://doi.org/10.1109/LACLO49268.2019.00052
19. Llerena J, Ayala-Carabajo R (2020) Significant learning activities (ASA) in the modality of face-to-face studies with integration of virtual educational media in Engineering careers. In: 2020 XV Conferencia Latinoamericana de Tecnologias de Aprendizaje (LACLO), pp 1–9. https://doi.org/10.1109/LACLO50806.2020.9381134
20. Bukeikhanov NR, Gvozdkova SI, Nikishechkin AP (2017) Automation of the technological system as nitriding - recycling processes. In: 2017 international conference on industrial engineering, applications and manufacturing, ICIEAM 2017 - proceedings, pp 1–5. https://doi.org/10.1109/ICIEAM.2017.8076447
21. Llerena Izquierdo J (2021) El Trabajo en Equipo Aligera la Carga. https://pure.ups.edu.ec/es/publications/el-trabajo-en-equipo-aligera-la-carga
22. Gautam A, Shankar R, Vrat P (2021) End-of-life solar photovoltaic e-waste assessment in India: a step towards a circular economy. Sustain Prod Consum 26:65–77. https://doi.org/10.1016/j.spc.2020.09.011
23. Koltsov VB, Sevrukova EA, Slesarev SA (2017) The allocation of gallium from a multicomponent eutectic in the disposal of technological waste recycle. In: Proceedings of the 2017 IEEE Russia section young researchers in electrical and electronic engineering conference, ElConRus 2017, pp 1164–1168. https://doi.org/10.1109/EIConRus.2017.7910765
24. Aguilar JA, López KE, Aguilar JN (2016) Diagnóstico de generación y manejo de los residuos eléctricos y electrónicos en instituciones educativas: un caso de estudio. Ingeniería 20(2):115–126
25. Aristizábal-Alzate CE, González-Manosalva JL, Vargas AF (2021) Revalorización de residuos de equipos eléctricos y electrónicos en Colombia: una alternativa para la obtención de metales preciosos y metales para la industria. TecnoLógicas 24(51):e1740–e1740
26. Franco-Urquiza E, Ferrando HE, Luis DP, Maspoch MLI (2016) Reciclado mecánico de residuos plásticos. Caso práctico: Poliestireno de alto impacto para la fabricación de componentes de TV. Afinidad 73(575):227–236

27. Llerena Izquierdo JF (2019) El Reciclaje Como Alternativa Al Desperdicio Electrónico: Compromiso Ciudadano Como Elemento Básico En La Preparación Profesional De Los Estudiantes De Computación De La Universidad Politécnica Salesiana. In: El Reciclaje Como Alternativa Al Desperdicio Electrónico: Compromiso Ciudadano Como Elemento Básico En La Preparación Profesional De Los Estudiantes De Computación De La Universidad Politécnica Salesiana. Editorial Abya-Yala. https://pure.ups.edu.ec/en/publications/recycling-as-an-alternative-to-electronic-waste-citizen-commitmen
28. Mudila H, Prasher P, Kumar A, Sharma M, Verma A, Verma S, et al (2021) E-waste and its hazard management by specific microbial bioremediation processes. In: Microbial rejuvenation of polluted environment. Springer, pp 139–166
29. (2019) In: 5 biotechnology for metal mechanic industrial wastes. River Publishers, pp 99–108
30. Masuduzzaman M, Amit SKS, Alauddin M (2018) Utilization of E-waste in concrete and its environmental impact - a review. In: 2018 international conference on smart city and emerging technology (ICSCET), pp 1–4
31. Thakur S, Verma A, Sharma B, Chaudhary J, Tamulevicius S, Thakur VK (2018) Recent developments in recycling of polystyrene based plastics. Curr Opin Green Sustain Chem 13:32–38. https://doi.org/10.1016/j.cogsc.2018.03.011
32. Li L, Zuo J, Duan X, Wang S, Hu K, Chang R (2021) Impacts and mitigation measures of plastic waste: a critical review. Environ Impact Assess Rev 90:106642. https://doi.org/10.1016/j.eiar.2021.106642
33. Pillain B, Loubet P, Pestalozzi F, Woidasky J, Erriguible A, Aymonier C, et al (2019) Positioning supercritical solvolysis among innovative recycling and current waste management scenarios for carbon fiber reinforced plastics thanks to comparative life cycle assessment. J Supercrit Fluids 154:104607. https://doi.org/10.1016/j.supflu.2019.104607
34. Sanchez Q, Jones M, Monir S, Vagapov Y, Lupin Junior S (2020) Evaluation of methodology for the carbon fibre recycling. In: Proceedings of the 2020 IEEE conference of Russian young researchers in electrical and electronic engineering, EIConRus 2020, pp 2178–2183. https://doi.org/10.1109/EIConRus49466.2020.9039382
35. Zhao X, Boruah B, Chin KF, Dokic M, Modak JM, Soo HS (2021) Upcycling to sustainably reuse plastics. Adv Mater 1–77. https://doi.org/10.1002/adma.202100843
36. Choubey S, Goswami P, Gautam S (2020) AI based sustainable approach for metal extraction from e-waste: a comprehensive literature review. In: Proceedings of the 3rd international conference on intelligent sustainable systems, ICISS 2020, pp 1445–1449. https://doi.org/10.1109/ICISS49785.2020.9316121
37. Colmenares AZ, Salaverría JD, Delvasto P (2018) Characterization of the chemical compounds obtained after using acetic acid as leaching agent in the hydrometallurgical treatment of spent Ni-MH batteries. Produccion y Limpia 13(1):19–29. https://doi.org/10.22507/pml.v13n1a2
38. Ghiga SC, David M, Minuţ M, Comanita-Ungureanu ED, Cozma P, Gavrilescu M (2020) Reducing health and ecological risks by using WEEE as secondary sources for critical raw materials. In: 2020 international conference on e-health and bioengineering (EHB), pp 1–4
39. Bermeo-Paucar J, Rea-Sánchez V, López-Bermúdez R, Pico-Yépez M (2006) El Reciclaje, La Industria Del Futuro. Ciencia en su PC 21(3):29–36
40. Mosbah MB, Eleon C, Passard C, Loridon J, Perot B, Barbot L, et al (2020) Performance assessment of amplification and discrimination electronic devices for passive neutron measurements. IEEE Trans Nucl Sci 67(4):662–668. https://doi.org/10.1109/TNS.2020.2965008
41. Semenov VK, Polyakov Y (2019) Current recycling: new results. IEEE Trans Appl Supercond 29(5). https://doi.org/10.1109/TASC.2019.2904961
42. Amankwah-Amoah J, Ahen F (2021) Editorial: sustainable waste management innovations: developing new ventures for improved health and environmental wellbeing. Sustainability 13(13):7132. https://doi.org/10.3390/su13137132
43. Bilici M, Călin L, Cătinean A, Samuilă A (2021) Increasing the recovery rate of metals from WEEE by corona-electrostatic separation. In: 2021 12th international symposium on advanced topics in electrical engineering (ATEE). IEEE, pp 1–6

44. Catinean A, Dascalescu L, Lungu M, Dumitran LM, Samuila A (2021) Improving the recovery of copper from electric cable waste derived from automotive industry by corona-electrostatic separation. Part Sci Technol 39(4):449–456. https://doi.org/10.1080/02726351.2020.1756545
45. Gorshenin A (2019) Toward modern educational IT-ecosystems: from learning management systems to digital platforms. In: International congress on ultra modern telecommunications and control systems and workshops, November 2018, pp 0–4. https://doi.org/10.1109/ICUMT.2018.8631229
46. Luong J, Arnold R (2020) Creating a digital learning community: four key considerations for remote learning during a pandemic. In: Proceedings of 2020 IEEE international conference on teaching, assessment, and learning for engineering, TALE 2020, pp 786–791. https://doi.org/10.1109/TALE48869.2020.9368385
47. Dieu HDT, Kim ODT, Bich HNV (2018) Sustainable development of collaborative problem solving competency for technical students through experiential learning (a case study in planning skills subject at Ho Chi Minh city University of Technology and Education). In: Proceedings 2018 4th international conference on green technology and sustainable development, GTSD 2018, pp 505–510. https://doi.org/10.1109/GTSD.2018.8595682
48. Garcia CA, Naranjo JE, Zambrano TP, Lanas D, Garcia MV (2018) Low-cost cyber-physical production systems for industrial control robots under IEC 61499. In: 2018 IEEE 23rd international conference on emerging technologies and factory automation (ETFA), vol 1, pp 1281–1284
49. Garcia CA, Castellanos EX, Buele J, Espinoza J, Lanas D, Beltran C, et al (2018) MPC under IEC-61499 using low-cost devices for oil pipeline system. In: 2018 IEEE 16th international conference on industrial informatics (INDIN), pp 659–664
50. Zea JA, Benalcázar ME (2020) Real-time hand gesture recognition: a long short-term memory approach with electromyography. In: Advances and applications in computer science, electronics and industrial engineering. Springer International Publishing, Cham, pp 155–167
51. Alvarez K, Reyes J (2021) Reconfigurable manufacturing system based on the holonic paradigm for the die-cutting process in a sports shoes company. In: Advances and applications in computer science, electronics and industrial engineering. Springer, Singapore, pp 19–36
52. Hu H, Liang YC, Zhang H, Soong BH (2017) Cognitive radio with self-power recycling. IEEE Trans Veh Technol 66(7):6201–6214. https://doi.org/10.1109/TVT.2016.2635150

Smart IoT Watering Platform Based on Orchestration: A Case Study

Thalía Gualpa, Paulina Ayala, Javier Cáceres, Edmundo Llango, and Marcelo Garcia

Abstract When we talk about a smart city we refer to a city that has changed its traditional management to a smart one being able to respond to its citizens to every-thing that is part of their daily life making it more sustainable and improving their quality of life with the implementation of information and communication technologies, an example is the present smart irrigation system for parks that with its application helps to optimize water consumption by irrigating the right amount, according to the climatic and soil conditions and depending to the needs of the plants, achieving that the exact water is consumed, at the right time and where it is required through the acquisition of information from field sensors, visualization and control in real time that allow the efficient use and management of resources and that through orchestration can coordinate and sequence the different processes present in a smart city. The system uses an ESP32 module to receive the environmental measurements of soil humidity and temperature which are stored in a database and later displayed on a website configured to supervise and control the status of irrigation. The physical configuration of the system allows the irrigation control actions to be carried out in both manual and automatic modes. The main water pump can be activated by an operator or when all the control conditions are met. The logical container decomposition of the architecture is done using the software Docker Host so that their administration can be done with the implementation of Kubernet containers.

E. Llango
Instituto Superior Tecnológico Cotopaxi, ISTX, Latacunga, Ecuador
e-mail: mellangop@istx.edu.ec

T. Gualpa · P. Ayala (✉) · J. Cáceres · M. Garcia
Universidad Técnica de Ambato, UTA, Ambato 180103, Ecuador
e-mail: ep.ayala@uta.edu.ec

T. Gualpa
e-mail: tgualpa9796@uta.edu.ec

J. Cáceres
e-mail: wcaceres0740@uta.edu.ec

M. Garcia
e-mail: mv.garcia@uta.edu.ec

© The Author(s), under exclusive license to Springer Nature Switzerland AG 2022
M. V. Garcia et al. (eds.), *Advances and Applications in Computer Science, Electronics, and Industrial Engineering*, Lecture Notes in Networks and Systems 433,
https://doi.org/10.1007/978-3-030-97719-1_11

Keywords Smart City · Orchestration · Automatic irrigation · Server · Temperature · Humidity

1 Introduction

In recent years Latin America and the Caribbean have presented a development that has caused several challenges for the quality of life of its inhabitants within cities, especially since more than 80% of its population lives in urban areas. This problem leads to the need of generating smart cities where technology and information are a fundamental pillar because their integration supports the fulfillment of the six fundamental pillars that are: Smart Governance, Smart People, Smart Living, Smart Mobility, Smart Economy and Smart Environment [1].

Cities are capable of consuming large amounts of natural resources, one of them being water. According to the Organization of Ibero-American States in the last fifty years the extraction of water in rivers and lakes has quadrupled. Taking into account that only 0,01% of the planet's water can be used for human activities, the rest is found in the oceans (97%) and in the form of ice or snow. The waste of this resource by cities is the result of several factors such as low water quality, floods, lack of irrigation systems and poor pumping systems. From the aforementioned, irrigation systems in parks represent a great waste of resources because it is generally done manually, that is, people go to the site to connect hoses to water intakes to water the plants without having adequate control of the quantity of liquid used [2].

Due to the number of pillars they handle, Smart cities involve a large number of processes and information that most of the time are routine. Some of this actions are the creation of new users, new sensors, actuators, updates, and assignment of permissions among others. At the beginning these tasks were carried out in a manual way but the appearance of software-defined networks (SDN) and network functions virtualization (NFV) allowed them to be virtualized, distributed and atomized. In order for this transition to be possible, flexible and dynamic tools are necessary at low cost, such as the orchestration of servers that allows coordinating and standardizing processes and sub-processes of a Smart city [3].

The development of the present research aims to the obtaining of an Intelligent Environment that allows optimizing natural resources, preserving and protecting the environment through intelligent irrigation of parks. In here the data collection will be done through the interconnection of sensor networks (WSN) which include sensing elements for the physical variables of: air humidity and temperature, soil humidity, temperature and ph, and flow of the water running inside the pipes. The former is integrated with Cloud Computing giving result to a structure called

Sensor Cloud, which can be applied to smart cities. This new structure highlights the role of IoT (Internet of Things) and server orchestration as fundamental elements in the conception and implementation of the project.

The project will take as a reference the "Palomino Flores" park in Baños de Agua Santa that is considered the most touristic city in Ecuador. The implementation of the intelligent system will improve water management, optimizing its consumption and preserving the life of the park's plants. The resulting benefits for the parks are diverse and the most remarkable are: i) the porosity of the pavements helps to recharge the aquifers and to manage rainwater [4], and ii) vegetation helps degrade noise pollution and local biodiversity by creating habitat and helping to improve air quality [5].

2 Related Works

The section analyzes various investigations and projects that have been developed in search of a high efficiency level at irrigation systems in parks, gardens or green areas where their maintenance requires various human resources that are expensive and water resources that are increasingly decreasing [1].

Boussarsar develops a ubiquitous park that seeks the self-sufficiency of electricity and water through a photovoltaic panel, wind turbines and an intelligent lighting system. It also integrates machines that allow generating electricity through physical activity. In the system, all devices are interconnected thanks to the Universal Plug and Play (UPnP) standard that allows the exchange of information by enabling devices and internet access to the system [6–9].

The systems developed by other studios are made up of several units including an Arduino UNO [1, 10] or an IoT platform that includes four parts: sensor, devices, gateway and cloud where it uses the E. Blynk application that allows controlling devices remotely [11]. For the information processing they use Raspberry pi [12] and NodeMCU ESP8266 that incorporates a 32-bit main microcontroller [11], for the information transmission they use a wifi module in [13, 14].

There are several sensors used in the work depicted in [11]. To obtain the temperature and humidity data they use the DHT22, a capacitive sensor that has a minimum cost [11] or the DHT11 sensor [10], both belong to the same family and provide an easy method to acquire the digital signal from an Arduino. To establish the amount of rainwater that is stored to irrigate and save water they use an ultrasonic sensor as presented in [1].

In order for the maintenance of the gardens to be automatically controlled, they developed a web server that allows consulting the value of humidity, the level of sun, water and batteries. WEBIOPI is also used as an ioT framework, which is built into RASpberry [1]. It is a mobile application that makes periodic requests and receives responses from the Arduino through a client-server architecture and allows manual control of irrigation [10].

The results of the research developed in [15] explain that traditional city management makes decision-making difficult due to the lack of data analysis. This affects the management of operations and the quality of the services it provides. Currently the development of technologies such as big data, cloud computing, IoT/CPS or others support the intelligent infrastructure of the platforms that can be developed for the construction of cities with transport, energy and intelligent services. The work presented in [15] has a common infrastructure shared between several cities that can identify and communicate different devices in a plug-and-play way to massively adopt solutions reducing costs [16].

The parks contribute to the sustainability of the urban ecosystem, but for the study of the characteristics we have two ways: to study case by case or a general study of all the parks, but the type of the later requires data that are consistent with the sample [17].

3 Study Case

In an irrigation system it is important to know when, how and how much water to irrigate, given the fact that it is a limited resource that directly influences the development of plants. Therefore, the environmental and soil parameters must be taken into account to irrigate on the right spot, at the right time and when the plants need it.

In this study, an irrigation system is proposed in order to allow the control and monitoring of parameters such as ambient temperature and humidity and soil temperature, humidity and pH through a website developed on a LAMP server, it also has both an automatic and a manual actuator for watering.

As can be seen in Fig. 1, at the monitoring stage the real data of soil temperature and humidity are obtained by means of the SHT10 sensor and of temperature and environmental humidity with the DHT22 sensor. Then those values are collected by the ESP32 module to be conditioned. And are sent to the database where they are collected and stored. After that, these values are displayed on a website to control the irrigation automatically or manually and finally the actuator is activated or deactivated according to system conditions or when determined by the operator.

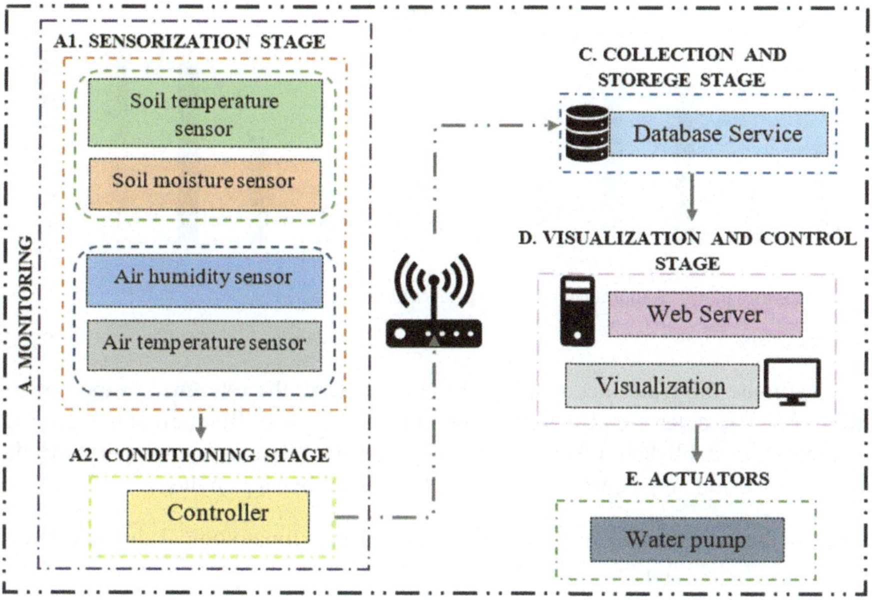

Fig. 1 System diagram

4 Solution Proposal

4.1 Plant Control Variables

In order to achieve an optimal development of the plants found in parks and gardens, climatic variables such as temperature and humidity of the environment must be considered given the fact that they intervene in the natural processes of evapotranspiration and photosynthesis of the plant. The "Palomino Flores" park is located in the city of Baños de Agua Santa and is characterized by its temperate climate and the presence of sandy-loam soil, making it an ideal place for the botanical growth and development.

The species available in here are low ornamental and trees having individuals such as Palm, Ficus Begonia, Higuerón Eugenia Calendula, Cypress, Petunia. For the aforementioned, soil conditions must meet the following requirements: temperature will be less than 35 °C, humidity equal to or less than 80% and neutral pH with values ranging between 6.5 and 7.5 pH. Also the following environmental requirements must be fulfilled: humidity higher than 50% and an oscilating temperature between 25 and 35 °C. It must be taken into account that with an ambient temperature higher than 25 °C the water can evaporate at the soil surface without reaching the roots, assuming a waste of water [18].

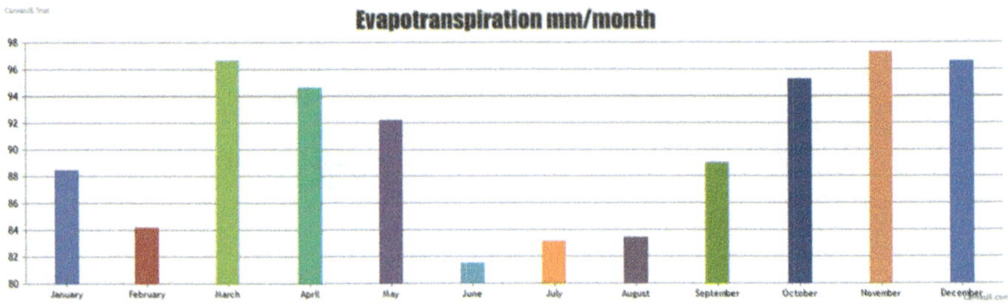

Fig. 2 Monthly plants evapotranspiration in $[mm/month]$

To estimate the water needs that the plants require, the reference evapotranspiration (ET_o) is determined using the method of Harry F. Blanney and Waine D. Criddle, who established a formula that uses the monthly average temperature, the daily photoperiod and a cultivation factor, following the steps below:

1. Relate the losses due to evapotranspiration with the climatological characteristics, given with the reference evapotranspiration (ET_o).

$$ETP_C = ET_o * kj \tag{1}$$

Where: ETP_C = potential evapotranspiration of the crop kj = garden coefficient
ET_o = evapotranspiration of reference plants
The value of ET_o is found using the Eq. 2

$$ET_o = f * kt \tag{2}$$

Where: f = consumption factor kt = temperature coefficient
Finally kt is found using 3

$$kt = 0.031144T + 0.2396 \tag{3}$$

Where T = the monthly average temperature in °C
2. Interlacing the characteristics of the crops with the evapotranspiration of the design.
3. Operate the effective monthly rains as depicted in [19] and [20].

According to the calculations made using Eqs. 1, 2 and 3 of evapotranspiration, it was possible to determine the amount of water that the plants lose month by month due to the evaporation processes of the water on the soil and the transpiration of the plants. As observed in the following graph the months of maximum loss correspond to the first and last months of the year (Fig. 2).

Once the evapotranspiration values of the park are obtained, the irrigation needs are calculated using the Eq. 4. They are obtained by balancing the losses produced by evapotranspiration and the gains of water by rain. Sometimes it happens that the

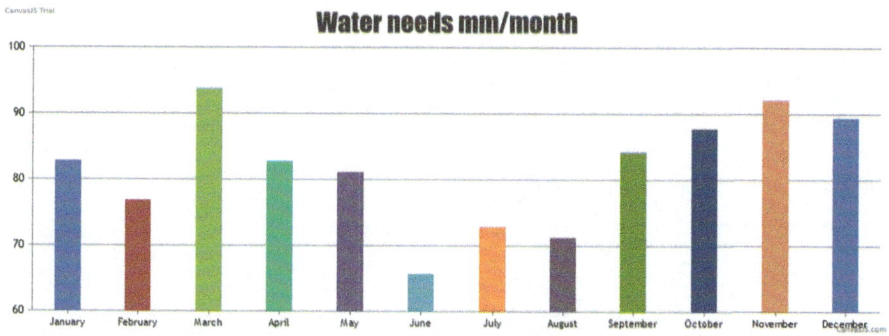

Fig. 3 Monthly water needs in $[mm/month]$

losses are greater than the gains, causing the accomplishment of water needs in the form of irrigation.

$$N_n = ETP_C - P_e[mm/month] \tag{4}$$

Where: ETP_C is the evapotranspiration of plants. P_e deals with effective precipitation.

The real water requirements are presented in Fig. 3. There are the overall needs, which are those that the plants require and use in an ideal system, and the real needs which are the water looses that occur during the flow of the liquid to the interior of the plants.

4.2 Web Site

A LAMP server was developed that uses the HTTPS protocol (HyperText Transfer Protocol Secure) which manages the exchange of information between user/server. The MySQL server is also used, which allows to store and manage the values obtained through the PHP language that is ideal for website development. It is the secure version of the HTTP (Hypertext Transfer Protocol) and uses the SSL/TLS protocol for encryption and authentication of websites. It comes with an SSL certificate that adds encryption, authentication and integrity to the HTTP protocol [21]. SST/TLS are cryptographic protocols that negotiate between two parts on a network, such as a browser and a web server. SSL provides insecure communications on the Internet, using encryption algorithms and certificates encrypted by a certified authority. TLS (Transport Layer Security), succeeds the SSL protocol, providing greater security and providing privacy and data integrity.

The System has a main page where the general information about the irrigation system is deployed, having as example: environmental information, soil information and the status of the water pump for each of the plots. Each of them has its own page

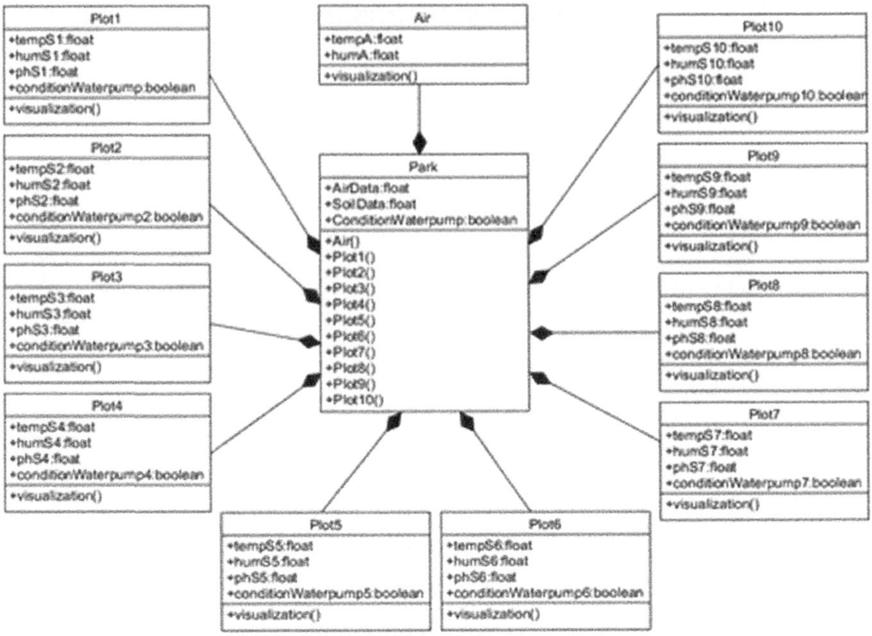

Fig. 4 Class diagram of the main page

of control and monitoring where the data of the soil and environment variables are shown in real time, a manual and automatic control to carry out the irrigation and a timer of the irrigation time.

The Apache server is in charge of acquiring the data sent by the controller, to visualize the sensors values. A script is required to select all the environmental and soil values from the database. Then this information is collected and stored and giving a way to control the system.

Figure 4 shows the classes of each of the plots of the main page that were used for the monitoring of the irrigation system, each one indicates their respective attributes and methods, and all of them have an identical composition since they monitor the variables of temperature and environmental and soil humidity, they are related to the main class "Park" and without it they cannot function independently.

Figure 5 indicates the classes used for monitoring and control of the individual plots, here we have the main class "Plot 1" composed of the classes of the data of temperature and humidity of the environment and soil and the class of Manual/Automatic Control, showing whether irrigation is being performed by environmental conditions, soil or by the actions of the user, it should be noted that it is the same scheme for the other plots.

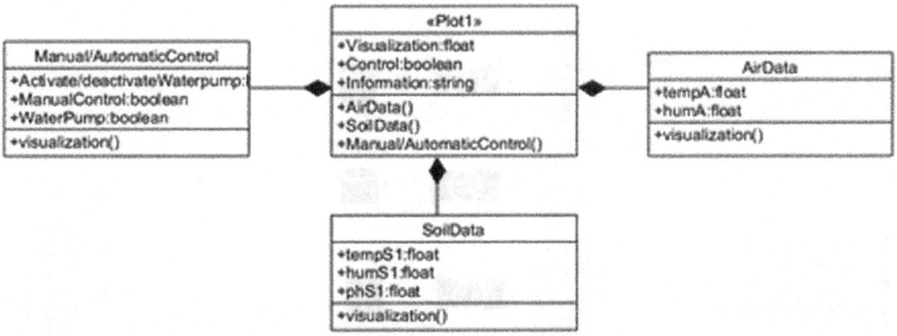

Fig. 5 Class diagram of the control and monitor page

To finalize the development of the project, there is documentation that has the history of the values obtained during the operation of the irrigation system extracted from the database, helping to know the behavior of the climate and soil. However, the data generated in the period covered by the project are not sufficient to draw decisive conclusions, at the same time the data are very useful for the optimization of the system and irrigation automation.

4.3 Orchestration

Orchestration tools allow systems to coordinate sequence services, processes and sub-processes, they can also manage data analysis considering service priorities and failures [22]. For this reason, TI can bring together different servers, storage units and network resources automatically as applications or the system loads increase or decrease. However, due to the heterogeneity of information and the existing requirements and configurations of current infrastructures it the automation and orchestration of servers and services get an increased difficulty [23].

For the development of the orchestration it was necessary to move the migrate system from an apolitical application (which over time loses control) to an application that has several software containers by means of the Docker Host, as shown in Fig. 6. These elements are managed with a Kubernetes container orchestrator, where each service is independent and encapsulates its own code, data, and dependencies; where some services require a large relational database.

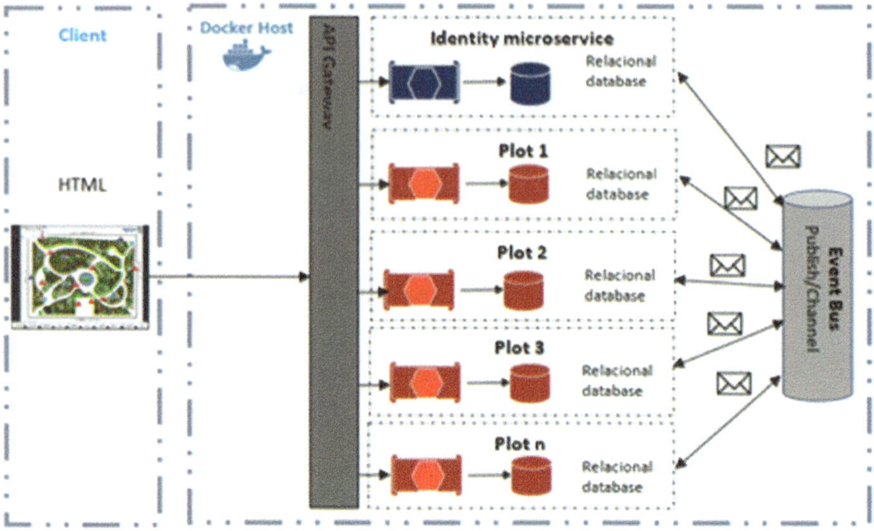

Fig. 6 Container diagram of the system

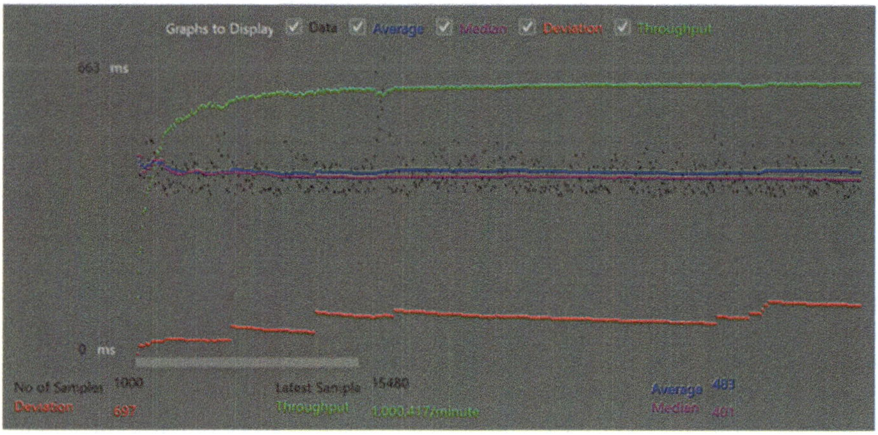

Fig. 7 1000 user test

5 Results

To verify the correct functioning of the website, stress tests were carried out with Jmeter for 100 and 1000 users. This type of test is used to determine the robustness of the website in a moment of extreme load, helping to indicate if the application is working sufficiently in cases where the actual load exceeds the expected load. The following server features were used for this test: PC with Windows 10 operating system, 2.50 GHz Intel® Core ™ i7 processor, 8 GB of RAM and 1 TB of hard disk.

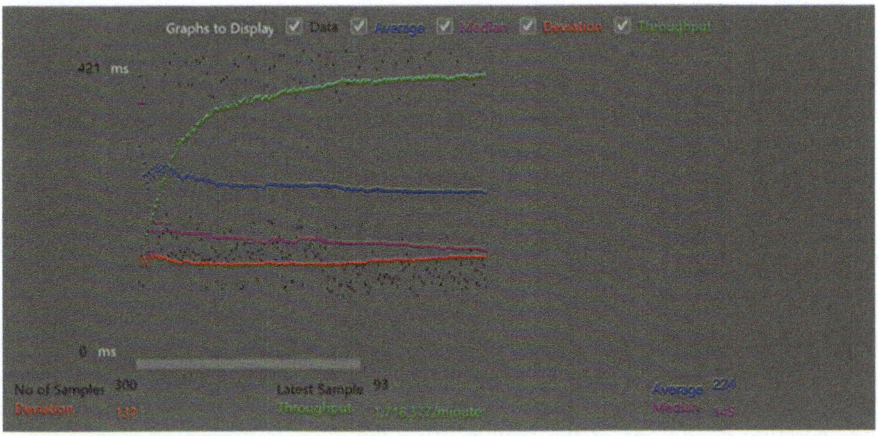

Fig. 8 100 user test

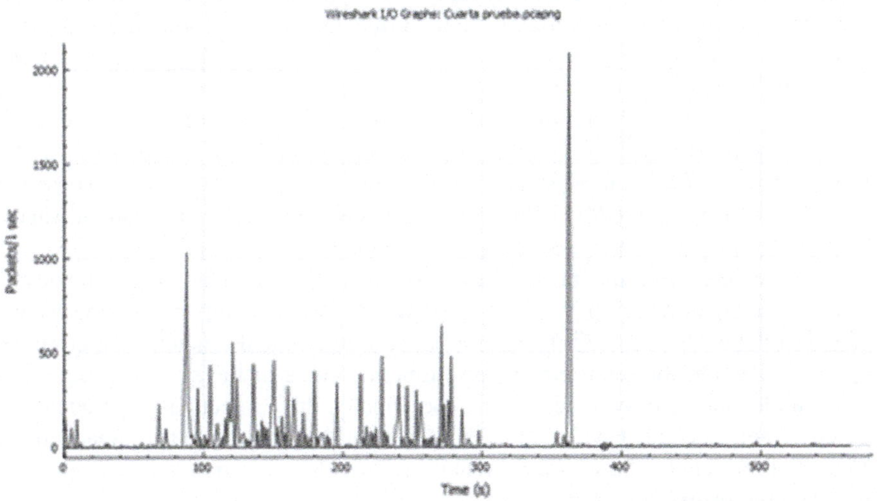

Fig. 9 Detailed packet input and output graph

To perform a stress test to the server HTTP requests are sent, creating views of results, therefore the most important data taken into account were the performance that indicates the number of requests per unit of time that have been sent to the server during the tests and their deviation. $Performance = requestnumber/totaltime$ Figure 7 shows the case of 1000 users where the throughput was 1,000.417/min and its deviation was 697 and Fig. 8 shows the graph for 100 users where the throughput was 1,718.377/min and its deviation was 137 in both cases, it is shown that the irrigation system server can support a high load and perform in an optimal way.

Finally, an analysis of the network traffic is carried out using the Wireshark tool, knowing that HTTPS works over TCP. Once all the server services have started, the first TCP and HTTP communications are recognized on the developed website. Figure 8 details the number of packets per second transmitted with an average of 381 packets per second (packets/sec) (Fig. 9).

6 Conclusions

This work presents the implementation of an intelligent irrigation system through the control and visualization of soil and environmental variables. Because of this, it was necessary to know the requirements of the plants grown at the park and thus establish the operating conditions of the system using the evapotranspiration measurement method, which allows irrigation to be more efficient and optimal.

The system allows real-time monitoring of the park soil and environmental variables from anywhere in the world, using both free hardware and software altogether with a server. This results in a safe browsing scheme, reducing its cost and with the capability of being decomposed into containers through the software Docker Host. The former are managed with the use of a Kubernetes Container Orchestrator.

The system sends a large number of data blocks, so it was decided to use the HTTPS protocol because it transfers the information only once. This makes it possible to discard the usage of the MQTT protocol, which is much better for small messages. When performing the server stress tests, low response times could be observed.

From the beginning, this project has served to lay the foundations for the development of a larger, more optimal and user-friendly system compared to the systems offered in the market, since it is oriented to be low cost and efficient, therefore we began by defining the requirements and functionalities of the system using well-known tools that are easy to access and use. Finally, this project has covered all the objectives set, since we have obtained a system for the integration and management of information sent by the sensors in real time, providing a response at the right time and when the plants require it.

Acknowledgements This work was financed by Universidad Tecnica de Ambato (UTA) and their Research and Development Department (DIDE) under project CONIN-P-0167-2017.

References

1. Boussarsar N, Kamoun Y, Ketatni M, Benhajji A, Jemal S (2015) Smart eco-urban cities ubiquitous public park. In: 2015 IEEE 29th international conference on advanced information networking and applications workshops, Gwangiu, South Korea, pp 638–643
2. Penha C, Campos-Rebelo R, Barros JP (2019) Self-configurable wireless automatic irrigation system. In: 2019 international young engineers forum (YEF-ECE), Costa da Caparica, Portugal, pp 52–58

3. Al-Fedaghi S, Alsharah M (2018) Modeling IT processes: a case study using Microsoft orchestrator. In: 2018 international conference on advances in computing and communication engineering (ICACCE), Paris, pp 394–401
4. Andersson-Sköld Y (2015) An integrated method for assessing climate-related risks and adaptation alternatives in urban areas. Clim Risk Manag 7:31–50. https://doi.org/10.1016/j.crm.2015.01.003
5. Janhäll S (2015) Review on urban vegetation and particle air pollution - deposition and dispersion. Atmos Environ 105:130–137. https://doi.org/10.1016/j.atmosenv.2015.01.052
6. Chiliquinga S, Manzano S, Córdova P, Garcia MV (2020) An approach of low-cost software-defined network (SDN) based internet of things. In: 2020 international conference of digital transformation and innovation technology (Incodtrin), pp 70–74
7. Caiza G, Bonilla-Vasconez P, Garcia CA, Garcia MV (2020) Augmented reality for robot control in low-cost automation context and IoT. In: 2020 25th IEEE international conference on emerging technologies and factory automation (ETFA), vol 1, pp 1461–1464
8. Caiza G, Garcia CA, Naranjo JE, Garcia MV (2020) Flexible robotic teleoperation architecture for intelligent oil fields. Heliyon 6(4):e03833. https://doi.org/10.1016/j.heliyon.2020.e03833
9. Naranjo JE, Sanchez DG, Robalino-Lopez A, Robalino-Lopez P, Alarcon-Ortiz A, Garcia MV (2020) A scoping review on virtual reality-based industrial training. Appl Sci 10(22). https://doi.org/10.3390/app10228224
10. Sukhdev N, Nahata N, Sridhara S, Swamy G (2018) IoT enabled smart gardening. In: 2018 fourteenth international conference on information processing (ICINPRO), Bangalore, India, pp 1–2
11. Sheth M, Rupani P (2019) Smart gardening automation using IoT with BLYNK app. In: 2019 3rd international conference on trends in electronics and informatics (ICOEI), Tirunelveli, India, pp 266–270
12. Caiza G, Alvarez-M E, Remache E, Ortiz A, Garcia MV (2020) Evaluation of AMQP and CoAP protocols for Shopfloor communication integration. RISTI - Revista Iberica de Sistemas e Tecnologias de Informacao 2020(E26):652–666
13. Leon CS, Ilvis DI, Remache EG, Villalba WR, Garcia CA, Garcia MV (2020) Evaluation of internet of things protocols for shopfloor communication integration. In: Advances and applications in computer science, electronics and industrial engineering. Springer International Publishing, Cham, pp 199–213
14. Manzano S, Granizo C, Velasteguí H, Guilcapi J, Benalcazar F, Gordón C (2020) Enabling electronic system for emergency alerts. In: Advances and applications in computer science, electronics and industrial engineering. Springer International Publishing, Cham, pp 184–198
15. Qingcai C, Wei Z, Rong Z (2017) Smart city construction practices in BFSP. In: 2017 29th Chinese control and decision conference (CCDC), Chongqing, China, pp 2714–2717
16. Rhee S (2016) Catalyzing the internet of things and smart cities. Global City Teams Challenge
17. Silva JP, Mestarehi M (2018) Bahrain parks cloud platform: live suit of tools for planning, designing, and managing public parks. In: 2018 international conference on innovation and intelligence for informatics, computing, and technologies (3ICT), Sakhier, Bahrain, pp 1–7
18. F A O (2020) Control del medio Ambiente. Accedido ago 20. http://www.fao.org/3/s8630s/s8630s06.htm
19. Pal L, Kumar A, Ojha CSP, Chandniha SK (2018) Sebal based evapotranspiration estimation for upper Tapi Basin (India). In: IGARSS 2018 - 2018 IEEE international geoscience and remote sensing symposium, Valencia, pp 7380–7382
20. Sowmya MR, Kumar MBS, Ambat SK (2020) Comparison of deep neural networks for reference evapotranspiration prediction using minimal meteorological data. In: 2020 advanced computing and communication technologies for high performance applications (ACCTHPA), Cochin, India, pp 27–33
21. Stanivuk I, Bjelic V, Samardzic T, Simic D (2017) Expanding Lua interface to support HTTP/HTTPS protocol. In: 2017 13th international conference on advanced technologies, systems and services in telecommunications (TELSIKS), Nis, pp 407–410

22. Kamal R, Agrawal S (2010) A design framework of Orchestrator for computing systems. In: 2010 international conference on computer information systems and industrial management applications (CISIM), Krackow, Poland, pp 410–413
23. Hwang J, Huang YW, Vukovic M, Anerousis N (2015) Enterprise-scale cloud migration orchestrator. In: 2015 IFIP/IEEE international symposium on integrated network management (IM), Ottawa, ON, Canada, pp 1002–1007

Gamification Resources Applied to Reading Comprehension: Projects of Connection with Society Case Study

Cristina Páez-Quinde, Sarah Iza-Pazmiño, Daniel Morocho-Lara, and Pablo Hernández-Domínguez

Abstract This research work focused on the use of gamification resources applied to virtual education, that is, the introduction of virtual resources based on games in the online education of the students of the Juan Léon Mera La Salle Educational Unit. The objective was to determine the use of ICTTAC tools in reading comprehension based on the Theory of the six readings. The methods used were the methodology for the development of ADDIE resources, which consists of providing learning experiences with the design of digital education, a survey was applied as a research technique with a structured questionnaire that served as a pretest to determine the use of web 3.0 tools and finally the Technological Acceptance Model (TAM) that served as a post-test, which allowed to check the level of acceptance of these tools; the reliability of the instruments was performed by means of the Cronbach's Alpha statistic, which had a reliability result of 0.854. For this research, it was possible to show that the vast majority of students accept this type of resource. In this way, it is concluded that the educational process based on the game improves children's learning; since they can be used synchronously and asynchronously; Web 3.0 tools such as Kahoot, Educaplay, and Quizizz are very useful since they facilitate the transmission of knowledge in an interactive, collaborative and fun way.

Keywords Learning and Knowledge Technologies (TAC) · Gamification · Web 3.0 tools · Reading comprehension · Virtual education

---

C. Páez-Quinde (✉) · S. Iza-Pazmiño · D. Morocho-Lara · P. Hernández-Domínguez
Faculty of Human and Educational Sciences, Research Group in Language, Education and Marketing, Universidad Técnica de Ambato, 180103 Ambato, Ecuador
e-mail: mc.paez@uta.edu.ec

S. Iza-Pazmiño
e-mail: sj.iza@uta.edu.ec

D. Morocho-Lara
e-mail: hd.morocho@uta.edu.ec

P. Hernández-Domínguez
e-mail: pe.hernandez@uta.edu.ec

© The Author(s), under exclusive license to Springer Nature Switzerland AG 2022

M. V. Garcia et al. (eds.), *Advances and Applications in Computer Science, Electronics, and Industrial Engineering*, Lecture Notes in Networks and Systems 433,
https://doi.org/10.1007/978-3-030-97719-1_12

1 Introduction

The objective of gamification is to motivate and, in turn, modify behaviors through the game, being this in an independent and spontaneous way, always making clear rules which must be followed in order to achieve that end.

Although gamification comes from the business world, education also seeks to motivate students to interact and generate positive learning experiences that touch the emotional aspect and give meaning to the teaching-learning process; therefore, it is necessary to acquire from early stages skills that develop critical thinking, creativity, self-confidence, perseverance, communication, and social skills, and enthusiasm, therefore it is sought that the teacher provides his students gamified didactic tools which will be focused clearly on the student increasing their motivation intrinsic in learning [1].

The teacher plays a leading role within the classrooms, as he is in charge of investigating different elements that are positively coupled to each subject, developing and designing activities that generate interest in the students that will make them a benchmark with the ability to acquire concepts and turn them into playful activities, where the teacher will use badges, rewards, norms, and challenges, which are ways of getting closer to the students; when handling creative stimuli that generate satisfaction and interest in the activities to be carried out [2–5].

The teacher must use digital media that surprise the student in each class, leaving the monotony and traditionality aside, leaving aside the digital divide and teaching stereotypes, which were focused on only sharing content; It may also be that the teacher is very good in their pedagogy, but yes, they do not use didactics, or they do not know how to integrate it, it will not help, since the two complements each other, therefore it is essential that the teacher properly handle Information Technologies and Communication (ICT) providing maximum benefit to your student and is formed as an active participant entity [6].

According to the Policies established in the current and prospective State of the Basic Education Career (2016) and the existing relationship with the Regional and National Development Plan, this project aims to: Promote the consolidation of a third and fourth level higher education system and of centers of excellence in applied research, linking the academy with the public and private sectors and social and productive organizations. Basic social protection services: emphasis on human settlements for indigenous people and women.

Affirmative action policies for the most excluded groups, articulated with proposals to reduce gaps. System of care for priority care groups. Coverage based on the circuits and districts model of the economic and social inclusion services.

In relation to detailed information, it is emphasized that this linking project called Reading Comprehension, based on the Theory of Six Readings (elementary level) through the use of ICTTAC tools, in Basic General Education students of the Juan León Mera La Salle Educational Unit of the city of Ambato [7], will be articulated to the research project Development of web 3.0 tools in education as support in collaborative learning; since the effective use of ICTTAC will allow us the first objective to diagnose reading comprehension in students of the levels that are taken as a sample, in turn, obtained the results of the diagnosis, the development of technological activities based on the web 3.0 will be carried out. For collaborative learning [8].

By implementing gamification in educational virtual spaces, they develop in the student the motivation to build self-learning, because when handling gamified technological tools from an early age, it will be easier for them to guide themselves to the formation of new skills in any subject of Basic General Education, but as we are in a digital age, it is necessary that it has the basic requirements such as a computer and broadband internet. The game converts the input content into an understandable product, which is directly related to the content, but in a more understandable way and should be used to arouse the curiosity and interest of students, thus impacting all training activities in the classroom. Therefore, systematic gamification techniques supported by recreational activities can be beneficial for student learning because they help stimulate children's interest and make them feel the need to learn; it should be emphasized that activities carried out with joy and pleasure usually be more effective than those considered unpleasant, thus reducing the feeling of fatigue, boredom, and obligation [9].

The Reading Comprehension, based on the Theory of the Six Readings, in the students of Basic General Education of the Educational Unit Juan León Mera "La Salle" of the City of Ambato, tries to diagnose the reading comprehension skills, design and execute a magazine digital reading for the development of reading comprehension, and evaluating the reading guide for the development of students' reading comprehension.

It is based on the Theory of the Six Readings devised by Miguel De Subiría, within the Conceptual Pedagogy Model, which supports the existence not of one but of six readings, ordered according to their level of complexity in elementary readings (phonetics, primary decoding, secondary decoding, and tertiary decoding) and complex readings (categorial and Metatextual). The author suggests that, as a general rule, during basic general education, students work with elementary readings and during high school and the first years of university with complex readings.

The coronavirus pandemic (COVID-19) has transformed several changes in different areas, even more so in education and at different levels, resulting in educational units in more than 190 countries closing their doors in order for the virus to spread and therefore mitigate its impact [7].

The Economic Commission for Latin America and the Caribbean (ECLAC) mentions that education has faced an overwhelming deterioration due to the impact of poverty and extreme poverty in the countries of this region; therefore, there is a crisis with discouraging effects on the social sectors, especially education.

The Ministry of Education of Ecuador (MINEDUC) through AGREEMENT No. MINEDUC-MINEDUC-2020-00014-A agrees to suspend face-to-face classes throughout the Ecuadorian territory, and therefore in all study modalities and days; It also provides the teachers of the National Education System as well as the administrative staff to continue working hours by teleworking [10].

The strategies that were implemented according to the website of the Ministry of Education www.mineduc.gob.ec were:

Access to the site https://recursos2.educacion.gob.ec in which around 840 digital resources are housed, the same ones that were developed as a COVID-19 Contingency Plan. A self-learning course is created for teachers called My Online Classroom by MINEDUC and the Central University of Ecuador [11].

The implementation of the Microsoft Teams platform, which allows interaction between teachers-students, teachers-authorities. Creation of several MOOCs for teachers on the website www.mecapacito.gob.ec. With the aforementioned information as a precedent, education in Ecuador currently faces several challenges focused on the technological era which must respond to changes within the knowledge society; therefore, the use of Information and Communication Technologies (ICT) is not enough, since it is not only knowing how to adapt a material or resources to virtual education, but also that this resource or activity is selected through Learning and Communication Technologies (TAC) so that this learning is relevant and adapted to the needs of each school year [12].

The correct implementation of ICTTAC allows teachers to minimize time and resources in the organization, registration, and storage of information, as well as students, to participate actively in the virtualization of education without the need for a specific academic program.

2 State of the Art

Virtual education refers to the development of teaching and learning based on ICT, and its main scenario is cyberspace, where new forms of teaching are developed that are integrated into didactic tools to stimulate teacher-student interaction.

In virtual education, gamification plays a fundamental role since it allows the student to pay voluntary attention due to its playful nature, thus increasing the spirit towards the objective of proposed goals and the processing of relevant stimuli such as concentration and memory, that is necessary for the cognitive development of the student producing positive results in skills such as problem-solving, reaction speed and multitasking thinking. Gamification is an active technique, as it allows the possibility of interacting between the student and the teacher, offering them an interactive exchange of knowledge and experiences, where specific actions will be rewarded, which generate positive experiences for the user [2].

It should be noted that to achieve good management of gamification in virtual classes, and the teacher must be properly trained in digital skills, technological tools, and educational portals to awakening interest in learning through digital resources used in their classes, which must be addressed to the search for playful strategies that are easy to understand and must vary according to the educational stage; The development of virtual gamification strategies not only clarifies the contents, they also strengthen interpersonal relationships since it manages to generate good communication with others, in addition, the teacher must handle a properly structured planning respecting the time of each gamified activity, in order to this, create study habits in their students through their example [13, 14].

Classroom education

- The teacher is the main source of questions.
- Students, being in a classroom, tend to become passive participants.
- Synchronous learning.
- The material is limited, having few reference sources such as books, dictionaries, etc.

Virtual Education

- The teacher is no longer the main source of questions, since students can investigate through various virtual sources.
- When students find themselves in virtual environments, they become active, self-taught, autonomous and digital natives.
- Asynchronous learning.
- The material is extensive, having many virtual libraries on hand.

Fig. 1 Differences between classroom and virtual education

Being in a globalized world and thanks to the advancement of technology, education has not been left aside because no longer, it is only face-to-face, but also becomes virtual; in this context [15], we will see below the main differences between education face-to-face and virtual (Fig. 1).

The integration of technological strategies used in pedagogy makes it easier to develop a class, thanks to its interaction between members of the classroom where knowledge is exchanged in a simpler and unusual way where real problems such as working skills and abilities can be solved. Self-esteem, in addition, it will be possible to encourage students to work autonomously in their free time and to build their own learning where the use of the internet will be purely educational (MinEduc, 2012).

Within education, emphasis is placed on what, in these times, the methodological strategies and the acquisition of knowledge, as well as the investigations that are carried out, is done digitally because we are in a globalized era where technology no longer it is a luxury, if not a necessity, to have a computer and internet in every place of study and work [16].

According to the [12], it mentions that videogame and gamification courses and contests were created that are available on the virtual educational platforms of the Ministry of Education and are aimed at students of Basic General Education and Baccalaureate; the purpose of creating These spaces is to motivate teachers and students to use and implement innovative virtual strategies, to improve the educational process. Also, thanks to the implementation of this space will allow students to develop skills in STEM (Science, technology, engineering, and mathematics) and thus to be able to function adequately in a technological universe; this project is promoted by MINEDUC, OEI, and CIESPAL.

It is planned to implement the Theory of the Six Readings of Miguel De Zubiría Samper, elementary readings through the use of ICTTAC tools, according to the following distributive: At the high school sublevel, work will be carried out in parallel, with phonetic reading (analytic perceptual, first synthesis, second synthesis). At the Basic Elementary sub-level, work will be carried out in parallel with primary decoding (lexical retrieval, contextualization, synonymy, radication) [17]. At the Basic Medium sublevel, work will be carried out in parallel with secondary decoding (pronominalization, chromatization and propositional inference); and, in the Upper Basic sublevel, work will be carried out in parallel with tertiary decoding (extraction of macropropositions, propositional relations, modeling) [15].

In the Instructional tests SER MAESTRO 2011, Language and Literature for Eighth of EGB to the third year of Baccalaureate, READING is defined as a process of apprehension of some type of information registered or stored in support and transmitted through certain verbal codes or not verbal reading is a human activity that allows, thanks to its implementation, for example, to interpret poetry, a story, a novel (this in terms of strictly literary matters). In addition, reading in a broad sense makes it possible to interpret signs, body movements, give or receive teaching. For the evaluation of reading, knowledge and skills are taken into account that allows understanding a written text on a literal, inferential and critical level [18].

In this same document, the following microskills are considered within the reading comprehension process:

- Identify explicit elements of the text: characters, objects, characteristics, and settings; narrator, time.
- Distinguish the main actions that make up the text and the order in which they occur.
- Distinguish cause-effect in the text.
- Recognize the type of text.
- Identify the parts of the text according to their type.

The central problem is the deficit in the development of reading comprehension skills in basic general education students. Possible causes are believed to be poor reading practices, poor reading habits, inadequate reading methodology. The potential effects can be a lack of interest in reading, a low level of general knowledge, poor academic performance, and school failure [19].

Ecuador's average in reading is 409, which places it at level 2, the minimum level of skills. Students performing at Level 2 in reading are able to locate one or more pieces of information. In addition, students can recognize the main idea of a text, understand relationships, or interpret its meaning within a limited part of the text when the information is not prominent, and the reader must make low level inferences. About 38% of students in Ecuador were at level 1st; this group of students is able to locate one or more independent pieces of explicit information, identify the main topic or the author's purpose in a text on a known topic or establish a simple connection, reflecting on the relationship between the information in the text and the common knowledge from day to day, according to PISA performance levels (OECD, 2017a). These students are the closest to reaching the basic level of all low-performing students [20].

In Ecuador, a higher proportion of students perform at level 1a than at any other level of PISA-D proficiency. Level 1st is the highest level of proficiency for approximately 33% of students in Ecuador; on the other hand, more than 49% of Ecuadorian students obtained a level higher than level 1a, which raised the Ecuadorian average in this field.

In OECD countries, about 14% of students can solve level 1a tasks, but not higher-level tasks. About 6.5% of OECD students do not even reach level 1a. It is important to mention that in Ecuador, there are students whose performance is even below level 1a. The proportion of students in Ecuador who reach level 1b at most is 15.5%. At level 1b, students can only solve the easier reading comprehension tasks of the PISA-D assessment, such as searching for a single specifically indicated piece of information, for example, in the title of a simple and familiar text or in a simple list [21].

3 Methodology

This research was experimental-exploratory, in which the following hypothesis was raised: gamification resources improve reading comprehension in students of the La Salle Educational Unit.

For the verification of the hypotheses of this research, the instrument used in the research project "Development of web 3.0 tools in education as support in collaborative learning" was taken into consideration, which shows a degree of reliability of 0.854. Consequently, the questions of the instrument are reliable for the investigation according to the Cronbach's Alpha statistic (Table 1).

It has an approach by parts, that is, it began with the qualitative part, which was to measure the perceptions of use and acceptability of the technology in the study population, and then obtain the quantitative data by applying the two instruments (questionnaire structured and TAM model), the same ones that allowed the quantification of the data obtained.

The research had a documentary bibliographic modality, in which it was possible to gather theoretical foundations of investigative antecedents of the study variables, as well as previous scientific documentation, information on gamification, virtual education, and the development of adequate tools, were used. For gamification in order to obtain truthful and precise information that I contribute significantly to the development of this research work based on the application of the project of connection with society.

Table 1 Reliability statistics

Cronbach's alpha	No. of elements
,854	19

Table 2 Stadistical contrast

	Do you consider that the teacher should generate his own resources based on web 3.0 tools. Do you consider that the development of web 3.0 resources by the teacher is important to improve teaching
Z	$-3,102^{b}$
Sig. asintót.	,002

In decision making by means of the null (H0) and the alternate (H1) hypothesis; The verification analysis is using the Wilcoxon non-parametric statistic for two related samples: Do you consider that the teacher should generate their own resources based on web 3.0 tools and Do you consider that the development of web 3.0 resources by the teacher is important to improve teaching; These two samples allowed us to identify that the use of gamification through the development of author's resources of web 3.0 improves the teaching process in virtual education (Table 2).

Having a p (value) less than 0.05, the null hypothesis (H0) is rejected, and the alternative hypothesis (H1) is accepted.

- H0: Gamification resources do not improve reading comprehension in La Salle Educational Unit students.
- H1: Gamification resources improve reading comprehension in La Salle Educational Unit students.

As a final decision, it is established that at the time the teacher develops their own resources (author resources), students improve their activities in virtual classes; at the same time, factors such as the use of devices, the type of tools, and the quality of the resources that are developed are considered. It should be mentioned that for this research, an experiment was carried out which was the development of author resources based on web 3.0, which were applied in the virtual classes of children of the La Salle educational unit; as a result, this research, there is a change in children's reading comprehension.

After the experimentation, we proceeded to use the TAM model (Technology Acceptance Model), where there was evidence that the vast majority of students accept this type of resource for their learning (Fig. 2).

By means of the normality test, the variables sex and generation of author's resources were taken into account by the teacher based on web 3.0 tools; When the data is tabulated, there is an asymmetry to the right, that is, both male and female students consider it important and very important for the teacher to develop their own resources for better learning.

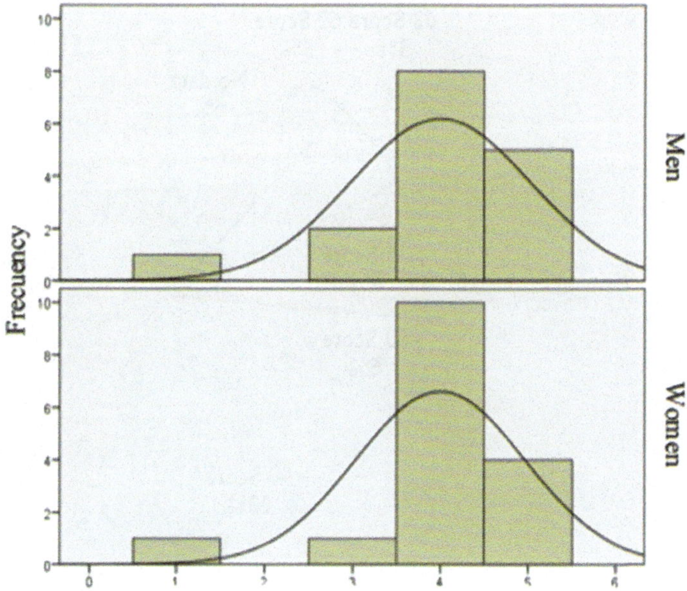

Fig. 2 Normality test

4 Results

The results obtained through the application of gamification tools to improve reading comprehension in the students of the La Salle Educational Unit are shown below, taking into account the different levels to which the readings were applied, based on the Theory of the Six readings.

4.1 Lexical Recovery

Homophone word completion exercise in order to make sense of a story and identify the meaning of said words.

Using Kahoot, the pretest of the lexical recovery text stage was developed, of which the following results were obtained (Fig. 3):

Of the 34 EGB students of the linking project of the six readings, 15 students representing 44% correctly developed the lexical recovery text stage obtaining 100 points, 8 students representing 23% obtained a score of 90 points, 2 students representing 6% obtained a score of 80 points, 2 students obtained different scores each with 3% of graphic representation and 2 students representing 6% had some error or did not complete the exercise correctly.

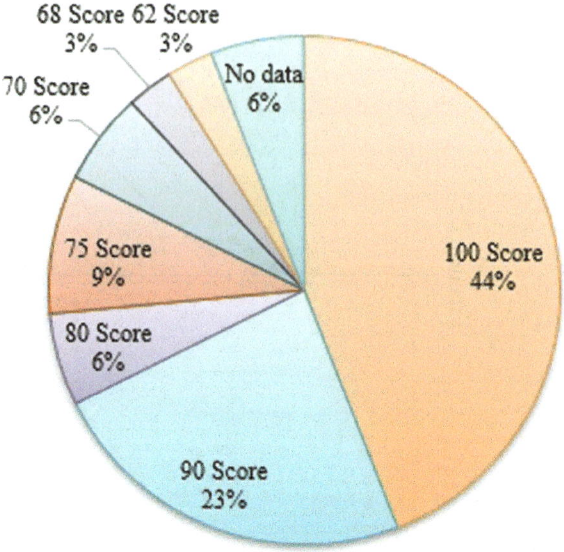

Fig. 3 Students score

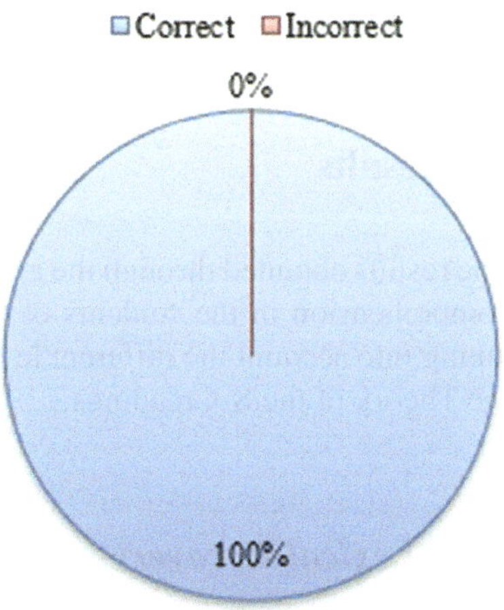

Fig. 4 Students score

4.2 Secondary Decoding

In question 1; 38 students from fifth grade of EGB answered correctly. This is equivalent to 100% correct answers; this means that students know how to use punctuation marks (Fig. 4).

For the development of the analysis and discussion of the results of the post-test, the results of the TAM evaluation were considered, applied to fifth grade EGB students.

After having applied the gamification tools, the surveyed students answered the TAM evaluation questionnaire. The most representative questions of the model were taken. The Question 2 was: **Using web 3.0 tools allows me to do my jobs faster?** (Table 3).

Of the 100% of respondents, 50% agree that using web 3.0 tools allows them to carry out work faster, 34.4% totally agree, 12.5% undecided, and 13.1 disagree. Most students agree that they totally agree and agree that using web 3.0 tools allows them to do their work faster; this is because it is easier to search for information on the web, and there are a variety of tools to edit and convert the work to suit the user (Fig. 5).

The Question 3 was: **Technological tools improve my initiative in class?** The answers are presented in the Table 4 and Fig. 6.

Of the 100% of respondents, 65.6% agree that technological tools allow them to improve initiative in class, 18.8% totally agree, 12.5% undecided, and 3.1 totally disagree. The majority of respondents state that they agree that technological tools improve initiative in class because having graphical interfaces and having easy accessibility to design tools adapt to any learning style and therefore draws attention to users.

Table 3 Using web 3.0 tools allows to work quickly

Option	Response	Percentage (%)
Strongly disagree	0	0,0
Disagree	1	3,1
Undecided	4	12,5
Agree	16	50,0
Total agree	11	34,4

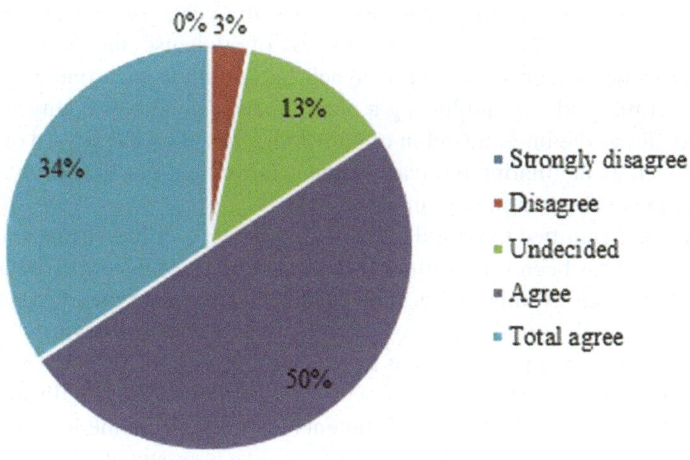

Fig. 5 Using web 3.0 tools allows to work quickly

Table 4 Best in-class initiative for web 3.0 tools

Option	Response	Percentage (%)
Strongly disagree	1	3,1
Disagree	0	0,0
Undecided	4	12,5
Agree	21	65,6
Total agree	6	18,8

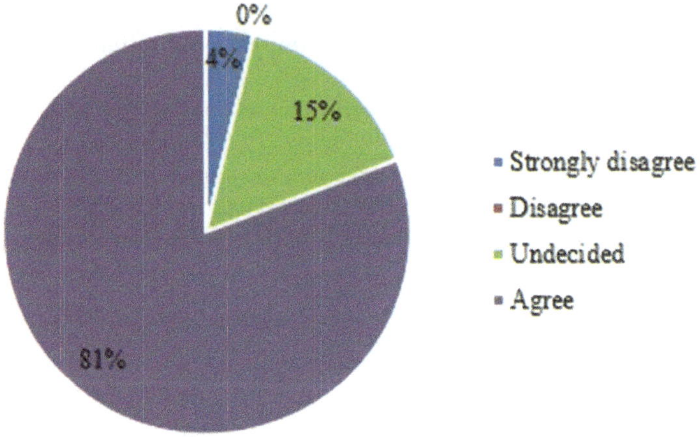

Fig. 6 Best initiative in class by web 3.0 tools

5 Discussion

The gamification tools in virtual education, over the years, have become a fundamental pillar for the online educational process of students, since when presented through the game, students can handle synchronous and asynchronous learning at the rhythm of the students and having a playful nature generates enthusiasm in the student and allows the internalization of knowledge based on the reward of concrete actions, as well as facilitating motivation, the achievement of objectives, activation of learning and evaluation in any subject.

The complete gamified tool for the use of gamification within virtual education is Kahoot; this tool has been very useful to remember the learning and associate it with the new one, as well as, this tool is very useful to apply any type of evaluation, In this case, it was applied in the diagnostic evaluation; Another necessary tool for the presentation of content and resources is Educaplay since when presented with a series of graphic interfaces, it allows the student to acquire knowledge in an illustrated way; and finally, Quizizz, this tool allowed students to demonstrate the level of learning acquired at the end of the class, based on a formative assessment.

Acknowledgements Thanks to the Technical University of Ambato, to the Directorate of Research and Development (DIDE acronym in Spanish) for supporting our research project Development of Web 3.0 Tools in education as support in collaborative learning and being part of the research groups: Research in Language and Education and Marketing Consumption and Society.

References

1. Ramos FJV (2021) Una propuesta para gamificar paso a paso sin olvidar el currículum: modelo Edu-Game. Retos: nuevas tendencias en educación física, deporte y recreación (39):2
2. Zapata Vega ZM (2019) Estrategias metodológicas de la gamificacion en el aprendizaje [B.S. thesis]. Universidad de Guayaquil. Facultad de Filosofía, Letras y Ciencias de la ...
3. Rico-Bautista D, Medina-Cardenas Y, Coronel-Rojas LA, Cuesta-Quintero F, Maestre-Gongora G, Guerrero CD (2021) Smart university: key factors for an artificial intelligence adoption model. In: Advances and applications in computer science, electronics and industrial engineering. Springer, Singapore, pp 153–166
4. Montalvo W, Escobar-Naranjo J, Garcia CA, Garcia MV (2020) Low-cost automation for gravity compensation of robotic arm, vol 10
5. Caiza G, Saeterosa M, Onate W, Garcia MV (2020) Fog computing at industrial level, architecture, latency, energy, and security: a review, vol 6
6. Guevara VC (2018) Estrategias de gamificación aplicadas al desarrollo de competencias digitales docentes. Universidad Casa Grande, Departamento de Posgrado
7. Gerdenitsch C, Sellitsch D, Besser M, Burger S, Stegmann C, Tscheligi M, et al (2020) Work gamification: effects on enjoyment, productivity and the role of leadership 43:100994. https://doi.org/10.1016/j.elerap.2020.100994
8. Cáceres-Piñaloza KF (2020) Educación virtual: Creando espacios afectivos, de convivencia y aprendizaje en tiempos de COVID-19. CienciAmérica 9(2):38–44
9. Espín Chuqui DK (2019) Aplicación de una estrategia lúdica para desarrollar las capacidades lectoras en niños de educación básica elemental. Pontificia Universidad Católica del Ecuador
10. Ministerio de Educacion del Ecuador: Curriculo de niveles de educacion. https://educacion.gob.ec/wp-content/uploads/downloads/2016/08/Curriculov2.pdf
11. Ministerio de Educacion del Ecuador: TICs Aplicadas. https://educacion.gob.ec/wp-content/uploads/downloads/2013/03/SiProfe-TIC-aplicadas.pdf
12. Ministerio de Educacon del Ecuador: Concurso de Videojuegos. https://educacion.gob.ec/lanzamiento-del-curso-y-concurso-virtual-de-videojuegos-para-el-aula/
13. Nogales R, Benalcazar ME, Toalumbo B, Palate A, Martinez R, Vargas J (2021) Construction of a dataset for static and dynamic hand tracking using a non-invasive environment. In: Advances and applications in computer science, electronics and industrial engineering. Springer, Singapore, pp 185–197
14. Comas Rodríguez R, Simón-Cuevas A, Sánchez Fleitas N, García Lorenzo MM (2020) An ontology-based data management model applied to a real information system. In: Advances and applications in computer science, electronics and industrial engineering. Springer International Publishing, Cham, pp 36–50
15. Sanchez DR, Langer M, Kaur R (2020) Gamification in the classroom: examining the impact of gamified quizzes on student learning 144. https://doi.org/10.1016/j.compedu.2019.103666
16. Mullins JK, Sabherwal R (2020) Gamification: a cognitive-emotional view 106:304–314. https://doi.org/10.1016/j.jbusres.2018.09.023
17. Mishra S, Malhotra G (2021) The gamification of in-game advertising: examining the role of psychological ownership and advertisement intrusiveness 61:102245. https://doi.org/10.1016/j.ijinfomgt.2020.102245
18. Xi N, Hamari J (2020) Does gamification affect brand engagement and equity? A study in online brand communities 109:449–460. https://doi.org/10.1016/j.jbusres.2019.11.058

19. Werbach K, Hunter D (2015) The gamification toolkit: dynamics, mechanics, and components for the win. University of Pennsylvania Press
20. Maliza Manobanda GW (2020) La gamificación y su relación en el aprendizaje [B.S. thesis]. Universidad Técnica de Ambato. Facultad de Ciencias Humanas y de la ...
21. Holguín RMV (2020) Retos de las universidades latinoamericanas en la educación virtual. Revista Virtual Universidad Católica del Norte 59:1–3

Conditioning System for an Electromagnetic Energy Collection Device

Darío Balarezo, Carlos Gordón, Julio Cuji, and Fabian Salazar

Abstract Power storage and conditioning today is very important for IOT devices, achieving high efficiency of conversion of RF energy into DC energy is essential to obtain a reliable voltage source that can be used to power low-power electronic devices. In this work it is shown that a higher output voltage is obtained by using a coupling stage between the source and the voltage multiplier. 1—and 4-stage voltage multipliers are designed and implemented with Schottky diodes in FR4 material, the same ones that are evaluated with various models of diodes and capacitances to select the one with the best performance. A considerable increase in voltage is obtained from −5 dBm, managing to accumulate up to 4.5 Vdc quickly in an electrolytic capacitor with the 4-stage circuit.

Keywords Optimization · Radio Frequency · Energy harvesting · Schottky diode · Voltage multiplier

1 Introduction

In recent years, energy demand has increased exponentially worldwide, the use of batteries has been recognized as a limiting factor for applications where devices are required to operate unattended for long periods of time [1].

There are numerous power sources from which energy harvesting can benefit, worldwide the number of transmitting antennas, base stations, mobile phones, WiFi

D. Balarezo · C. Gordón (✉) · J. Cuji · F. Salazar
GITED Research Group, Facultad de Ingeniería en Sistemas, Electrónica e Industrial, Universidad Técnica de Ambato, UTA, 180150 Ambato, Ecuador
e-mail: cd.gordon@uta.edu.ec

D. Balarezo
e-mail: abalarezo7769@uta.edu.ec

J. Cuji
e-mail: julioecuji@uta.edu.ec

F. Salazar
e-mail: fr.salazar@uta.edu.ec

© The Author(s), under exclusive license to Springer Nature Switzerland AG 2022

M. V. Garcia et al. (eds.), *Advances and Applications in Computer Science, Electronics, and Industrial Engineering*, Lecture Notes in Networks and Systems 433,
https://doi.org/10.1007/978-3-030-97719-1_13

devices, continues to increase exponentially, the RF signals found in the environment are essentially "free" energy that can be collected and converted into electrical energy [2, 3]. Of the different existing sources, the harvesting of energy from radio frequency bands (RF, Radio Frequency) has been the least exploited, although it has begun to gain a lot of attention due to the advantages it offers [4, 5]. The efficiency of a collection system depends mainly on the design of the conditioning circuit, which allows impedance matching with the antenna and is capable of receiving all the RF power available in free space in the desired frequency band [6].

The fundamental concept of RF energy harvesting involves the use of RF signals to generate DC power, several methods have been used in recent years, in [2, 7] a modified form of the CMOS-based voltage doubler circuit is presented, obtaining a 160% increase in output power over traditional circuits at 0 dBm. In [8] it is mentioned the importance of minimizing the power loss in the circuit through the appropriate selection of diodes and impedance matching, this approach is considered in [9–11] and [12–14], with the aim of maximizing efficiency, in [15] an RF converter that is composed of a limiter, a rectifier and a control loop was used to provide a stabilized DC output, while in [16] a two-stage energy harvesting circuit is proposed, the former is more responsive in low input power regions, while the latter is more suitable for a higher power range.

The objective of this work is the optimization of the conditioning system, which is based on the proper selection of the components and on the adaptation of the impedances between the source and the harvester circuit at a frequency of 2.4 GHz.

In the proposed system it can be verified that the coupling stage allows better use of the collected energy as suggested in the works presented in [17] and [18], it has been possible to improve the power transmission between the antenna and the voltage multiplier, obtaining a higher level of output voltage.

The organization of this document is as follows: in Sect. 2, we describe the methodology used for the design of the proposed system; in Sect. 3, we show the results obtained in the tests carried out with the prototype; finally, in Sect. 4 we mention the conclusions of this work.

2 Methodology

In this section an RF energy converter circuit is proposed that allows obtaining a higher DC output voltage, for this, simulations are carried out that allow determining the components and coupler circuits to obtain the greatest amount of energy at a frequency of 2.4 GHz.

The process used has been the design, simulation, manufacture and characterization of the conditioning system.

In the design, the multiplier type to be used and the elements to be tested are determined, the electrical diagram and PCB design are carried out in the Proteus system. For the simulation, the ADS (Advance Design System) program is used, through which the capacitance and model values of the diodes that present the best

Fig. 1 Layout for 4-stage circuit

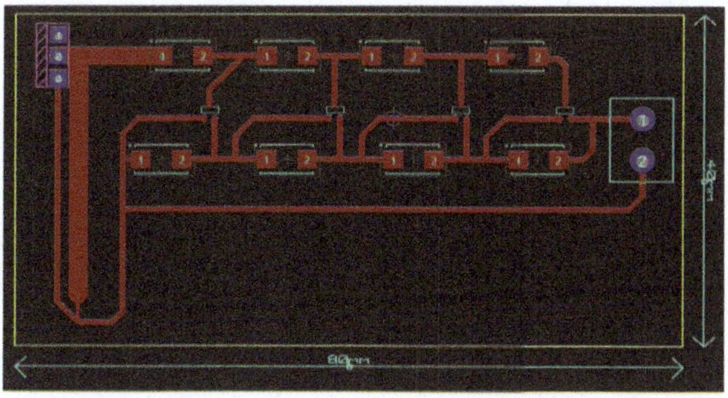

performance at a frequency of 2.4 GHz are defined. The Smith3.1 tool is used to determine the impedance of the circuits in order to define the distance and length of the coupling stubs for each of the stages. In manufacturing, FR4 material plates are used with electrical characteristics ϵ_r of 4.4, and a substrate height of 1.5 mm, for the diode welding process, suitable equipment must be used due to their size. The characterization uses the antennas developed in [19] as capture devices, as storage elements a 100 uF electrolytic capacitor and a 1.2 V 1800 mAh NiMH rechargeable battery are used.

2.1 *Design*

Component Selection: The components determine the frequency of operation and the efficiency in the conversion of energy, the efficiency of an energy harvesting system depends on the efficiency of each block of it. Due to the complex interaction between the blocks, the system design must be considered as a whole.

Layout Design. For the design of the tracks that will be printed on our boards, the PROTEUS software is used, in Fig. 1 the design made for the 4-stage circuit is shown.

2.2 *Simulation*

The simulation was carried out in the ADS program (Advance Design System), to determine the low voltage performance of the Schottky diodes 1N5819, BAT43 and HSMS286C. The circuits were fed with a source that produces an input power ranging from −30 to 10 dBm, with an impedance of 50 Ω at a frequency of 2.4 GHz. The value of the load resistance and the capacitors was varied, the Harmonic Balance simulation control was used, which allows an analysis in the frequency domain to be carried out.

The results obtained are presented in Fig. 2. As can be seen, the BAT43 and 1N5819 diodes present a low performance in the entire test power range, the HSMS286C diode

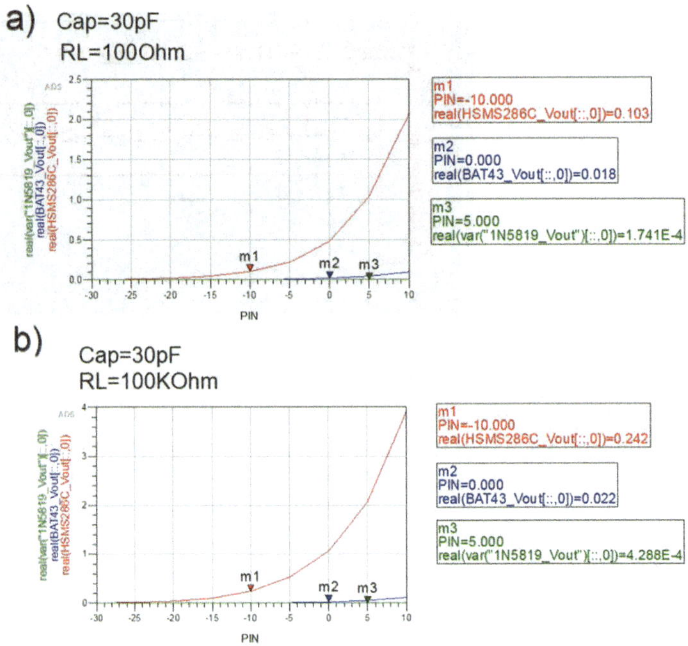

Fig. 2 Diode voltage output m1: HSMS286C, m2: BAT43 and m3: 1N5819 with 30 pF capacitor and different load resistance, **a** 100 Ω and **b** 100 KΩ

presents a better performance in all cases, obtaining a better response from −10 dBm and a resistance of 100 KΩ; This test does not reflect significant changes with the variation of the value of the capacitors.

With the results obtained, the use of the HSMS286C diode is determined. Several simulations are carried out with the selected diode, to find the optimal value of the capacitors, in Fig. 3 it can be observed that, when using a capacitor with a very small value, the value of the output voltage is not the desired one. And this is due to the rapid charge and discharge of the capacitor. It has been decided to use 33 pF capacitors as they are considered the most optimal for both practical implementation and simulations.

Since the capacitors to be used have been determined, the simulation is carried out by varying the value of the load resistance, in order to determine the effect on the voltage output of the circuit, Fig. 4 shows the results obtained, as shown You can check, the load resistance does not greatly influence the voltage output.

Once the elements that are part of the circuit have been determined, we proceed to the simulation of voltage multiplier with 1 and 4 stages. In Fig. 5, you can see the circuits implemented in ADS.

Figure 6 shows the result obtained from the simulation, the multiplier effect of the circuits can be seen; however, the output voltage at low powers does not reach the

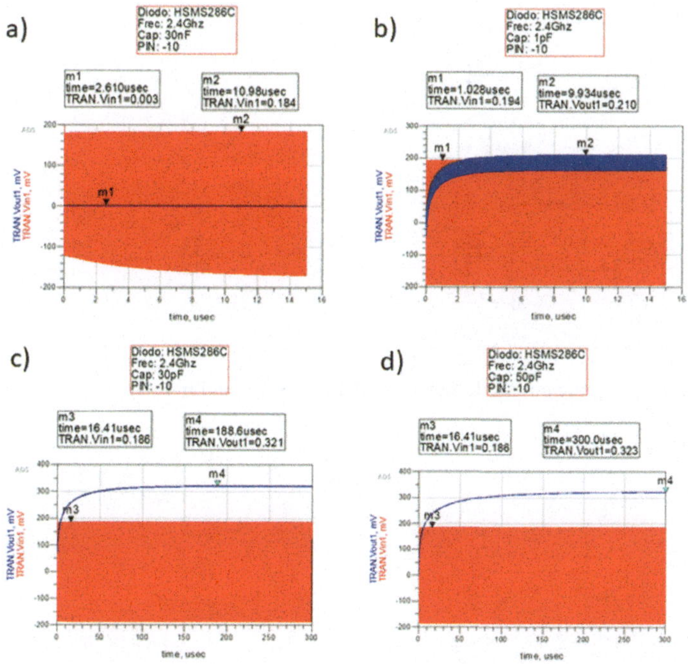

Fig. 3 Results with variation of the capacitor value, **a** 30 nF, **b** 1 pF, **c** 30 pF and **d** 50 pF

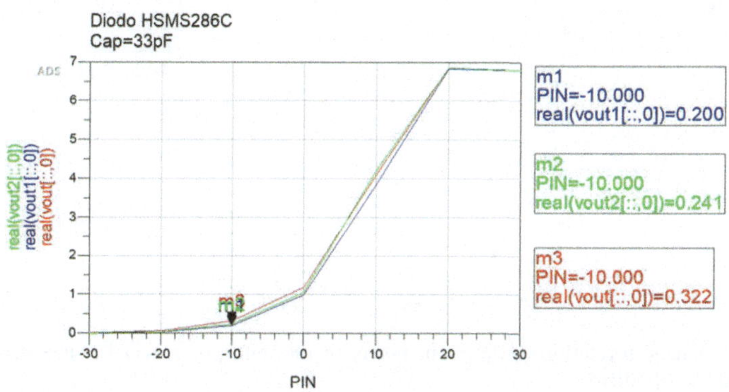

Fig. 4 Output voltage with different load resistance m1): 10 KΩ, m2): 100 KΩ and m3): no load resistance

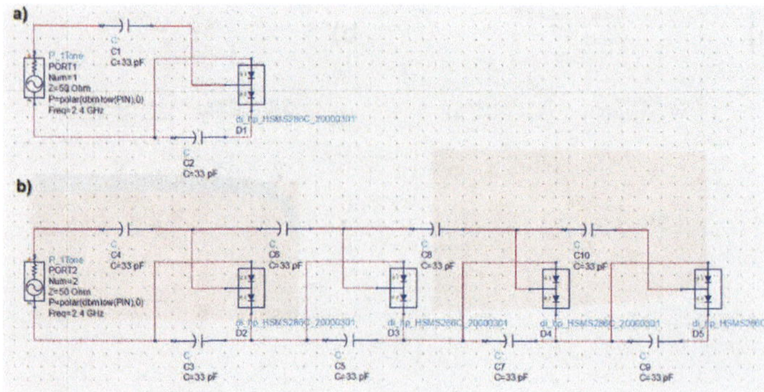

Fig. 5 RF energy capture circuit designed in ADS, **a** 1 stage and **b** 4 stages

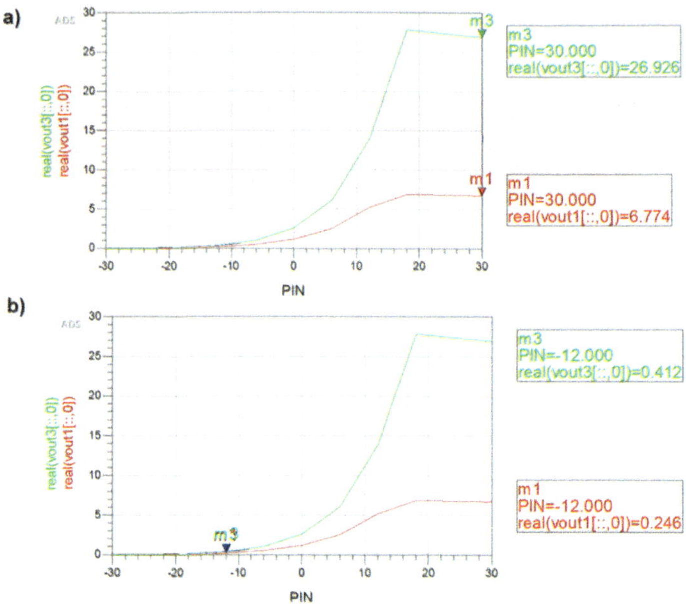

Fig. 6 Vout values in relation to the input power of the multipliers m1) 1 stage, m3) 4 stages; **a** 30 dBm and **b** −12 dBm

ideal value, the results obtained are considered without the adaptation between the source impedance and the voltage multiplier impedance.

The impedances of the antenna and of the RF-DC conversion stage are initially decoupled from each other, which causes a flow of reflected power in the circuit that reduces its efficiency. To avoid this energy leakage, we must calculate and design the coupling circuit or impedance matching network, in order to ensure that the maximum power is transferred between the RF source and the load.

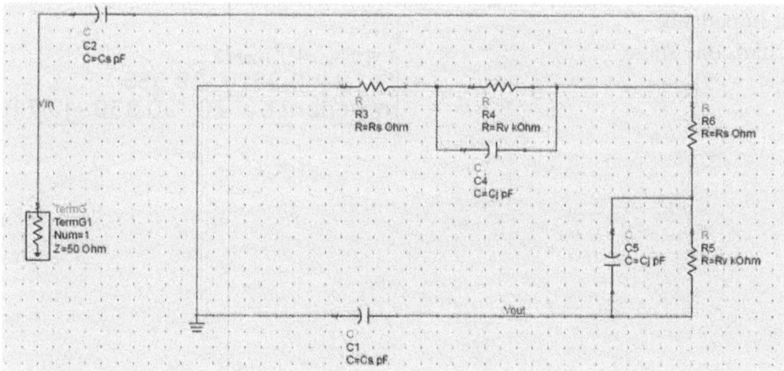

Fig. 7 HSMS286C diode equivalent circuit, one stage

Table 1 1 and 4 stage multiplier impedance

Number of stages	Z	Zin
1	Z0*(0.330 – j3.719)	16.5 – j185.95
4	Z0*(0.083 – j1.054)	4.15 – j52.7

To find the impedance of the multiplier circuit, the equivalent circuit of the HSMS286x diode is used, which is found in the Schottky diode datasheet provided by the manufacturer, in Fig. 7 we can see the one-stage multiplier circuit where the diodes are replaced by its equivalent circuit. The TermG component is used at the input of the circuit that allows to designate the 50 Ω adaptation between the source and the multiplier, which will allow us to simulate the S parameter.

The S-Parameter simulator is used, which allows to visualize the $S11$ parameter through the Smith graph, in Fig. 8 we can see that the impedance of the one-stage circuit is $Z0 * (0.330 - j3.719)$, with a $Z0 = 50\,\Omega$ the input impedance $Zin = (16.5 - j185.95)\Omega$.

To determine the impedance of the 4-stage circuit, the same procedure is used. Table 1 shows that as the number of stages increases, the resistance and reactance values decrease.

The Smith Chart tool is used to determine the distance and length of the Stubs that will allow our circuits to be coupled to 50 Ω. Figure 9 shows the method used.

With the values provided by the Smith chart, we proceed to calculate the distance of the coupling stubs using the formula (1), in the formula the permittivity $\epsilon_r = 4.4$ of the FR4 is considered, since it is the material to be use to manufacture the circuits.

$$\lambda = \frac{C}{\sqrt{4.4} * f} \tag{1}$$

$$\lambda = 0.0595914$$

Where, C is Speed of Light and f correspond to the Working frequency

Fig. 8 Input impedance with Smith chart, one stage

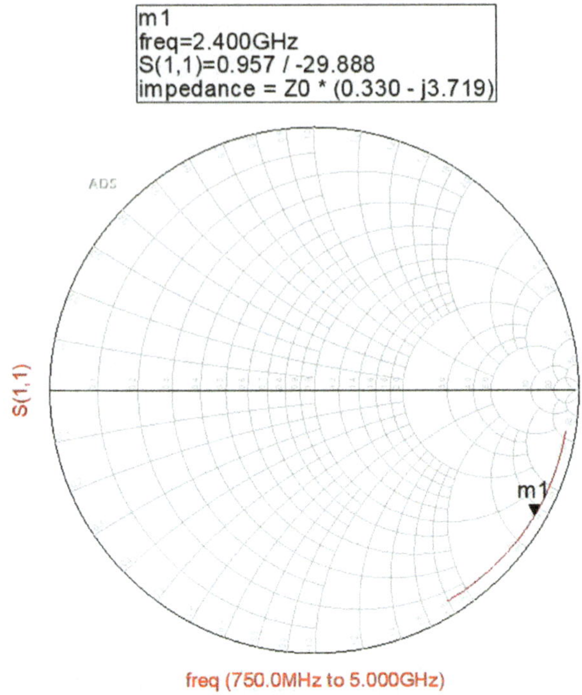

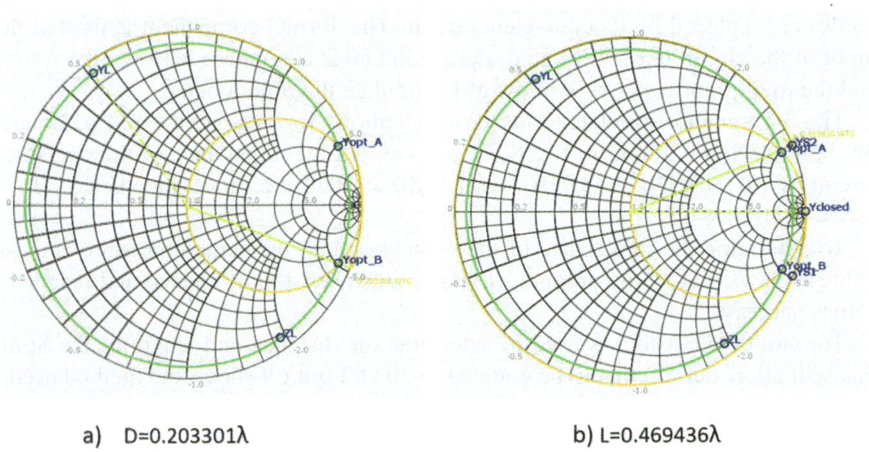

Fig. 9 **a** Coupling stub distance for 4-stage circuit. **b** Coupling stub length for 4-stage circuit

Table 2 Values of D and L of the coupling stubs of the circuits of 1 and 4 stages

Number of stages	D(mm)	L(mm)
1	13.82	28.36
4	9.56	27.87

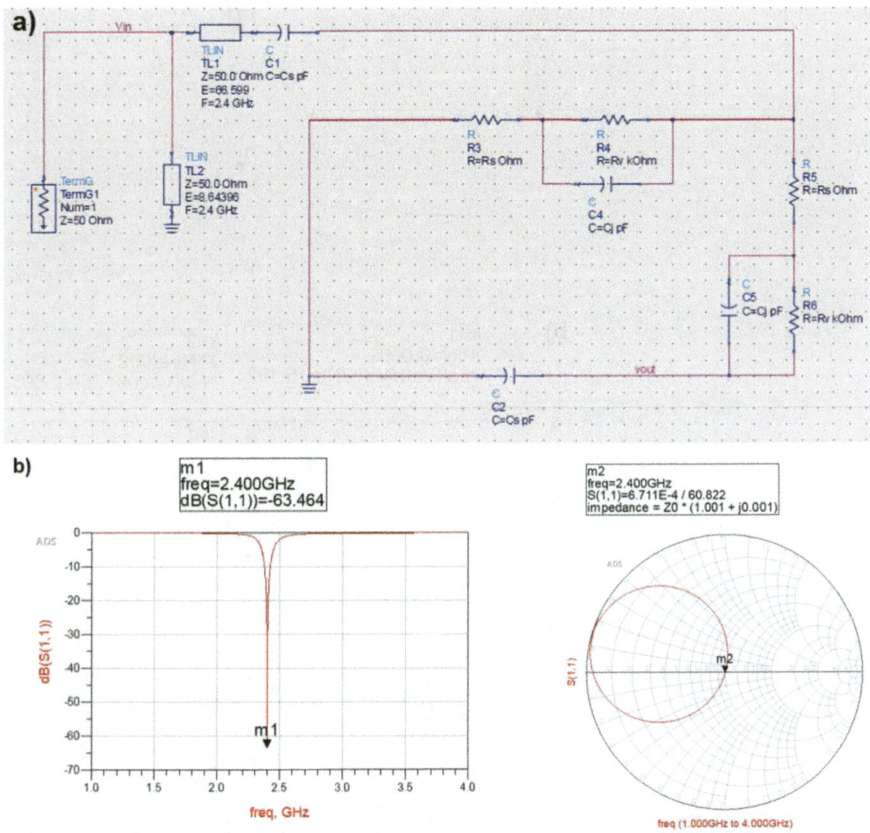

Fig. 10 **a** 1-stage RF energy harvester circuit with coupling stubs, **b** Coupling verification at 2.4 Ghz and 50 Ω

Using the same procedure, the distance and length of the stubs are calculated to achieve the 50 Ω coupling for the 1-stage circuit, the results are detailed in Table 2.

Once the values of the coupling stubs have been determined, they are incorporated into the circuits to carry out the simulations in the ADS program, as can be seen in Fig. 10 a) the coupling is included for the 1-stage circuit and in b) the coupling is checked at 50 Ω with 2.4 Ghz.

Figure 11 shows the comparison of the voltage obtained from the 1 and 4-stage circuits with and without coupling for a power range from −30 to 30 dB. You can

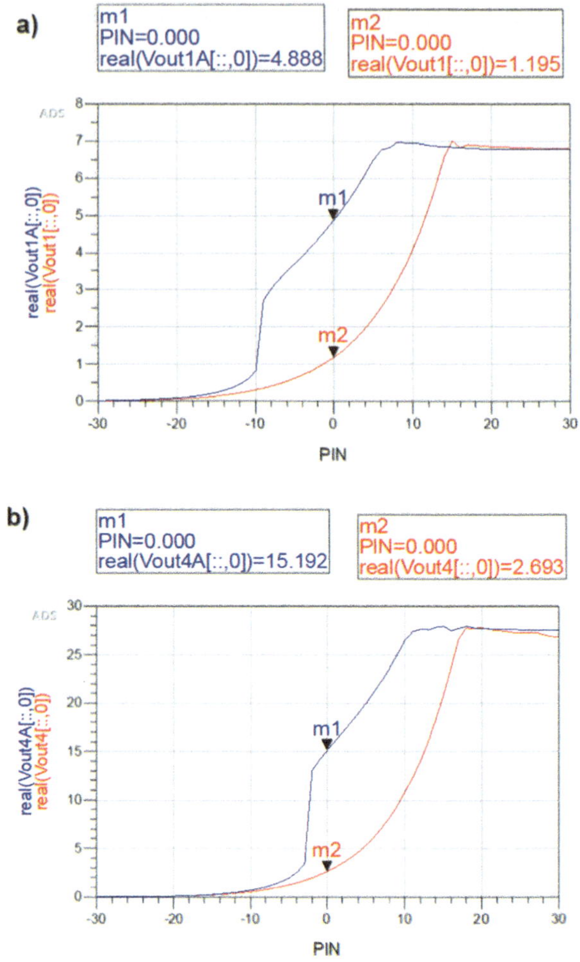

Fig. 11 Graph Vout vs PIN m1) with coupling stubs, m2) without coupling stubs; **a** 1 stage, **b** 4 stages

see the voltage gain with the coupling circuit is important from −10 dBm for 1 stage and −15 dBm for 4 stages, because the impedance matching produces the maximum energy transfer from the source to the multiplier circuit, reaching its maximum peak at approximately 8 dBm for 1 stage and 15 dBm for 4 stages.

Table 3 shows the voltage values obtained in each of the system stages with respect to the PIN value (−30 to 30 dBm).

Table 3 Output voltage obtained with and without coupling stubs for 1—and 4-stage circuits

PIN	*Vout* without coupling		*Vout* with coupling	
(dBm)	1 stage	4 stage	1 stage	4 stage
−30	0.0008	0.009	0.011	0.011
−25	0.023	0.029	0.033	0.033
−20	0.063	0.086	0.094	0.1
−15	0.151	0.237	0.252	0.286
−10	0.322	0.581	0.853	0.767
−5	0.638	1.287	3.706	2.201
0	1.193	2.673	4.873	14.772
5	2.245	5.33	6.488	19.523
10	4.11	10.84	6.959	25.969
15	7.016	20.692	6.842	27.837
20	6.865	27.889	6.788	27.749
25	6.822	27.344	6.781	27.639
30	6.801	26.926	6.779	27.581

2.3 Manufacturing

Track Printing and Element Soldier: Once the plates are printed, the elements that make up the circuit are soldered, due to the size of the HSMS286C diode it is required to use precision tools, in Fig. 12 the plate designed Vs fabricated plate.

2.4 Characterization

For the tests of the 1 and 4-stage circuits, the antennas manufactured in [19] are used, a periodic loc and a spiral, designed to receive 2.4 GHz signals. As a frequency source, an Aruba Networks brand Access Point, model APIN0225 and a Fluke multimeter are used as measurement equipment.

The voltage collected by each of the antennas was measured, tests are carried out by varying the distance from the source in relation to the antenna to check the effect of the output voltage with respect to the power received by the antenna. In Fig. 13, part of the tests and values obtained with the 4-stage circuit and the periodic loc antenna can be observed. The access point of our Aruba team is identified as "Los Aztecas", as you can see the power varies considerably with distance.

To measure the storage capacity, a 100 uF electrolytic capacitor is placed and tests are carried out for a certain time with each of the circuits and antennas. The same test is performed with a 1.2 V NiMA battery, the values obtained are recorded in Table 5.

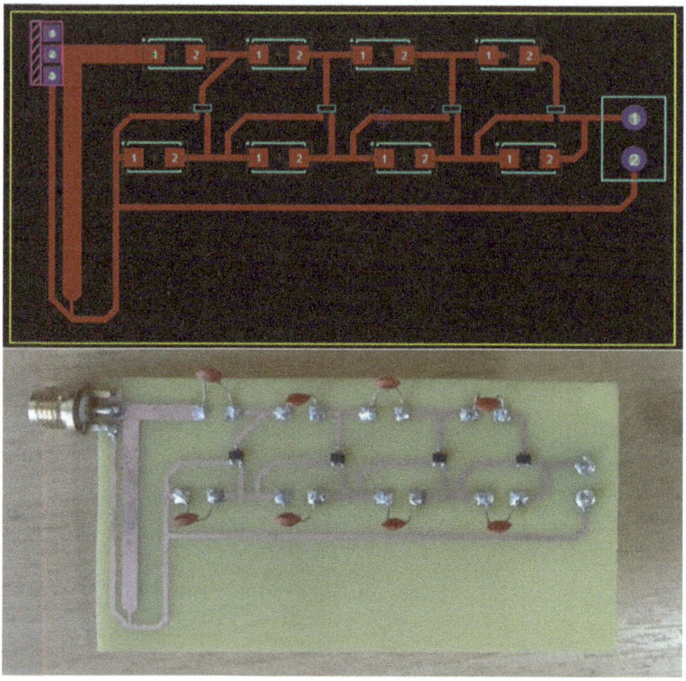

Fig. 12 Designed plate vs finished plate

Fig. 13 Received power and voltage obtained at 3.5 m from the source

3 Discussion of Results

As can be seen in Table 4, the received power varies considerably in relation to the distance at which the 2.4 GHz source is located. Higher output voltage output is obtained with the periodic Log antenna in all tests, the increase in voltage is evidenced as the number of stages increases.

Table 5 records the values of the voltage stored with each of the circuits, as can be seen, by using the capacitor it is possible to store a greater amount of voltage in a shorter period of time. The 4-stage circuit is the one that provides the greatest load with both the capacitor and the battery, managing to store 109.4 mV with the battery in 90 min; and, 4 V in 30 min with the capacitor.

Figures 14 and 15 show the exponential trend of the voltage stored in the rechargeable battery with each 1 of the antennas and stages.

Table 4 Received power and voltage obtained by antenna and in stages

Distance (m)	Power received (dBm)	Stages	Log-antenna $Vout$	Spiral antenna $Vout$
3.5	−51	1	0.0005	0.0004
3.5	−51	4	0.0061	0.0034
0.1	−18	1	0.1269	0.0740
0.1	−18	4	0.4005	0.3514
0	−12	1	0.3524	0.2288
0	−12	4	2.8562	1.3616

Table 5 Received power and voltage obtained by antenna and in stages

Storage element	Stage	Test time (min)	Initial voltage VDC	Final voltage log antenna $Vout$	Initial voltage VDC	Final voltage spiral antenna $Vout$
Electrolitic capacitor	1	30	0	2.2084	0	1.2604
100 uF	4	30	0	4.0212	0	2.5058
Battery	1	30	1.0065	1.0233	0.9953	1.0099
NiMA	1	60	1.0233	1.0417	1.0099	1.0201
1.2 V	1	90	1.0417	1.0995	1.0201	1.0309
Battery	4	30	0.9735	1.0097	1.0031	1.0251
NiMA	4	60	1.0097	1.0457	1.0251	1.0494
1.2 V	4	90	1.0457	1.0829	1.0494	1.0712

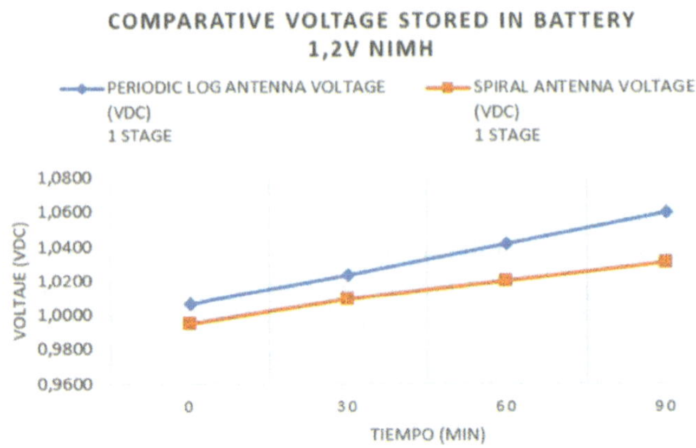

Fig. 14 1.2 V battery-stored voltage with 1-stage circuit

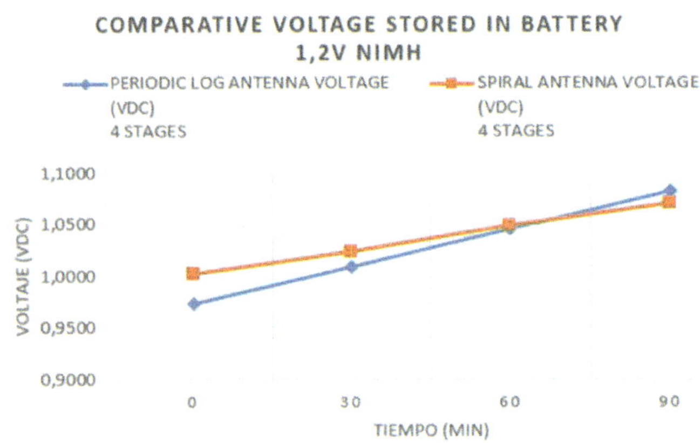

Fig. 15 1.2 V battery-stored voltage with 4-stage circuit

4 Conclusions

In this article, it is shown that the coupling stage has a high impact on the output voltage, the behavior of the system depends on the received power, which varies depending on the distance at which the source is located. The amount of energy that can be collected is very low, on the order of milivolts, however the voltage stored with the coupling circuit and HSMS286C diodes carried out in this work is higher compared to the voltage stored in the work carried out in [20] and [21] where BAT43 diodes were used and the coupling step is not performed. The 4-stage circuit presents better performance, managing to accumulate more voltage in less time. The main limitation for carrying out this work was in the HSMS286C Schottky diodes, because this type of elements is not marketed in our country, another limitation is that due to its size it is required to use special implements for handling and welding. Since they were found to be sensitive to the heat generated when placed on the plate.

Acknowledgements The authors thank the Technical University of Ambato and the "Dirección de Investigación y Desarrollo" (DIDE) for their support in carrying out this research, in the execution of the project "Sistema de Captación de Energía Electromagnética para Abastecimiento de Energía en Terminales de Internet de las Cosas (IoT) en entornos de Quinta Generación (5G).", project code: SFFISEI 04.

References

1. Nintanavongsa P, Muncuk U, Lewis DR, Chowdhury KR (2012) Design optimization and implementation for RF energy harvesting circuits. IEEE J Emerg Sel Topics Circ Syst 2(1):24–33
2. Jabbar H, Song YS, Jeong TT (2010) RF energy harvesting system and circuits for charging of mobile devices. IEEE Trans Consum Electron 56(1):247–253
3. Salazar-Moya A, Garcia MV (2021) Lot streaming in different types of production processes: a PRISMA systematic review 5:67. https://doi.org/10.3390/designs5040067. MDPI AG
4. Villalonga DAU, Gómez SJT (2018) Estudio de las políticas de optimización de energía para la transmisión de información en sistemas recolectores de energía de Radio Frecuencia. Tono, Revista Técnica de la Empresa de Telecomunicaciones de Cuba SA 14(2):23–34
5. Riofrio-Morales M, Garcia MV (2021) Training virtual reality-based system for detection and simulation of motors failures, vol 1983. IOP Publishing, p. 012099. https://doi.org/10.1088/1742-6596/1983/1/012099
6. Devi KKA, Din NM, Chakrabarty CK (2012) Optimization of the voltage doubler stages in an RF-DC convertor module for energy harvesting
7. Caiza G, Torres FI, Garcia MV (2021). Identification of patterns in the involvement of novice software developers in software testing processes. https://doi.org/10.1109/icaica52286.2021.9498137
8. Kim JH, Bito J, Tentzeris MM (2015) Design optimization of an energy harvesting RF-DC conversion circuit operating at 2.45 GHz. In: IEEE international symposium on antennas and propagation & USNC/URSI national radio science meeting. IEEE, pp 1280–1281
9. Yan H, Montero JM, Akhnoukh A, De Vreede LC, Burghartz J (2005) An integration scheme for RF power harvesting. In: Proceedings of the STW annual workshop on semiconductor advances for future electronics and sensors, vol 2005, pp 64–66
10. Montalvo-Lopez W, Catota P, Garcia CA, Garcia MV (2021) Development of a virtual reality environment based on the CoAP protocol for teaching pneumatic systems, pp 528–543. https://doi.org/10.1007/978-3-030-87595-4_39
11. García MV, Irisarri E, Pérez F, Estévez E, Marcos M (2017) An open CPPS automation architecture based on IEC-61499 over OPC-UA for flexible manufacturing in oil&gas industry. IFAC-PapersOnLine 50(1):1231–1238. https://doi.org/10.1016/j.ifacol.2017.08.347
12. Harrist DW (2004) Wireless battery charging system using radio frequency energy harvesting. University of Pittsburgh
13. Morales F, Mogrovejo D, González E, Toasa RM (2021) Monitoring of industrial variables based on LoRa communication protocols. In: Advances and applications in computer science, electronics and industrial engineering. Springer, Singapore, pp 201–214
14. Hojas-Mazo W, Simón-Cuevas A, de la Iglesia Campos M, Ruíz-Carrera JC (2020) Semantic processing method to improve a query-based approach for mining concept maps. In: Advances and applications in computer science, electronics and industrial engineering. Springer International Publishing, Cham, pp 22–35

15. Lhermet H, Condemine C, Plissonnier M, Salot R, Audebert P, Rosset M (2008) Efficient power management circuit: from thermal energy harvesting to above-IC microbattery energy storage. IEEE J Solid-State Circuits 43(1):246–255
16. Prusayon N, Muncuk U, Lewis DR, Chowdhury KR (2012) Design optimization and implementation for RF energy harvesting circuits. IEEE J Emerg Sel Topics Circ Syst 2(1):24–33
17. Barrionuevo Ortiz NI, Cruz Hurtado JC (2014) Análisis de la eficiencia de un Multiplicador de Tensión de baja potencia de entrada en cuanto al número de etapas. Ingeniería Electrónica, Automática y Comunicaciones 35(3):90–101
18. Longoria DMB, Navarro ES, Casas AA, Juárez MAC, Rico UP (2020) Comparación de Circuitos Multiplicadores de Voltaje para Cosecha de Energía en RF (Comparison of Voltage Multiplier Circuits for RF Energy Harvest). Pistas Educativas 42(136)
19. Sucozhañay G, Cabrera F, Sucozhañay D, Guaman R, Siguenza-Guzman L, Vanegas P (2021) Toward a sustainability balanced scorecard for managing corporate social responsibility: a conceptual model. In: Advances and applications in computer science, electronics and industrial engineering. Springer, Singapore, pp 279–298
20. Mendoza Chicaiza JL (2019) Sistema electrónico portátil para la recarga eléctrica de dispositivos móviles mediante la captación de energía electromagnética [B.S. thesis]. Universidad Técnica de Ambato. Facultad de Ingeniería en Sistemas ...
21. Freire Cárdenas EL (2021) Antenas Logarítmicas Para Aplicaciones De Captación De Energía Electromagnética [B.S. thesis]. Universidad Técnica de Ambato. Facultad de Ingeniería en Sistemas ...

Electronics Engineering

Predictive Dynamic Matrix Control (DMC) for Ball and Plate System Used in a Stewart Robot

Javier Montenegro, Paúl Ayala, and Alexander Ibarra

Abstract This paper presents the development of a predictive dynamic matrix controller (DMC) on a Stewart platform with 6 degrees of freedom for the ball and plate application that consists of controlling the position of a ball on a rotating plane surface. Prior to implementation, the construction characteristics of the system and its kinematic analysis are established. Obtaining the position of the ball on the platform is carried out by means of digital signal processing of the images from a camera by means of algorithms implemented in a Raspberry Pi 3B microcomputer. It also includes the description of the identification process of the mathematical model of the system, which starts from an excitation of the closed-loop system with a pseudo-random binary signal (PRBS). The qualities of this type of controllers are detailed, as well as the considerations necessary for their application and design. The device that runs the control algorithm is a STM32F4 Discovery development board. The results show that the proposed controller stabilizes the ball in an average time of less than 6 s and reacts to disturbances by re-stabilizing the system in a similar time with an expected error of 10%.

Keywords Nonlinear control · Dynamic matrix control · Ball and plate system · Stewart platform

1 Introduction

The Stewart Platform is a parallel robot with six degrees of freedom, proposed by D. Stewart in 1965, who conceived it as a means to contribute to the development of

J. Montenegro · P. Ayala (✉) · A. Ibarra
Departamento de Eléctrica, Electrónica y Telecomunicaciones, Universidad de las Fuerzas Armadas ESPE, Av. General Rumiñahui s/n, Sangolquí 171103, Ecuador
e-mail: jpayala@espe.edu.ec

J. Montenegro
e-mail: ejmontenegro@espe.edu.ec

A. Ibarra
e-mail: oaibarra@espe.edu.ec

© The Author(s), under exclusive license to Springer Nature Switzerland AG 2022

M. V. Garcia et al. (eds.), *Advances and Applications in Computer Science, Electronics, and Industrial Engineering*, Lecture Notes in Networks and Systems 433,
https://doi.org/10.1007/978-3-030-97719-1_14

transport systems [1]. The platform has been useful in all types of aviation simulators and space docking applications, and has recently been used in docking devices for submarine rescue ships, robot assemblers [2], and even in the automotive and industrial areas [3]. Other fields of application of the platform are biomedical, in rehabilitation therapies [4], in entertainment, as motion simulators for video games [5], in radiotelescopes such as the AMiBA in Hawaii [6] or in panel systems for solar tracking [7].

On the other hand, DMC is not a relatively new control technique, but it has gained notoriety over classical control techniques. Recent works cite the use of predictive controllers in the industrial sector [3], the chemical sector [8], in continuous stirring reactors and mixing tanks [9], in temperature control [10–12], in sweetening plants gas [13]. There are also works that focus on developing tools for DMC control [14] or improving the dependency of software in the implementation of explicit DMC controllers because the boom in the application of this type of controllers comes with the sophistication of the means to carry out control in discrete time [15]. The ball and plate application is an extension of ball and beam systems and is a typical problem in control engineering [16]. Several researchers have used various classical control strategies such as PD [17], PID [18, 19], and more modern ones such as adaptive PID [20], Fuzzy, LQ, LQR [21] or SMC [16] for this application.

In this work, the use of a dynamic matrix controller (DMC) is proposed, which is a control technique that handles non-linearities well, taking into account considerations prior to applying this type of control strategy, such as the identification of the system and the solution of the stability problem.

2 System Description

2.1 System Parts

The Stewart platform is a closed kinematic chain, motorized by six rotary servo motors with their respective links and joints that join the base with the rotating surface (Fig. 1).

On the structure, lies the acrylic platform 6 mm thick and 35 cm per side. The servomotors have a torque of 9.4 kgf.cm at 4.8 V, with 24 mm aluminum arms connected to aluminum body links and plastic ball joints at their ends. A 6-channel Pololu Micro Maestro USB controller is used to command the motors. From the base extends a rail that holds a Logitech C170 camera, with 720p resolution at 30 fps, pointed at the platform.

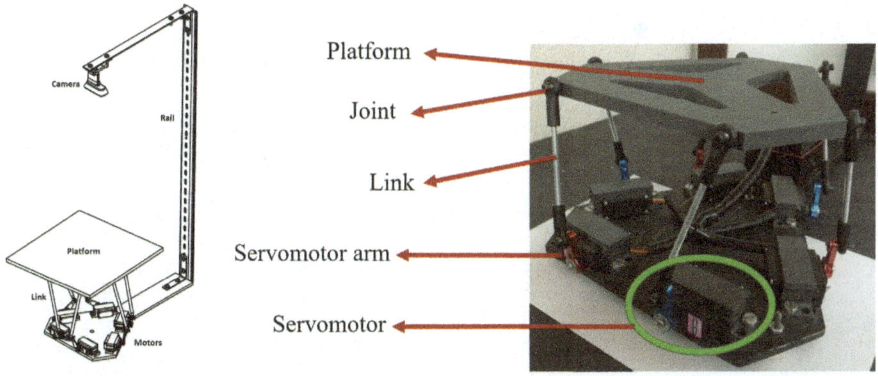

Fig. 1 On the left show the description of the parts of the Stewart robot, on the right show the description of the parts from the base of the Stewart robot

Fig. 2 Schematic of the Stewart platform for explanatory purposes of the calculation of the inverse kinematics

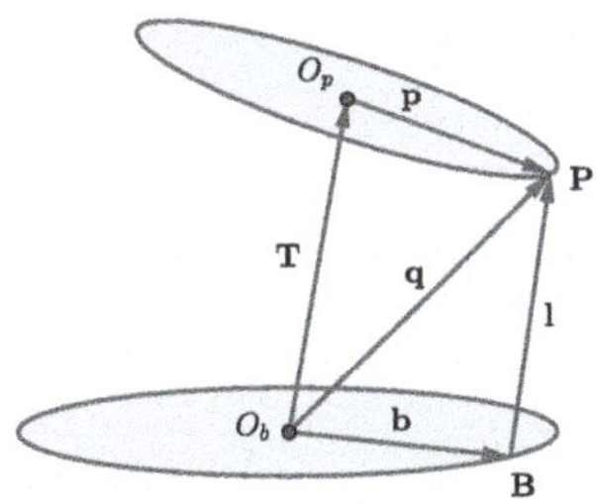

2.2 *Kinematic Analysis*

The inverse kinematics of a robot manipulator determines the actions that the actuators must perform together to achieve a desired position and orientation [22]. Assuming that the platform works with extendable arms whose length is represented by the vector **l** (Fig. 2), the relationship of how much each arm must measure to reach a certain position and angle is given by:

$$\mathbf{l} = \mathbf{T} + \mathbf{R}\mathbf{p} - \mathbf{b} \tag{1}$$

where **T** and **R** are the translation and rotation matrices of the platform respectively, **p** is the vector that joins the center of the platform with the joint and **b** is the vector that joins the center of the platform base with the actuator location. By doing this calculation for each actuator, the distances **l** under which the desired position and rotation are obtained are calculated.

Given that we have rotary servomotors, the angle α of rotation of each servomotor must be known in such a way that the vector sum of the lengths of the arm **h** and the link **d** results in the distance **l**. Eisele [23] gives a solution to this problem, arriving at the following expressions:

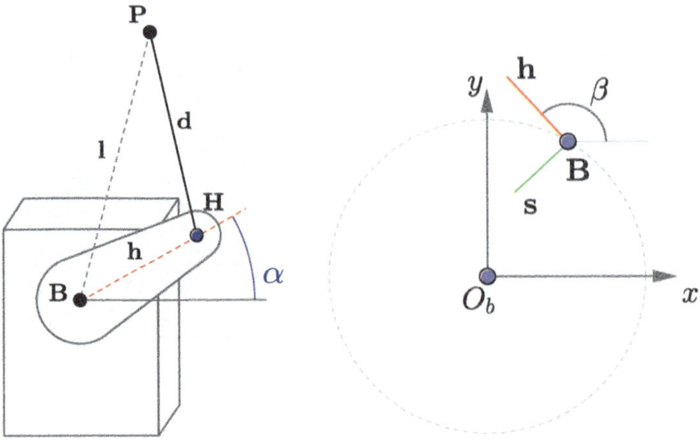

Fig. 3 Schematics of the servomotors for calculating the inverse kinematics

$$q_1 = 2l_z|\mathbf{h}| \quad (2)$$

$$q_2 = 2|\mathbf{h}|\left(l_x \cos\beta + l_y \sin\beta\right) \quad (3)$$

$$q_3 = |\mathbf{l}|^2 - \left(|\mathbf{d}|^2 - |\mathbf{h}|^2\right) \quad (4)$$

$$\alpha = \sin^{-1}\left(\frac{q_3}{\sqrt{q_1^2 + q_2^2}}\right) - \operatorname{atan2}(q_2, q_1) \quad (5)$$

where q_1, q_2 and q_3 are auxiliary variables that relate all the vectors involved pertinent to the real system. Figure 3 on the left shows the side view of a servomotor, where α is the angle that the arm forms with the horizontal plane. The right figure is a top view of the system, where O_b is the center of the platform base and β is the angle that the servomotor arm makes with respect to the x axis in the x-y plane.

2.3 System Assembly

A Raspberry Pi 3B microcomputer and a STM32F4 Discovery microcontrolled development board are available for signal processing and subsequent calculations. The microcomputer takes care of the tasks of acquiring images from the camera and processing them to obtain the position of the ball. The development board performs

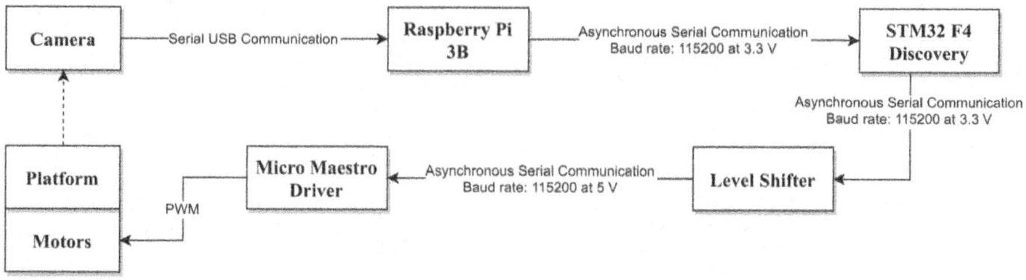

Fig. 4 Scheme of the system where the elements, connections and communication between them are detailed

calculations the inverse kinematics and controller (Fig. 4). To determine the position of the ball on the platform, artificial vision algorithms are used, mainly contour detection.

3 Identification

The application of the DMC requires knowing the mathematical model of the system. For this, you can take two paths, model the system analytically, or obtain this model through an empirical approach based on the analysis of the observations made, which is known as identification [24]. Due to the construction of the system, obtaining an analytical model is complicated, so the latter is chosen. The system to be identified has as inputs the *roll* and *pitch* signals to which the platform must tilt, measured in degrees. The outputs are the positions of the ball in x and y measured in pixels.

3.1 Input Signal

The input signal for identification must have limited amplitude when considering the physical restrictions of the actuators [25], in addition, it must have a wide spectrum of frequencies [26], and for this system it should be periodic and symmetrical, therefore, a pseudo-random binary signal (PRBS) is chosen. This signal is generated by linear feedback shift registers or LFSR, and the number of bits in this register must be such that the shortest time that the signal is active must be at least half the characteristic time of the system, and the longest time should be longer than the stabilization time [25–27].

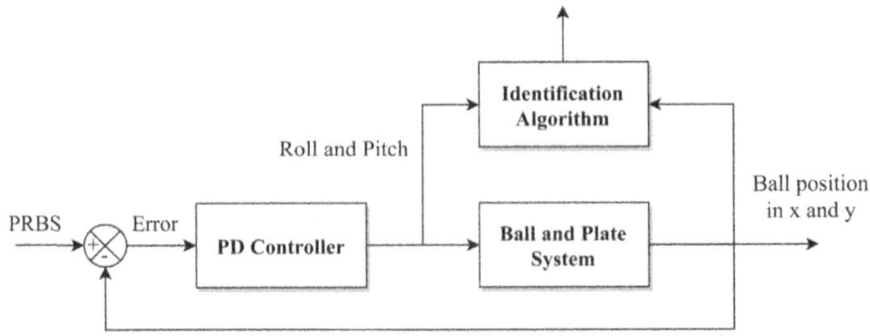

Fig. 5 Scheme of the system for data collection during identification

3.2 Closed Loop Identification

For this system, if a constant excitation is provided in *roll* or *pitch*, the ball will fall, if the platform were infinite the ball would continue to roll, it is concluded that the system is unstable, and it is necessary to stabilize it.

Although it is recommended to avoid it when possible, for unstable systems the most suitable alternative is to use closed-loop identification [28]. The use of this method is proposed in unstable plants and when the open loop could involve problems or unwanted behaviors, although it is noted that the closed loop can trigger inaccuracy in the result [29]. There are several methods for closed-loop identification: the direct approach, where the input and output are measured directly from the process, and the indirect approach, in which the reference and the output are measured, and where we must know the controller [30]. As shown in Fig. 5, the direct method is used, since PRBS will be used to excite the system providing a good noise model [31], and also because the configuration of the system allows us to take measurements directly from different stages of the process.

To stabilize the system, a PD controller is used, whose parameters were obtained taking as reference the methods described in [32], a system that, although it is much simpler in construction, uses methods that serve as a starting point to tune the controller parameters. These parameters are k_p 0.0075 and 0.0026 for x and y respectively, and k_d 0.023 and 0.020 for x and y respectively.

3.3 Models Resulting from Identification

The system will be excited with a PRBS of order 7, with this configuration, the ball can roll for a considerable time without falling. The system sampling time is 0.133 s. The identification algorithm used is the non-linear least squares method. For the x-axis, 2290 samples were taken, of which 1700 were used for identification and 590 for validation. In the y-axis, 525 samples were taken, 350 for identification and

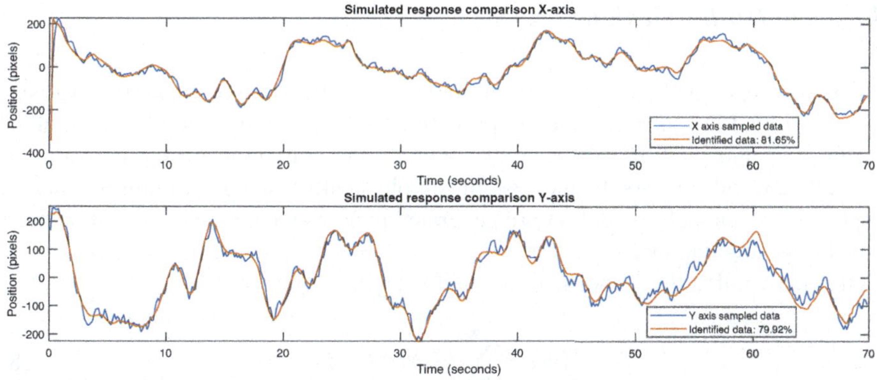

Fig. 6 Comparison between sampled and identified data

175 for validation. The transfer functions with the best percentage of coincidence (81.65% and 79.92% for x and y respectively) were:

$$\frac{pos_x}{roll} = \frac{6.568s^2 + 335.7s - 2.1}{s^3 + 6.295s^2 + 1.827s + 0.542} \tag{6}$$

$$\frac{pos_y}{pitch} = \frac{5.056s^3 + 63.46s^2 + 3.492s + 0.4599}{s^3 + 0.360s^2 + 0.1312s + 0.002924} \tag{7}$$

The comparison between sampled and identified data is shown in Fig. 6.

4 Dynamic Matrix Control Design

4.1 Dynamic Matrix Control

Dynamic matrix control or DMC is a type of model-based predictive control (MPC) [33] since they use an explicit model to predict the output of the process in future time instants [34].

DMC uses as a model the step-response of the system [35] and predicts the dynamics of the process in a time window known as the prediction horizon, estimates the movements of the control action in another time window known as the control horizon based on the minimization of a cost function [36]. DMC is applicable to asymptotically stable linear systems; for unstable systems, this algorithm can be applied after system stabilizing [37].

4.2 Controller Design

The following considerations are taken for the controller design: the system was stabilized with a PD controller; the sampling time is 0.133 s; the manipulated variables are the *roll* and *pitch* in degrees and the controlled variables are the positions of the ball in x and y in pixels; the system can physically reach maximum inclination angles of 15°, in addition delays are generated in cases where the engine arms must travel large angular arcs.

It is known that the response of a system can be expressed as:

$$y(t) = \sum_{i=1}^{\infty} g_i \Delta U(t-i) \tag{8}$$

A prediction can be estimated with **k** samples in the future, considering a disturbance $\hat{\boldsymbol{n}}$, with:

$$\hat{y}(t+k|t) = y_0 + \sum_{i=1}^{\infty} g_i \Delta U(t+k-i) + \hat{n}(t+k|t) \tag{9}$$

Considering a constant disturbance, being the difference between the estimated and the measured output, the free response of the system can be defined as **f**, and the estimate as:

$$\hat{y}(t+k|t) = y_0 + \sum_{i=1}^{k} g_i \Delta U(t+k-i) + f(t+k|t) \tag{10}$$

In matrix form, the prediction can be expressed as:

$$\hat{\mathbf{y}} = \mathbf{Gu} + \mathbf{f} \tag{11}$$

where **G** is the dynamic matrix, which is a lower triangular matrix with the coefficients of the step response. If the sample taken from the estimate is subtracted and squared, the least squares of the error is obtained as a cost function that can be optimized; in addition, a penalty will be included in the effort of the control action [35]. The cost function is:

$$J = \sum_{i=1}^{p} e_i^2 + \sum_{j=1}^{m} \lambda \left[\Delta u(t+j-1)\right]^2 \tag{12}$$

where e is the prediction error for each instant in the prediction horizon, Δu is the variation of the control signal, p is the prediction horizon, m is the control horizon and λ is the penalty factor known as performance index. The cost function expressed in matrix form is:

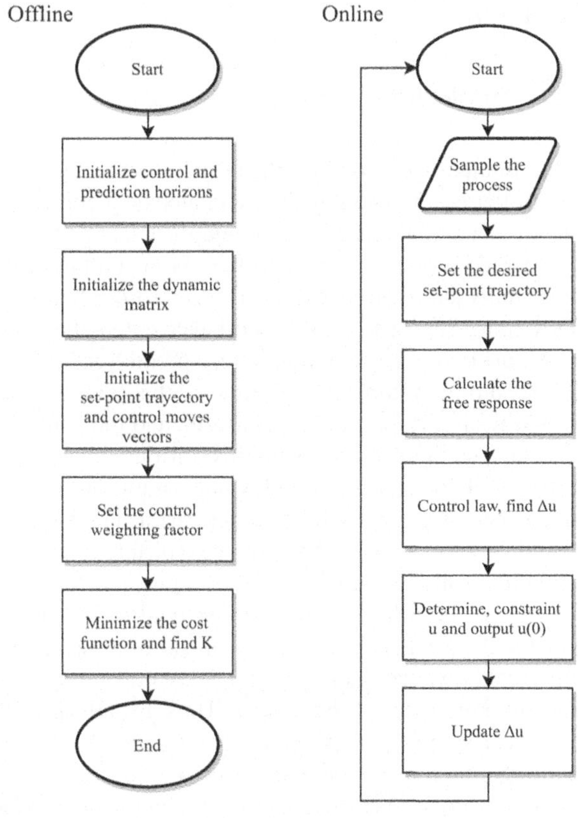

Fig. 7 Flow chart of programmed DMC

$$J = \mathbf{e}\mathbf{e}^{\mathrm{T}} + \lambda \mathbf{u}\mathbf{u}^{\mathrm{T}} \tag{13}$$

Optimizing (13), the control law is obtained:

$$\mathbf{u} = \left(\mathbf{G}^{\mathrm{T}}\mathbf{G} + \lambda \mathbf{I}^{-1}\right) \mathbf{G}^{\mathrm{T}} \left(\mathbf{y}_{\mathbf{sp}} - \mathbf{f}\right) \tag{14}$$

where $\mathbf{y}_{\mathbf{sp}}$ is the vector set point. The optimization is only done at the beginning, so for $t > 0$ a vector **K** can be defined, which is the first row of the expression $\left(\mathbf{G}^{\mathrm{T}}\mathbf{G} + \lambda \mathbf{I}^{-1}\right) \mathbf{G}^{\mathrm{T}}$ [34].

The DMC algorithm programmed in the microcontroller has two stages, an offline in which all the parameters are defined and started, and an online stage (Fig. 7).

5 Results

5.1 Simulation

The simulation diagram for the DMCs is shown in Fig. 8. The presence of the digital controller is simulated by the ZOH blocks or zero order holders [25]. The set-point, the output signal of the system and the execution time are introduced to a function that would represent the controller. At the output of this block, there is a saturation that limits the rotation angles. The zero order holders (ZOH) simulate the A/D and D/A converters, a step-like disturbance enters the system input.

Considering the sampling time, a prediction horizon of 10 will give a perspective 1.33 s into the future, enough time so that with a certain inclination the ball travels a significant distance on the platform, and the control movements are allowed to be executed on the platform; with this, various values of λ are tested. You can notice the effect of the performance index and saturation on the signal (Fig. 9). In the y-axis, it was necessary to establish larger performance index because the physical actions that the motors must perform are very different when trying to modify the pitch than when the roll is modified due to the physical asymmetry of the system, therefore, to change the roll, significant movement of only two motors is required, while to change the pitch, significant movement of at least three motors is required and in a different way, even generating a delay depending on the angles to be reached and the previous states of the motors. Thus, for all cases, the overshoot does not exceed 5%, the rise time is approximately 0.6 s, the steady-state error tends to zero and the settling time remains between 1.0 s to 1.5 s.

Tests were also carried out by introducing disturbances in cases where the highest performance index was used, to which the controller reacts and controls the position of the ball (Fig. 10).

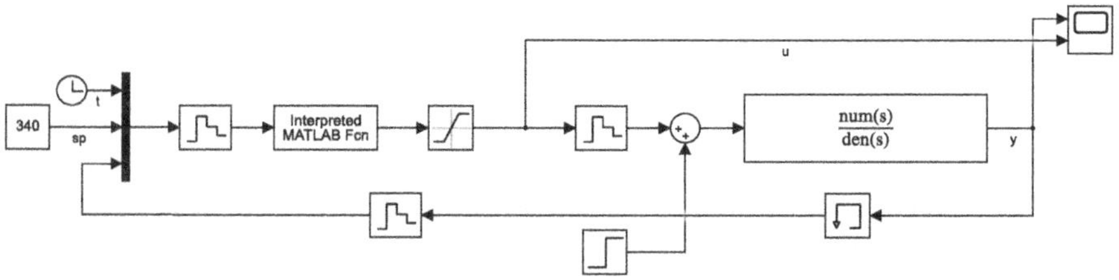

Fig. 8 DMC simulation scheme

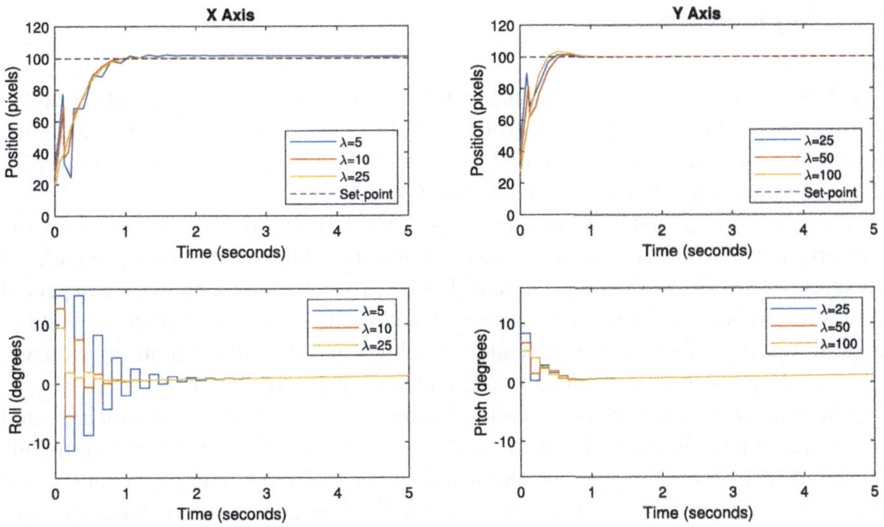

Fig. 9 Simulation of the DMC for various values of λ. Above is the performance of the system. Below is the behavior of the control signal

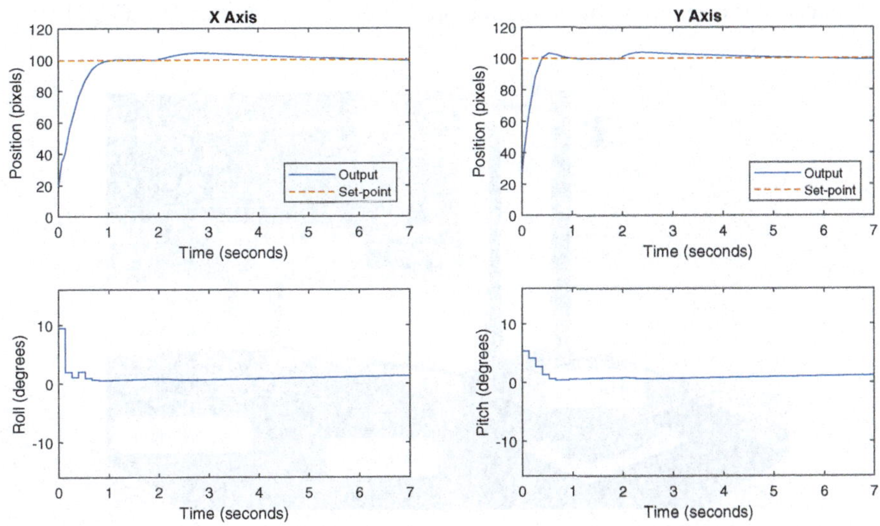

Fig. 10 DMC response to step-type disturbances

5.2 Implementation

The DMC was programmed starting from the most successful parameters of the simulation and then tuning the controller. Finally, a prediction of approximately one second is established, this is detailed in the Table 1. The physical configuration for the execution of the experiments is shown in Fig. 11.

The tests carried out consisted of placing the ball in one of the corners of the platform and releasing it without giving it any impulse. The results that are considered most relevant for analysis are presented (Fig. 12, 13, 14). Table 2 shows the expected values in terms of performance parameters and their respective standard deviation. In addition, one more case is presented in which two disturbances are introduced, when pushing the ball after having reached the set-point (Fig. 15).

The controller makes the system set to average times less than 6 s and with expected steady state errors between 8 and 11%. The standard deviation of the measurements is around 15% and 6% for the overshoot and error parameters. These percentages are due to the wear of the platform, since its material has scratches that act as grooves where the ball tends to go; this is clearly seen in some tests where close to reaching the set point the control signal varies substantially due to having to act more aggressively as there is a lot of difference between the prediction and what is actually measured, especially in the central parts doing that prediction at these points fails (Fig. 13).

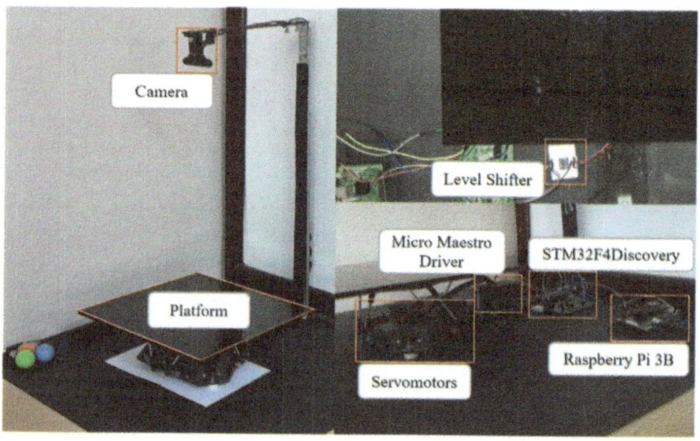

Fig. 11 Physical implementation of the system

Table 1 Parameters of the implemented DMCs

Parameter	X Axis	Y Axis
λ	13	25
p	7	7
m	5	5

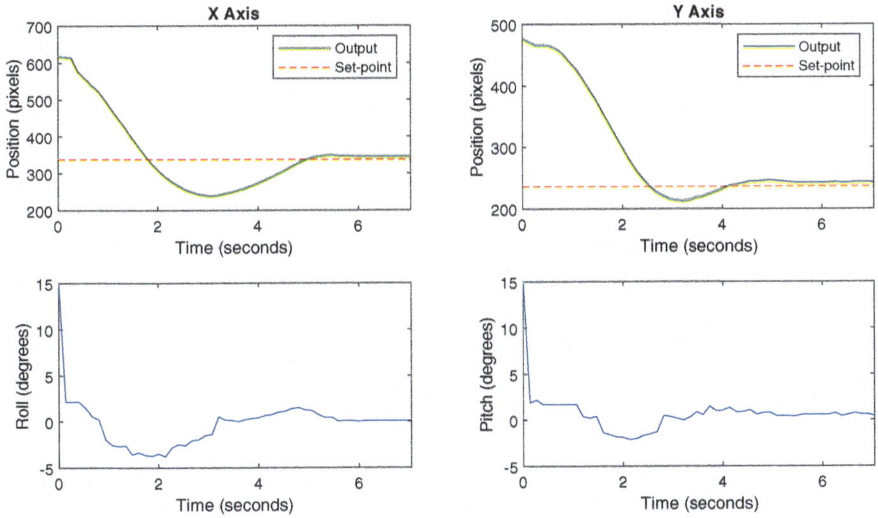

Fig. 12 DMC test closest to expected performance

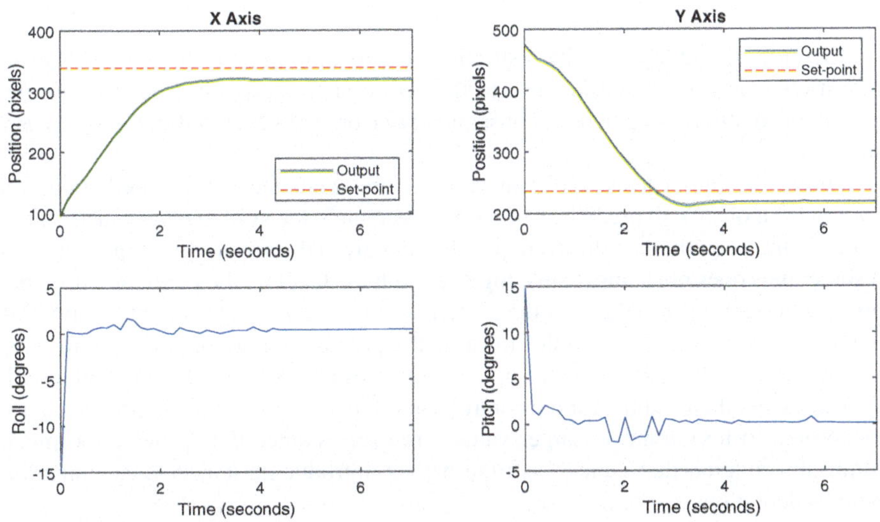

Fig. 13 Trial of the DMC with the best establishment time

Table 2 Summary of the DMC performance tests implemented

	X Axis				*Y* Axis			
Parameter	Mp [%]	$e_s s$ [%]	t_r [s]	t_s [s]	Mp [%]	$e_s s$ [%]	t_r [s]	t_s [s]
$\bar{x}$	19.05	10.26	1.67	5.37	14.35	8.92	1.85	5.10
m	17.45	6.32	0.41	1.65	14.19	5.46	0.73	1.86

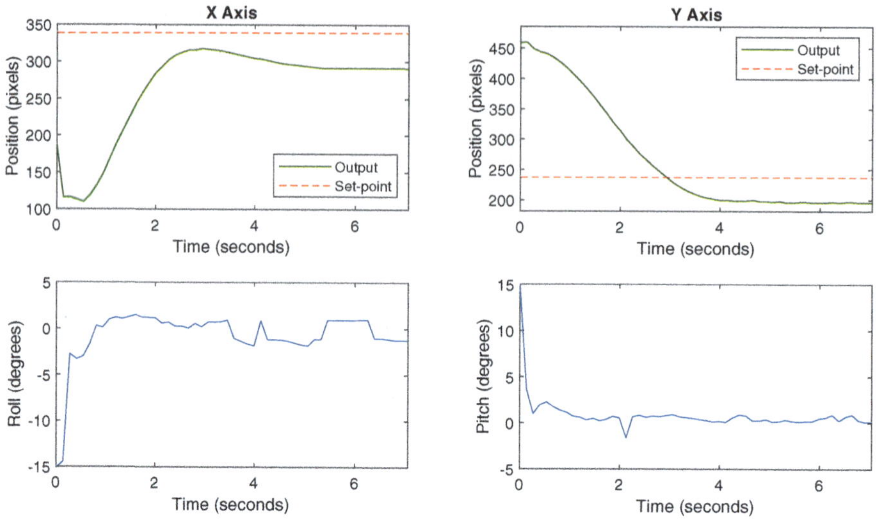

Fig. 14 DMC test with the highest steady-state error

Therefore, in addition to the restrictions on the amplitude of the control signal, a restriction when the ball is close to the set-point was programmed, causing the controller to almost stop acting, consequently errors are generated in the stationary state (Fig. 14).

During the initial tests, different values were evaluated for the prediction and control horizons. With prediction horizons less than 5, the system takes much longer to reach the set-point than the average value and given the sensitivity of the actuators it sometimes does not achieve this, together with the fact that the prediction does not provide enough information on the system so that it reacts. With control horizons that approach less than 3 samples towards the prediction horizon, the system fails, as it generates aggressive changes in the control signal which causes the ball to fall.

Values less than 2 and 4 for the control and prediction horizons respectively cause the system to lose control. Larger values are useless since they provide too much future information that change as time passes, introducing unnecessary computational calculations.

Another remarkable trend is that the more the performance index is reduced, the more the system becomes unstable, this is mainly since the system demands actions that are physically impossible for real actuators, such as inclinations out of range, or drastic changes in the direction of rotation, producing delays in the system, both to reach the required angle of inclination, as well as to change the direction of the ball. On the contrary, small performance index values cause the platform not to rotate enough and the ball to fall.

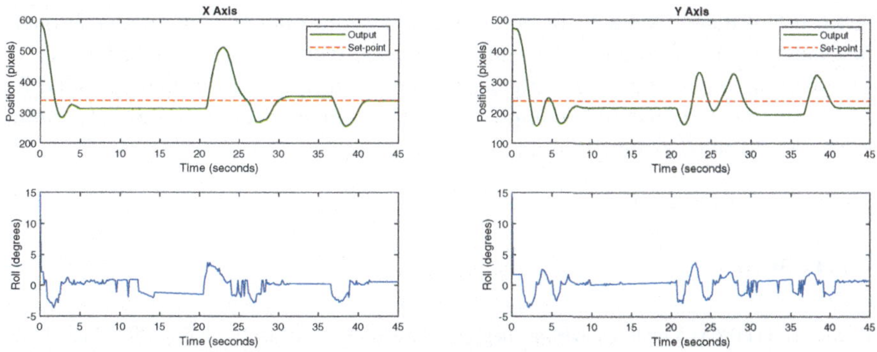

Fig. 15 Test of the behavior of the DMC against disturbances

In the scenario where steady state is reached when pushing the ball, it is noted that the controller acts to return the ball to the desired position. Faced with this type of disturbance, it is expected that the system will take around 6 s to return to the stable state, which may be a longer time if the ball is influenced by the wear problems of the platform (Fig. 15). A solution to this problem could be to use a ball with a greater mass, although if so, the system should be re-identified and the controller tuned for this case, which is a disadvantage compared to other strategies of control which are more flexible being able to a certain extent, dispense with a mathematical model of the system.

Regarding execution times, within the microcontroller, which works at 168 MHz, the inverse kinematics calculations are executed in 820 μs, the stabilization PID is executed in 4 μs and the DMC calculations are carried out online performed at 101 μs and 103 μs for the x and y axis, respectively. On the other hand, the digital acquisition and processing of the camera image takes 115.2 ms to be executed by the Raspberry Pi, while the serial sending of the data takes 17.8 ms longer, resulting in 133 ms corresponding to the sampling time.

6 Conclusion

In this paper, the implementation of a dynamic matrix control strategy was proposed in a system consisting of a Stewart platform for the ball and plate application. All the components of the system were detailed, in addition to the methodology to arrive at the implemented design of the controller, including an identification stage, which was done in a closed loop given the nature of the system. The results show that after stabilizing the system, DMC is applicable for this application. With the DMC, stabilization times of less than six seconds were achieved. It was also shown that the controller reacts to disturbances by returning the ball to the desired position. There was a need to consider the physical constraints of the system, and to program extra routines that affected the performance of the controller by generating stationary pulses and errors of around 10% and 15% respectively. Future work will focus on implementing other control techniques and comparing their performance with DMC. In addition, the system could be improved, not only in terms of the control algorithm, but also in its physical construction, in order to improve the sampling time.

References

1. Stewart D (1965) A platform with six degrees of freedom. Proc Inst Mech Eng 180(15):371–386
2. Bo Y, Zhongcai P, Zhiyong T (2011) Fuzzy PID control of stewart platform. In: Proceedings of 2011 International Conference on Fluid Power and Mechatronics, pp 763–768
3. Moriones E (2016) Diseño, implementación y control de sistema de balance ball on plate. Revista SENNOVA 2(2):134–149
4. Boian RF, Bouzit M, Burdea GC, Lewis J, Deutsch JE (2005) Dual Stewart platform mobility simulator. In: 9th International Conference on Rehabilitation Robotics, pp 550–555
5. Full Motion Dynamics: About the Project. http://fullmotiondynamics.com/about-the-project. Accessed 30 May 2021
6. Koch PM, Kesteven M, Nishioka H, Jiang H, Lin KY, Umetsu K et al (2009) The AmiBA hexapod telescope. Astrophys J 694(2):1670–1684
7. Basim N, Sharma RN, Vignesh A, Dinesh P, Ajith I (2018) Design of modified stewart platform for solar tracing applications. Int J Comput Appl 180(38):33–40
8. Ramdani A, Grouni S (2017) Dynamic matrix control and generalized predictive control, comparison study with IMC-PID. Int J Hydrog Energy 42(28):17561–17570
9. Campoverde M, Guayasamín R (2018) Diseño, simulación y comparación de tres controladores predictivos: control predictivo generalizado, control por matriz dinámica y predictor de Smith robusto, aplicados a un reactor de agitación continua y un tanque de mezclado con retardo [Bachelor's thesis]. EPN. Quito
10. Rodriguez M (2017) Desarrollo de un sistema de control predictivo de la temperatura en un reactor de transesterificación. PUCP, Lima
11. Garcia CA, Naranjo JE, Gallardo-Cardenas F, Garcia MV (2019) Virtual environment for training oil & gas industry workers. In: De Paolis LT, Bourdot P (eds) AVR 2019, vol 11614. LNCS. Springer, Cham, pp 379–392. https://doi.org/10.1007/978-3-030-25999-0_32
12. Garcia CA, Naranjo JE, Alvarez-M. E, Garcia MV (2019) Training virtual environment for teaching simulation and control of pneumatic systems. In: De Paolis LT, Bourdot P (eds) AVR 2019, vol 11613. LNCS. Springer, Cham, pp 91–104. https://doi.org/10.1007/978-3-030-25965-5_8
13. Arteaga F, Contreras J (2003) Identificación de sistemas y control de matriz dinámica para la optimización de una planta de endulzamiento de gas. Revista Ingeniería UC 10(2):79–86
14. Oravec M, Jadlovska A (2017) Intelligent positioning plate predictive control and concept of diagnosis system design. J Manuf Ind Eng 15(1–2):1–9
15. Gawkowski P, Ławryńczuk M, Marusak P, Tatjewski P, Sonowski J (2008) Software implementation of explicit DMC algorithm with improved dependability. In: Proceedings of Novel Algorithms and Techniques In Telecommunications, Automation and Industrial Electronics. Dordrecht, pp 215–219
16. Bang H, Lee YS (2018) Implementation of a ball and plate control system using sliding mode control. IEEE Access 6:32401–32408
17. Spacek L, Vojtesek J, Gazdos F, Kadavy T (2018) Ball & plate model for robotic system. In: Proceedings of 32nd European Conference on Modelling and Simulation. Wilhelmshaven, pp 226–231

18. Ham C, Taufiq MM (2015) Development of a ball and plate system. In: Proceedings of 2015 ASEE Annual Conference & Exposition. Seattle, Washington: ASEE Conferences, pp 26.518.1–26.518.13
19. Ojo-Gonzalez K, Prosper-Heredia R, Dominguez-Quintero L, Vargas-Lombardo M (2021) A model DevOps framework for SaaS in the cloud. In: García MV, Fernández-Peña F, Gordón-Gallegos C (eds) Advances and Applications in Computer Science, Electronics and Industrial Engineering, vol 1307. AISC. Springer, Singapore, pp 37–51. https://doi.org/10.1007/978-981-33-4565-2_3
20. Xiao J, Buttazzo G (2016) Adaptive embedded control for a ball and plate system. In: Proceedings of The Eighth International Conference on Adaptive and Self-Adaptive Systems and Applications, pp 40–45
21. Kassem A, Haddad H, Albitar C (2015) Commparison between different methods of control of ball and plate system with 6dof stewart platform. In: IFAC-PapersOnLine. 1st IFAC Conference on Modelling, Identification and Control of Nonlinear Systems-MICNON 2015, vol 48, no 11, pp 47–52
22. Barrientos A, Peñín LF, Balaguer C, Aracil R (2007) Fundamentos de Robótica, 2nd edn. McGraw-Hill, Madrid
23. Eisele R: Inverse Kinematics of a Stewart Platform–Computer Science and Machine Learning. https://www.xarg.org/paper/inverse-kinematics-of-a-stewart-platform
24. Tangirala AK (2015) Principles of System Identification: Theory and Practice. CRC Press, Boca Raton
25. Landau ID, Zito G (2006) Digital Control Systems: Design, Identification and Implementation, 1st edn. Springer-Verlag, London
26. Ayala JP, Bernal CP, Guerra S, Fernández A (2014) Identificación de sistemas basado en remuestreo de señales periodicas implementado en FPGA. Energética 43:23–29
27. Loya H, Enríquez V, Salazar FW, Sánchez C, Urrutia F, Buele J (2020) Analysis and determination of minimum requirements of an autopilot for the control of unmanned aerial vehicles (UAV). In: Nummenmaa J, Pérez-González F, Domenech-Lega B, Vaunat J, Oscar Fernández-Peña F (eds) CSEI 2019, vol 1078. AISC. Springer, Cham, pp 129–142. https://doi.org/10.1007/978-3-030-33614-1_9
28. Poulsen NK (2019) Closed loop identification. In: Proceedings of Stochastic Adaptive Control, pp 1–28
29. Forssell U (1999) Closed-loop Identification: Methods, Theory, and Applications. Linköping University, Automatic Control, The Institute of Technology
30. Mandloi R, Shah P (2015) Methods for closed loop system identification in industry. J Chem Pharm Res 7(1):892–896
31. Jianhong W, Ramirez-Mendoza RA (2020) The practical analysis for closed-loop system identification. Cogent Eng 7(1):1796895
32. Arruga S (2011) Control De Posicionamiento De La Plataforma Hexapod. Universitat Politècnica de Catalunya (UPC), Barcelona
33. Ruchika NR (2013) Model predictive control: history and development. Int J Eng Trends Technol 4(6):2600–2602
34. Camacho EF, Bordons C (2007) Model Predictive Control, 2nd edn. Springer-Verlag, London
35. Cutler C.: Dynamic Controller. https://www.controlglobal.com/articles/2003/159/
36. Bao-Cang D (2010) Modern Predictive Control, 1st edn. CRC Press, Boca Raton
37. Xi Y, Li D (2019) Predictive Control: Fundamentals and Developments, 1st edn. Wiley, Shanghai, Hoboken

Coordination of Two Robots for Manipulating Heavy and Large Payloads Collaboratively: SOFOCLES Project Case Use

Pablo Bengoa, Itzel de Jesus González-Ojeda, Aitor Ibarguren, Borja Goenaga, Sandra Martínez-De-Lahidalga, Christos Gkournelos, Konstantinos Lotsaris, Panagiotis Angelakis, Sotiris Makris, and Juan Carlos Antolín-Urbaneja

Abstract During the last two decades a great impulse in the development of applications to improve the collaboration between human and robots has been achieved in several sectors. This paper presents a new research carried out for coordinating two robots with the aim to manipulate heavy and large payloads up to 250 kg. This

P. Bengoa (✉) · I. de J. González-Ojeda · A. Ibarguren · B. Goenaga · S. Martínez-De-Lahidalga · J. C. Antolín-Urbaneja
Advanced Manufacturing Unit, TECNALIA, TECNALIA, Basque Research and Technology Alliance, (BRTA), 20009 Donostia-San Sebastián (Gipuzkoa), Spain
e-mail: pablo.bengoa@tecnalia.com

I. de J. González-Ojeda
e-mail: itzel.gonzalez@tecnalia.com

A. Ibarguren
e-mail: aitor.ibarguren@tecnalia.com

B. Goenaga
e-mail: borja.goenaga@tecnalia.com

S. Martínez-De-Lahidalga
e-mail: sandra.martinezdelahidalga@tecnalia.com

J. C. Antolín-Urbaneja
e-mail: juancarlos.antolin@tecnalia.com

C. Gkournelos · K. Lotsaris · P. Angelakis · S. Makris
Laboratory for Manufacturing Systems and Automation, Department of Mechanical Engineering and Aeronautics, University of Patras, 26504 Patras, Greece
e-mail: gkournelos@lms.mech.upatras.gr

K. Lotsaris
e-mail: lotsaris@lms.mech.upatras.gr

P. Angelakis
e-mail: angelakis@lms.mech.upatras.gr

S. Makris
e-mail: makris@lms.mech.upatras.gr

© The Author(s), under exclusive license to Springer Nature Switzerland AG 2022
M. V. Garcia et al. (eds.), *Advances and Applications in Computer Science, Electronics, and Industrial Engineering*, Lecture Notes in Networks and Systems 433,
https://doi.org/10.1007/978-3-030-97719-1_15

process is performed with the collaboration of the final user who clamps each robot to one point of the part by means of a flexible and reconfigurable tool. The architecture of the system and the followed methodology is envisaged. Also, information about the generation of the trajectory is explained. The work is focused on the coordination problem of two collaboratives robots used in a pick and place application, synchronized with a second process (e.g., sealing, painting, welding). The test has been performed using an 80 kg payload, handling the part synchronously using both collaborative robots, showing good articular coordinates tracking. This research has been done within SOFOCLES project, funded by the European Institute of Innovation and Technology (EIT), a body of the European Union.

Keywords Cobots · Dual-Arm Robots · Collaboratively · Heavy and large payloads · Digital Twin · Coordination

1 Introduction

During the last decades the use of robots in industrial environment is continuously increasing. In fact, according to Zanchettin et al. [1], only in 2016 nearly 300,000 units of industrial robots were installed worldwide. Moreover, this kind of devices have been introduced in a wide variety of industrial sectors like packaging, assembling, welding or inspection, among others. This intensive use of robots in such different sectors is due to the possibility of carrying out repetitive tasks, with a repeatability error. In this sense, industrial robots capable of supporting small and medium payloads are being used to pick and place parts with significant dimensions and weight without any support of a labour. However, one of the drawbacks is the lack of flexibility due to the fact they need to be provided with some security systems, like barriers, fences or electronic devices. These systems prevent risk situations when the access of a person inside the working area of the robot is required. This essential interaction in such a shared workspace introduces novel mechanical risks with respect to "traditional" industrial robots [2].

Nevertheless, many applications need a collaborative work between robots and persons, specifically when humans provide know-how and cognitive skills qualifying the activity. Thanks to the development of the Industry 4.0, during the last two decades the robotic industry has been developing robots to be used in presence of humans. This kind of robots, also called collaborative robots (hereinafter, cobots), provides increased versatility, process adaptability, and some kind of intrinsic safety compared to the industrial robots. Cobots are used to complete tasks together with the human workforce while sharing the same working space, offering greater mobility and flexibility [3]. However, the safety issues concerning these systems continues worrying researches when human-robot collaboration is putting into practice in real situations [4].

In the literature, other researches have been led to investigate and work out problems related to simultaneous coordination between robots which overlap workspaces ensuring smooth and collision free transitions [5]. Others are working into planning and execution of trajectory generation for industrial applications using collaborative systems [2, 6–8]. Rojas et al. have been working in multicriteria trajectory planning approach with the aim to provide a trade-off between smoothness and the disadvantage of reducing the jerk through a composite cost function [2]. Other studies are focused on the integration of this kind of robots in assembly systems with the aim to handle small and light parts, limiting the maximum payload up to 10 kg [7, 9, 10] in a collaborative mode. After looking for researches in dual collaborative robotics for handling payloads higher than 15 kg, only the Fanuc RC-35iA cobot is able to operate at maximum payload of 35 kg [11]. Nevertheless, during the last years new cobots with higher capabilities has been reached to the market in order to handle bigger payloads up to 170 kg [12]. In spite of that, nowadays, new boosts must be carried out in order to overcome limitations in handling heavy and voluminous payload using only one cobot when sequence of movements are really complex. These complex trajectories can tackle movements turning the part 180° in different planes. One solution can be based on the utilization of two cobots handling large and weighty parts synchronously.

This paper describes the work performed under SOFOCLES project (https://eit-mfg-sofocles.eu/) which develops a flexible, reconfigurable and smart collaborative robotic system for handling and clamping large, heavy, bulky and complex aeronautical components, but not limited, to be used in different processes, e.g. sealing and welding, using two collaborative robots at the same time. The final goal is to integrate this robotic system in a productive work cell to coordinate this dual-robot system for high payloads, up to 250 kg, with a high degree of autonomy, but collaborating with humans, safely. This solution minimizes the limitation of handling weighty and bulky parts. Also, the system has been prepared to be managed collaboratively by an operator who is interacting with the two cobots by means of a Digital Twin (DT) interface taking advantages of Robot Operating System (ROS) tools [13]. The main innovation under this research corresponds to the availability of handling heavy and bulky payloads, turning the part 180° in different planes. This fact makes easy the labour to be performed, nowadays, by the process end user. This solution can be easily implemented in other processes apart from sealing and welding, like additive manufacturing, drilling, sanding and painting among others. The work described in this paper is only delimited to the coordination of two robots, from the grasping position to the ungrasping position, driving out of the study the integration of a third-part process.

The rest of the paper is structured as follows. Section 2 introduces the description of the complete system consisting in the hardware architecture and the DT of the system. Section 3 explains the methodology carried out to develop the application based on three software main layers: the path generator, robot trajectory generator and trajectory executor. Section 4 is focused on the description of the performed tests and the obtained preliminary results using a representative part. Finally, conclusions and future work are presented in Sect. 5.

2 Description of the System

As indicated in the introduction, the aim of the SOFOCLES project is the development of a flexible, reconfigurable and collaborative robotic system for handling and clamping large, heavy, bulky and complex parts reaching up to 250 kg. As there is not a commercial, flexible and collaborative robotic system able to handle those high payloads (up to nine different parts), with dimensions around 2800 × 900 mm (length x width), a solution of two robots working co-ordinately has been envisaged. The goal is to move, to turn and to support those parts in a specific position meanwhile the second process (e.g. sealing, welding or painting) is starting its work. The cobots must follow their trajectories synchronously in order to keep the part in the correct position in the world frame minimising additional stresses.

To fulfil the specification of being able to hold such weight and size of the parts, two AURA robots were selected. This kind of robots are able to hold payloads up to 170 kg each one. Each robot has coupled a reconfigurable and flexible tool to clamp part by means of a guidance tool, managed by an end user. Each guidance tool drives by hands the cobot in order to place the clamping device in front of the part. The drawback of these cobots comes from the unavailability of controlling two of them with the same controller. Therefore, a High Level Controller (HLC) upstream of robot controllers is needed to command them. This HLC generates the required trajectory off-line, checking that the generated trajectory is able to be executed by the robots. In next two paragraph the description of the hardware architecture and the DT are included.

2.1 Hardware Architecture Description

The hardware architecture of the system is illustrated in Fig. 1. The robotic system is based on two collaborative AURA robots manufactured by COMAU. These robots are provided with sensorized skins, stopping completely the robot if pressure contact is detected. In that case, a manual reset is needed. If sensorized skins detects the presence of an object or a person in its proximity, the robot stops and, after a while, it rearms itself automatically. At the same time, as laser barriers scan the working area, the speed is reduced in the proximity zone.

This functionality provides additional safe degree in non-collaborative situations, the hardware architecture includes a security barrier by means of two laser sensors (SLS-SA5-08) properly located in the working area to limit the collaboration zone and covering the area of interest. These two sensors are connected to a safety PLC, CPU1214F-DC, which interchanges some variables with third-party robot controller and two AURA controllers in order to stop the operation. The threshold limits to differentiate between both zone and the speed limits can be modified by software. This functionality reduces risks in case of any collisions with parts or operators. In addition, it allows to perform manual operations in safe conditions, without the need

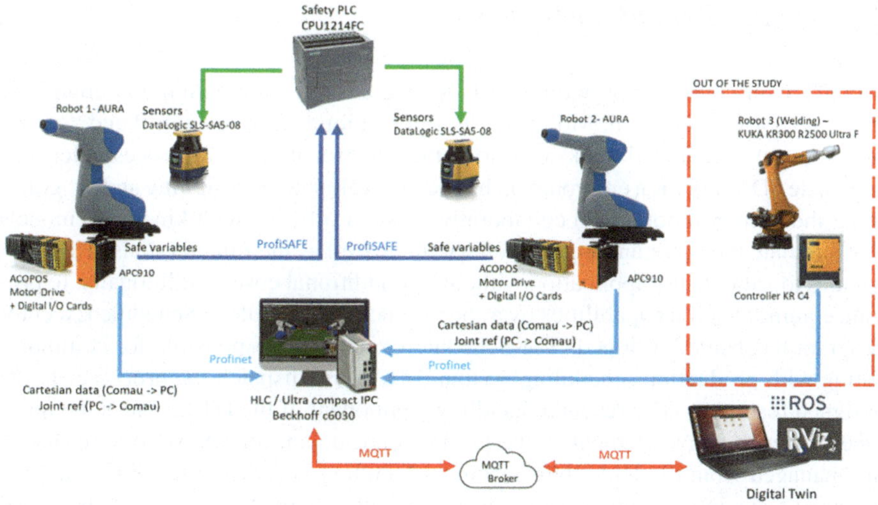

Fig. 1 Architecture of the complete system

for fences [12], guiding the robot inside its reach. Despite of that, these robots can be operated in non-collaborative way, as an industrial robot. To finish the description of Fig. 1, the system includes a third-party non-collaborative robot (out of the scope of this work) to apply a final operation, e.g., sealing or welding.

The main innovation underlies in coordinating two robots to hold, transport and change the orientation of an unique bulky and heavy part at the same time. Nowadays, the coordination of two robots is performed using the same controller to manipulate all axes at the same time. However, the controller of the AURA robots, APC910, is connected via Powerlink synchronous communication to the ACOPOS motor drive system. The former system is only able to manipulate up to 8 axes. Hence, in order to control up to 16 axes at the same time a HLC is required, being the master of the application. The selected HLC for generating the trajectories to be executed in each robot is an ultracompact c6030 Industrial PC (IPC). This device is connected to each robot controller using PROFINET network. It is in charge of sending the commands to each robot, following the implemented criteria in the state machine structure. The other main task of this HLC is the generation of the trajectory. This controller receives the commands from the Digital Twin (DT) that monitors the whole production system. This connection is established over the MQTT protocol [14]. DT offers the easy integration of smart Human Machine Interface (HMI), such as tablets or smartphones, for assisting the human operators.

2.2 Digital Twin of the System

The DT is in charge of providing tools for the accurate simulation and control of the SOFOCLES work cell. Moreover, the application is able to control and "smart" monitor of the simulated work cell during the real execution of the process. Hence, a full-scale 3D simulation environment has been developed for designing and programming the human-robot (H-R) collaborative system off-line, with kinematic models that simulate the robot and human behaviour, as well as their interactions. This contributes to validate the application without any additional costs. Additionally, to provide communication capabilities over the production schedule, a Schedule Executor program has been developed. The Schedule Executor is responsible for communicating with production scheduling system (MES) and dispatch the processes to the assigned resources. The resource handling component of the DT provides the interfaces for each active element of the production (human, robot). Robotic resources are managed from the HLC thus, resources handling is required for translating the assigned tasks into specific operations. On the other hand, the operator is informed about the progress of the production and the tasks that are assigned to them throughout smart interfaces. A high-level architecture of the DT is illustrated in Fig. 2.

One of the requirements of the DT is to be used as HMI, allowing easy programming, control and monitoring of the system. For this purpose, the Monitoring Environment of the DT gathers the data from the sensors and the resources and provides: a) online monitoring of the schedule execution and the status of the work cell's devices, b) basic schedule controlling capabilities, and c) information on the operator's tasks. Taking advantages of the MQTT communication protocol between the real and digital world, all described functionalities are available also remotely.

The DT of the SOFOCLES system is used for the accurate simulation of the physical cell, both from a structural and process perspective. The online reconstruction of the work cell provides valuable information during the execution, for tracking

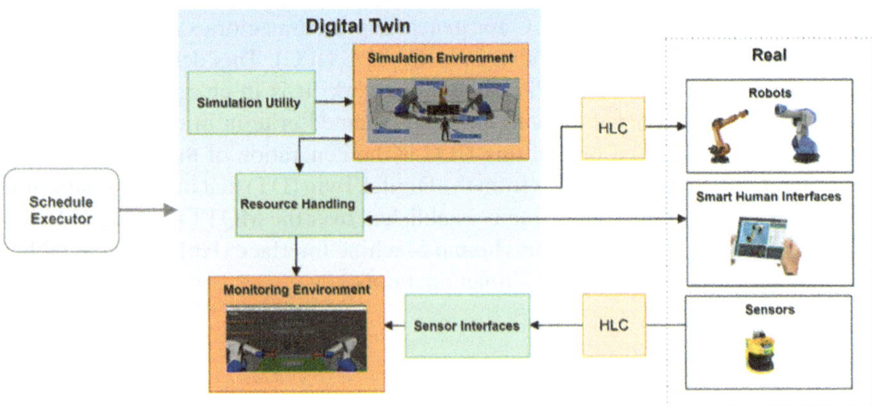

Fig. 2 Component diagram of the Digital Twin

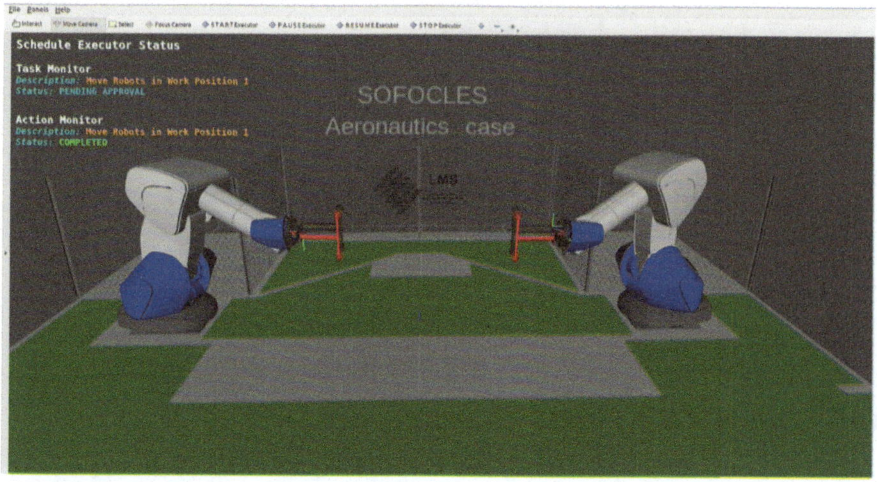

Fig. 3 Digital Twin monitoring environment

the status and the behaviour of the resources and components, collecting data as shown in Fig. 3. This data can be used by the operator for quick decision-making during the execution, as well as for the analysis of the quality and the effectiveness of the production procedure. The update frequency of the DT depends on the linked data sources. In this case of study, the DT exchanges information with the HLC over MQTT. Therefore, this communication channel defines the final monitoring frequency, being in our case 20 kHz.

As seen in Fig. 3 and Fig. 4, DT simulation environment is the component in which the offline design and programming for the H-R collaborative system can be tested and validated. In the first figure, both robots are executing tasks and after placing the part in the correct position, they are waiting to receive the approval in order to follow next tasks. This has been built with the use of the Gazebo [14] simulation software.

Figure 4 shows the outcome of the simulated world for the SOFOCLES sealing case. For this case, it is required to simulate the behaviour of three robots and one operator that transfer the product to the station. ROS is used as a middleware for the interconnection of DT's components.

ROS in combination with Gazebo, provides a set of tools that can facilitate the prototyping and testing of a concept set-up, while making available a handful of high-level controlling features, both for the simulated and the physical work cell. These characteristics, in combination with the hardware abstraction logic, that is inherent to ROS, create the described solution, highly versatile and powerful. In this sense, the schedule of the application is replicated inside Gazebo as seen in Fig. 3. All the needed human and cobots actions and interactions have been simulated for the achievements of the closes-to-real validation of the execution. Moreover, the

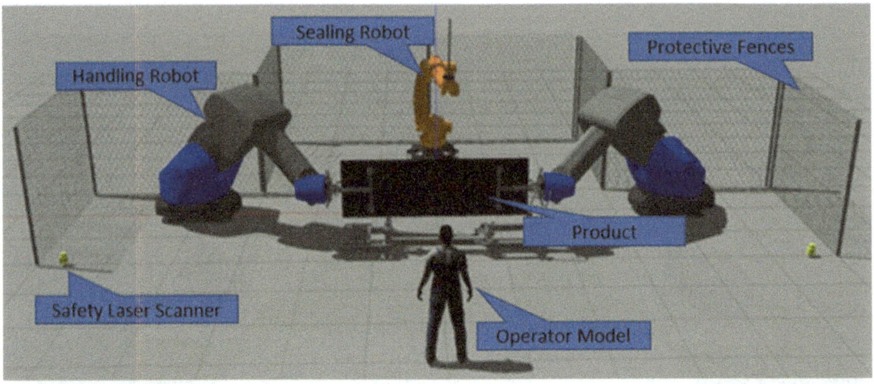

Fig. 4 View of the simulated work cell in Gazebo

components that are needed for the creation of the Gazebo simulation environment are included, as well as the back-end modules that integrate the Gazebo simulator in combination with the core modules of ROS for building the infrastructure of the system.

3 Methodology

The characteristics of the H-R collaborative process define the approach for developing the software application which must be running in the HLC. This fact is due to the necessity of executing similar trajectories at the same time but controlled by different robot controllers. In this sense, once the part is placed in a specific position in the work cell, the process must carry out the following main steps sequentially:

- Move the handling tool of each cobot closer to the part. The tool consists in a manual guidance system (MGS) and the reconfigurable clamping device
- Move the MGS of each robot manually in a collaboratively way to clamp the part using the flexible handling tool. This tool is fixed in the wrist of each AURA cobot. The operator grasps the part and rearm the system
- Once the operator acknowledges the part is correctly attached, execute the required trajectory to present and keep the part in a specific place of the world frame under the work cell
- Execute the required second process operation, performed by a third-party industrial robot (e.g. sealing)
- Change the orientation of the part to follow the process operation in the other side of the part
- Execute the second process operation on the other side of the part if it is required
- Execute the required trajectory to place the part in the ungrasping point
- Ungrasp the part by operator manual operation, using the MGS

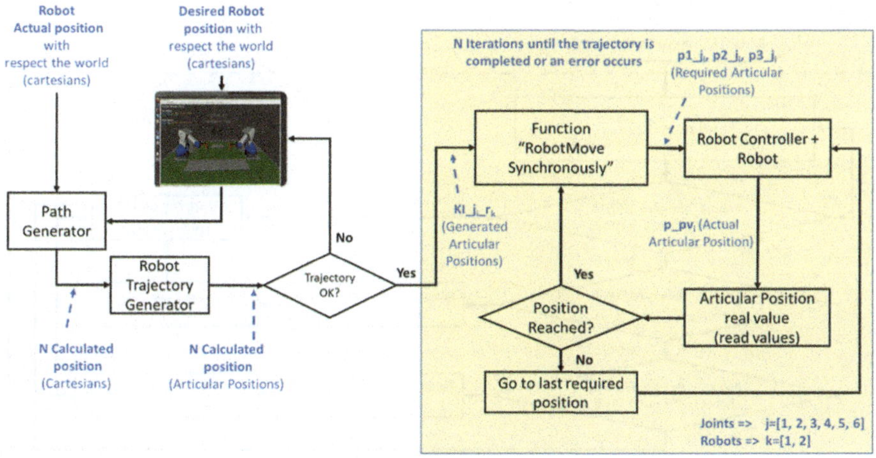

Fig. 5 Block diagram implemented in HLC program

- Remove the cobots to the initial point, if it is required, or repeat the process with another part

Apart from the clamping and ungrasping processes management, the software application must generate the trajectory to place each cobot from home to the vicinity of the part. Moreover, after the clamping and ungrasping processes, each cobot must execute some trajectories in order to place the part in the correct position before executing the second process operation. All these trajectories have been implemented following the trajectory block diagram detailed in Fig. 5. In this sense, k index represents the number of the robot (k = [1, 2]), meanwhile, j index represents the joints of each robots (j = [1,.., 6]). Under the cited block diagram, the function "Robot-MoveSynchronously" has the responsibility of sending the generated points of each robot and control the points flow with the aim to move both robot synchronously following the simplified flow diagram indicated in Fig. 6.

The previous described steps are commanded by the HLC main program. This program is connected to the MQTT server looking for messages to execute new motion commands. In this sense, once the operator introduces some data of the part by means the DT application, the different modules are able to generate the required trajectories depending on a defined State Machine. The executed action of the HLC program is sent to the DT in order to inform about the safety variables and the position of each robot, calculating the speed of the cobot's joints.

The execution of the movements is carried out by means of the Human Machine Interface (HMI) developed as a Digital Twin of the work cell. This HMI acts as the main program of the application. As it has been said before, the main program has been developed in TwinCAT3 and covers all the application flow: firstly, read commands from the HMI; secondly perform the connection with cobots; thirdly, run a C++ program, generating the trajectory and performing the Inverse Kinematic;

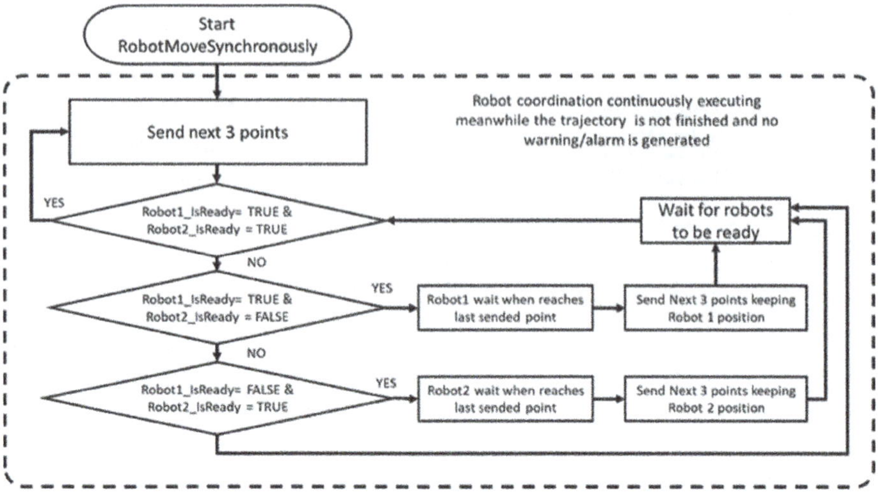

Fig. 6 Function "RobotMoveSynchronously" flow diagram

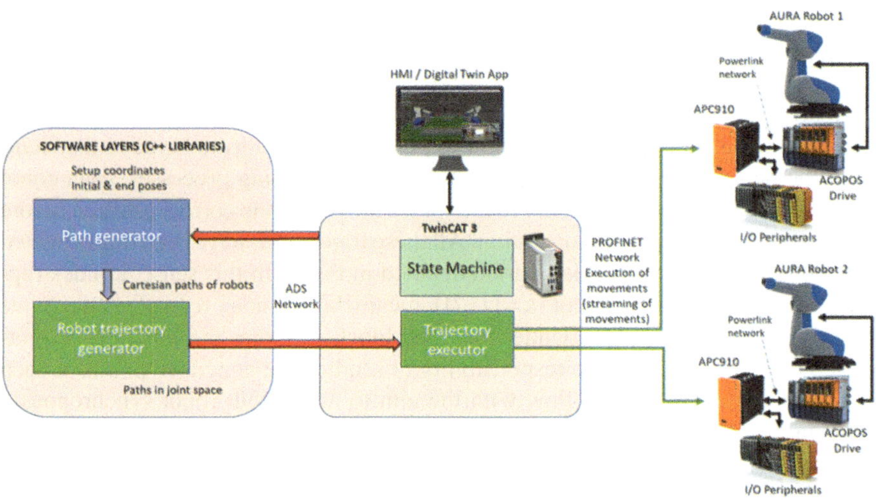

Fig. 7 Software architecture

fourth, execute the coordinated trajectory in the cobots; fifth, update the actual states and, finally, send the actual state to the HMI.

The key of the application rests in the generation and the implementation of the trajectory in both cobots. Therefore, a three-layer software architecture has been ordered to calculate and execute trajectories in both cobots, as indicated in Fig. 7. The initial module (Path generator) will calculate the trajectory to be carried out and a second module (Robot trajectory generator) will verify if the trajectory is achievable by the cobot. After that, the trajectory is executed in the third module, the Trajectory executor, sending the poses to the cobots controllers in order to carry out the synchronized movements by them. In next paragraphs, further details of the three-layers architecture are introduced.

3.1 Part Path Generator

This initial module allows generating the paths (in Cartesian space) that the cobots must travel from the initial to the final position of the manipulated part. The obtained paths provide a series of points in the coordinate system of each of the cobots that allow transporting the part.

Initially, the library allows defining a part trajectory on the world frame, the AURA cobots speed and trajectory to place the part in the desired position. From these data the movement of each cobot is calculated in order to ensure that the trajectories are coordinated to be able to transport large parts safely. The path generator consists of a C++ library based on Eigen library to perform geometric and linear algebra operations. Specifically, the developed C++ library allows:

- the definition of the grasping point of both robots on the part frame (transformation between part origin and grasping points) and the tools of each robot (transformation between flange and tool).
- the parametrization of the points of the trajectory to be executed by cobots, in such a way that the part will travel in the common world frame.
- the parametrization of the handling trajectory and speed to ensure the coordination of the movement of both cobots.
- the implementation of helper functions to change the frames of the calculated trajectories from the world (common frame between robots) to each robot base.

This module aims to generate cartesian trajectories for the two AURA robots in order to achieve the goal part path. The module is implementing a linear interpolator to generate high resolution discrete trajectories, enabling a better control of the manipulator behaviours. The generated robot paths will be sent to the Robot Trajectory Generator module for further kinematic calculations, as explained in the next section.

3.2 Robot Trajectory Generator

The module Robot Trajectory generator performs a kinematic calculation, allowing the description of bodies motion, without considering the external forces (loads, Coriolis forces, centre of mass, inertia moments).

So, this module allows to analyse, calculate and verify the movements that each cobot must follow, according to previously generated path. The goal of this module is the generation of the joints poses for every points of the generated trajectories, taking into a consideration, the space range and maximum speed of every axis.

The kinematics analysis of cobots, defined above, is performed with the Denavit-Harteberg Modified (DHM) parameters. Afterwards, taking into account these data and taking the advantage of Robotics Library (RL) the robot motion that each cobot must follow is calculated. This RL are programmed and operated using C++ language [15]. Also, the RL allows dynamic calculation, hardware abstraction, motion planning, collision detection and visualization, not used in this work.

In the module Part Path generator, the acquired information is in cartesian coordinates defined on the world frame. However, these (desired) positions must be decoded to robot coordinates (articular coordinates) on base frame of each robot. Then, the kinematics inverse is employed. Due to the use of the developed C++ library and the employed RL this module is able to:

- Calculate direct kinematics
- Calculate inverse kinematics
- Gets working range of each axis
- Gets the maximum speeds of each axis defined by the robot manufacture
- Verify the robots can reach every pose of trajectory
- Verify the singularities (in development)
- Motion simulation in 3D

3.3 Trajectory Executor

This last module will execute the trajectories in joint space, ensuring the synchronization of movements. In order to use this function, some mechanisms are needed with the aim to execute movements in the robots in real time to ensure coordination.

This feature is developed in TwinCAT3 inside the main state machine which controls the application. Whenever it is needed the TwinCAT3 program makes a call to the libraries detailed in previous paragraphs, obtaining the complete trajectory of both cobots. Afterwards, the downloaded TwinCAT3 program in HLC commands the COMAU cobots exchanging information using PROFINET network. Additionally, a PDL2 program has been developed in order to get the actual position information and to move the robot according the trajectory.

The complete application tackled in SOFOCLES is composed by two COMAU AURA robots operated by a PC901 B&R controller each cobot. It is worthy to mention that this article only deals with the manipulation part moved by two coordinated cobots, but a second process is expected to be done (e.g. sealing).

During the process, the HLC downloaded program must call the C++ program to obtain the trajectory of each cobot and communicate with both of them. Specifically, the DT will send the midpoints and the end poses. Afterwards, with the initial position obtained from the current position of each COMAU cobot and the data obtained from the DT, the HLC will run the C++ program. This program will return the whole trajectory between the poses of each cobot as well as the signal to start the movement of the third-party industrial robot. These trajectories will be executed in such a way that the coordination between both cobots is satisfied.

This solution allows communicating with both robots, performing a low-level control through a joint position streaming. However, to ensure a coordination between arms, it is highly recommended that both robots have the same time clock frequency in order to obtain the best performance in synchronization movements.

4 Tests and Results

To check the developed application some preliminary tests have been performed using a dummy part of 1125 × 2450 mm weighting about 21 kg. Also, the system has to support two flexible and configurable tools, used for clamping the part in both sides. In these preliminary tests, both robots must support a total weight of about 80 kg.

Figure 8 represents the completed sequence generated by the program. Instead of applying a subprocess in the middle of the operation, a waiting state has been simulated. Hence, after placing the dummy part in a specific position inside the work cell, the process carried out for these preliminary tests followed next steps sequentially:

– Move the handling tool of each robot closer to the dummy part using independent movements (Step 0)

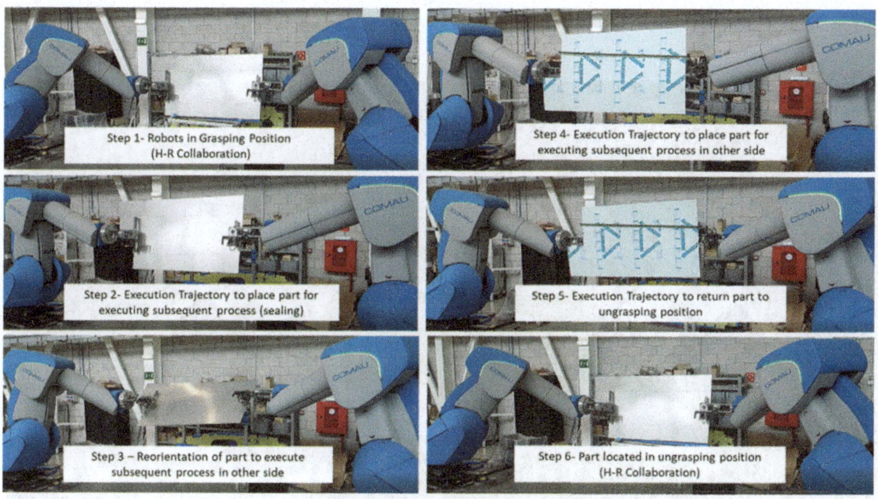

Fig. 8 Step by step handling process

- Move the MGS of each robot manually in a collaboratively way to clamp the dummy part using the flexible handling tool fixed in the wrist of each AURA robot (Step 1-Fig. 8)
- Execute the required coordinated trajectory in both cobots to place the dummy part for executing the second process (Step 2-Fig. 8)
- Keep the dummy part in the same position until the operator's acknowledgment
- Change the orientation of the dummy part to follow the process operation in the other side of the dummy part (Step 3-Fig. 8)
- Execute the required trajectory to place the dummy part for executing the second process in the other side (Step 4-Fig. 8)
- Keep the dummy part in the same position until the operator's acknowledgment
- Execute the required trajectory to place the dummy part in the ungrasping point (Step 5-Fig. 8)
- Ungrasp the dummy part by operator manual operation, using the MGS (Step 6-Fig. 8)

As the schedule of the application has been replicated inside Gazebo, it is possible to check the behaviour of the simulated and real system superimposing images. For the sake of simplicity, in Fig. 9, one photogram of the real execution, moving the dummy part, has been included at the same time that simulation is carried out, as seen at bottom -right of the figure.

Figure 10 illustrates the current position of each robot in joints coordinates. To acknowledge that the position has been reached and execute the next points, the implemented algorithm in HLC device checks continuously the difference between the required position and the current position in each joint of each robot. The coordination in joints coordinates starts at 5,276 milliseconds, being the previous movement of each cobot independents. This has been graphically explained in Fig. 5 and Fig. 6.

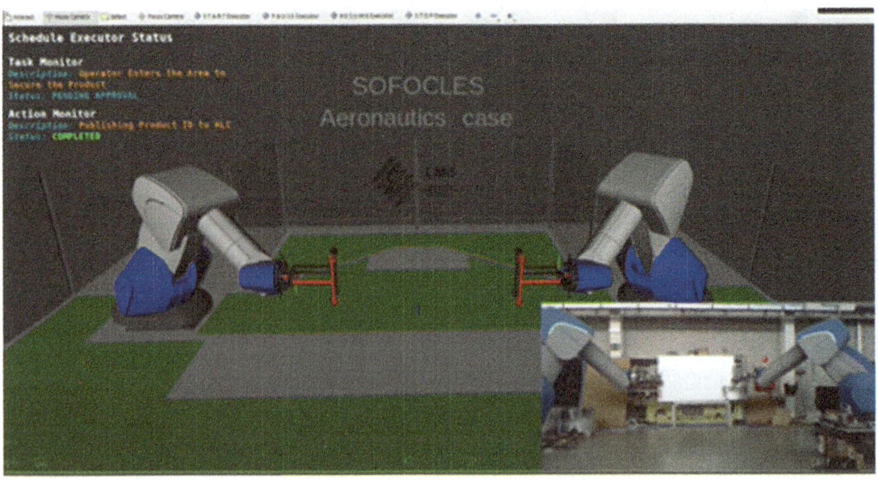

Fig. 9 Digital Twin commanding and monitoring the work cell

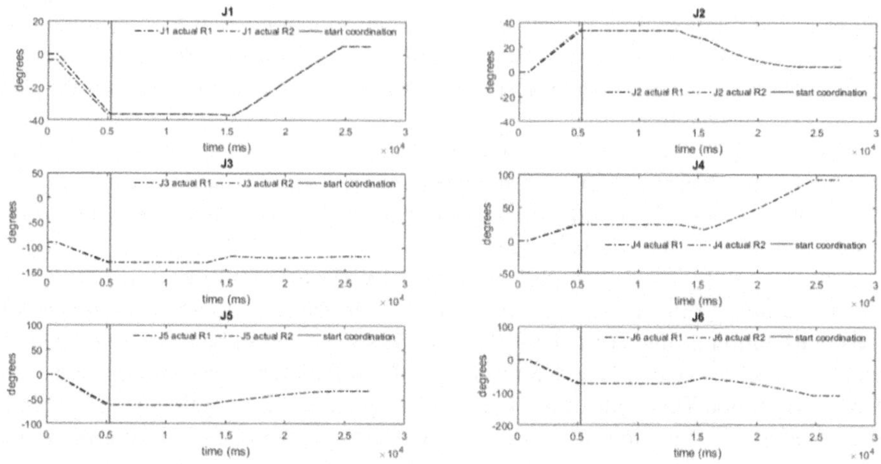

Fig. 10 Comparison of current position of the joints 1 (J1), 2 (J2), 3 (J3), 4 (J4), 5 (J5), 6 (J6)

In each robot, the law in (1) must be fulfilled, being p1i the desired point and pvi the current position of each joint. Therefore, the algorithm performs in every scan cycle the next calculation,

$$\sqrt{\sum_{i=1}^{6}(p1_i - p_v i)^2} < 0.35^\circ \tag{1}$$

Along the studied intervals the difference between the position of each robot's TCP is kept under 0.35°, following the required trajectory. As seen in figures, the first section of the movements up to vertical continuous grey line do not follow a synchronous movement as each robot works independently (Step 0). However, in the second section of the movement, both robots follow the same trajectory synchronously (Step 2). In these graphs, the corrections due to misalignment and the position of both robots respecting the world have been included.

This means that the required synchronization process has been achieved. However, some more tests must be performed in order to assure successful results with real parts, heavier than the used in these preliminary tests.

5 Conclusions

This paper presents a new research carried out for coordinating two robots with the aim to manipulate heavy and large payloads up to 250 kg. This process is performed with the collaboration of the end user who clamps each robot to one point of the part by means of a flexible and reconfigurable tool. The architecture of the system and the followed methodology is described. Also, information about the generation of the trajectory is detailed. The work has been focused on the coordination problem of two robots for handling and transporting the part to the area in which a third-part process is expected to be done.

The accomplished tests with a payload of 80 kg weight, indicate promising results to achieve the objective of moving nine different real parts synchronously with two robots able to work in a collaborative and coordinated way, reaching a precision lower than 0.35° in the whole system.

Future work is devoted to performing tests with different trajectories with the aim to be able to move nine real parts in an industrial environment by means of the coordination of the two cobots. This work is expected to be done at the end of the year. The last step will be led to include the collaboration communication with third-party robot. This industrial robot is in charge of performing the second process (e.g. sealing) on this kind of heavy and bulky parts in both sides during the same complex process.

Acknowledgements The work leading to this publication has been funded by EIT Manufacturing under the code 21055. EIT Manufacturing is supported by the European Institute of Innovation and Technology (EIT), a body of the European Union.

References

1. Zanchettin AM, Croft E, Ding H, Li M (2018) Collaborative robots in the workplace [from the guest editors]. IEEE Robot Autom Mag 25(2):16–17. https://doi.org/10.1109/mra.2018.2822083
2. Rojas RA, Wehrle E, Vidoni R (2020) A multicriteria motion planning approach for combining smoothness and speed in collaborative assembly systems. Appl Sci 10(15):5086. https://doi.org/10.3390/app10155086
3. Sherwani F, Asad MM, Ibrahim BSKK (2020) Collaborative robots and industrial revolution 4.0 (IR 4.0). In: 2020 International Conference on Emerging Trends in Smart Technologies (ICETST). IEEE. https://doi.org/10.1109/icetst49965.2020.9080724
4. Maurtua I, Ibarguren A, Kildal J, Susperregi L, Sierra B (2017) Human–robot collaboration in industrial applications. Int J Adv Rob Syst 14(4):172988141771601. https://doi.org/10.1177/1729881417716010
5. Ćurković P, Dubić K, Jerbić B, Stipančić T (2012) Simultaneous coordination of robots with overlapping workspaces. In: Annals of DAAAM for 2012 & Proceedings of the 23rd International DAAAM Symposium. DAAAM International, pp 549–552
6. Cesta A et al (2017) Planning and execution with robot trajectory generation in industrial human-robot collaboration. In: 4th Italian Workshop on Artificial Intelligence and Robotics, AIRO 2017, vol 2054. CEUR-WS, pp 47–52
7. Ragaglia M, Zanchettin AM, Rocco P (2018) Trajectory generation algorithm for safe human-robot collaboration based on multiple depth sensor measurements. Mechatronics 55:267–281. https://doi.org/10.1016/j.mechatronics.2017.12.009
8. Reyes J et al (2020) A study on modeling and simulation of automobile painting process based on flexsim. In: Nummenmaa J, Pérez-González F, Domenech-Lega B, Vaunat J, Oscar Fernández-Peña F (eds) CSEI 2019, vol 1078. AISC. Springer, Cham, pp 260–273. https://doi.org/10.1007/978-3-030-33614-1_18

9. Dimitropoulos N, Michalos G, Makris S (2021) An outlook on future hybrid assembly systems - the Sherlock approach. Procedia CIRP 97:441–446. https://doi.org/10.1016/j.procir.2020.08.004
10. Kousi N, Michalos G, Aivaliotis S, Makris S (2018) An outlook on future assembly systems introducing robotic mobile dual arm workers. Procedia CIRP 72:33–38. https://doi.org/10.1016/j.procir.2018.03.130
11. Ogenyi UE, Liu J, Yang C, Ju Z, Liu H (2021) Physical human-robot collaboration: robotic systems, learning methods, collaborative strategies, sensors, and actuators. IEEE Trans. Cybern. 51(4):1888–1901. https://doi.org/10.1109/tcyb.2019.2947532
12. COMAU, COMAU: AURA-Collaborative Robots. https://www.comau.com/en/media/news/2017/05/aura. Accessed 19 June 2021
13. Open Robotics: ROS.org. http://wiki.ros.org/. Accessed 07 Jan 2020
14. Organization for the Advancement of Structured Information Standards, MQTT: The Standard for IoT Messaging. https://mqtt.org/. Accessed 07 Jan 2020
15. Rickert M, Gaschler A (2017) Robotics library: an object-oriented approach to robot applications. In: 2017 IEEE/RSJ International Conference on Intelligent Robots and Systems (IROS). IEEE. https://doi.org/10.1109/iros.2017.8202232

A Brief Literature Review of Mathematical Models of EMG Signals Through Hierarchical Analytical Processing

Ruben Nogales, Jaime Guilcapi, Freddy Benalcazar, and Javier Vargas

Abstract This paper presents a systematic review of the literature for the selection of a mathematical model to determine the behavior and constitution of electromyography (EMG) signals. The selection of a mathematical model can be determined according to the search criteria related to obtaining the same EMG signal. In this context, determining the alternatives and most representative characteristics of a mathematical model is done through a hierarchical analytical process (AHP). This process allows determining the functional, modular, descriptive, and resulting characteristics of a proposed mathematical model. Subsequently, these criteria are correlated with quantitative values for each alternative by means of the Centroid Method, which allows for multiple criteria decision making. Finally, the study presenting the mathematical models proposed by the EMG composition with the highest acceptance weight for Hand Gesture Recognition using Electromyographic Signals has been performed.

Keywords Hierarchical analytical process · Centroid method · EMG · Mathematical model

1 Introduction

Electromyographic (EMG) signals are electrical impulses generated by the muscle [1]. These signals are taken by devices composed of electrodes that can be needle

R. Nogales (✉) · J. Guilcapi · F. Benalcazar · J. Vargas
Facultad de Ingeniería en Sistemas Electrónica e Industrial, Universidad Técnica de Ambato, Ambato, Ecuador
e-mail: re.nogales@uta.edu.ec

J. Guilcapi
e-mail: jr.guilcapi@uta.edu.ec

F. Benalcazar
e-mail: fg.benalcazar@uta.edu.ec

J. Vargas
e-mail: js.vargas@uta.edu.ec

© The Author(s), under exclusive license to Springer Nature Switzerland AG 2022

M. V. Garcia et al. (eds.), *Advances and Applications in Computer Science, Electronics, and Industrial Engineering*, Lecture Notes in Networks and Systems 433,
https://doi.org/10.1007/978-3-030-97719-1_16

or surface electrodes [2–4]. The accuracy of an electromyographic data or signal depends on the type of electrode used. At present, non-invasive devices are used in research, among them surface electrodes such as Myo Armband [5]. Because it is a non-invasive device with low precision but high acquisition because the signals contain a considerable amount of noise, according to the authors who used this type of device. The composition of the EMG signal in summary is based on Motor unit firings (contraction force or impulse), Motor unit potential (electrical potential for each firing), Motor unit potential train (producing the EMG signal) [6–9]; all these together are captured by the device or sensor.

Focusing on the search, a high-performance digital device with logical and mathematical operations will process the electromyographical signal. This reconstruction of the EMG signal composition is proposed with the representation of a mathematical model, either described in an equation, an algorithm or a block diagram [10]. For this purpose, articles or book chapters in which a mathematical model is applied or created to describe the behavior of electromyographic signals are analyzed.

From the following search information, the most representative works on the subject were established. What we need to know is for each of them: What did they do?, What did they propose?, If they used any method or mathematical model?, What results did they obtain?, if those who are there are not convinced of their results, if it is a "bad" article not suitable for our study.

The bibliographic search for articles or book chapters describing the behavior of electromyographic signals is carried out with a research method aimed at the systematic review of the literature [10–12]. The systematic review (SR) of the literature on the subject related to electromyographic signals covers general aspects of the mathematical models used, the obtaining of signals and the validation of results. For the search related to the research topic, the most used scientific databases were used, where key-words were parameterized as: "mathematical models", "EMG", "behavior of EMG", among other things.

On the other hand, a process of selection of a mathematical model is established to determine the functional, modular and successful characteristics valid for the articles. This process is carried out in the research project on Hand Gesture Recognition Using Electromyographic Signals.

To segment the research results, the Centroid Method is used, which is based on multi-criteria through the comparison between subsets of criteria for decision making [13]. Subsequently, the selection of these criteria is complemented by means of a basic Analytic Hierarchy Process (AHP) model which [14], through the selection of alternatives and criteria, weighs the matches or crosses of each characteristic with what is extracted from an article. This extracted information is given by the researcher of the subject in an arbitrary way that can determine if the article fits the research or not. Therefore, the Centroid Method and an AHP model allow the identification of measurable criteria so that the weights of article selection can be minimized for decision making by assigning evaluation criteria with respect to the project.

2 Context and State of the Art

Each surface electrode electromyography item is the most common one for recording electrical activity of muscles. In the field of procedural research, electromyography with non-invasive sensors is the most widely used, due to the majority of tests and data that can be obtained. These data undergo a pre-processing which, applying a mathematical model or simulation models, allows to determine the behavior of the electromyography signals.

The mathematical model of the EMG signal was born as a random sample process, to estimate the amplitude of the EMG signal due to the large data size and processing time with variability of each signal. Most of the articles point out their studies to find an applicable model that fits the evaluation needs.

The literature review presents different mathematical models, each of which tries to solve different problems. Among the main ones we can mention: the noise produced by the sensors, the measured force of the electric impulse of the motor units for a long time, and the propagation speed of the electric impulse.

The articles and book chapters with the most significant results are listed in Table 1. Each article that has a match is taken as a systematic mapping study, based on the following query, the search strings are determined by the criteria within the research project [15–17].

The search chain in scientific databases or personal studies is done by determination of the researchers. The keywords are determined according to the search indexes of the most cited articles. If an article is cited in different relevant researches it is taken into account, but this does not mean that the others are discarded. Each discarded article is analyzed to determine if the study is validated in other articles. The mathematical models studied are analyzed according to the search criteria within the systematic review, this allows filtering each algorithm, method or technique of selection of a mathematical model for EMG signals.

The selection of articles is an extraction method based on the needs of the initial EMG signal research project in [18]. Each article can be weighted according to the evaluation criteria of the authors. On the other hand, each reference is determined by the researchers, according to the areas determined for the evaluation criteria.

A mathematical model has its own characteristics, which were taken as a study in a systematic review of the literature within different references of the selected. The one that fits best to the research project will be with support and validation of the behavior of electromyographic signals [19].

The review of some adaptive models to electromyographic signals are mentioned in the articles or book chapters [26], which analyze the behavior of the EMG signal, also use mathematics as a basis for building equations or output blocks of the desired values for each signal.

In [20] proposed an alternative model to the one proposed by Dimitrov and [37], which for the potential of the motor unit calculates a unique convolution (convolution theorem - Fourier) between the input signal and the common impulse response for an electrode. The studies are validated by the authors in [22], presenting an

Table 1 Most influential primary studies present in different mathematical models

Article name	Reference
A Mathematical Model of Surface Electromyographic Measurement	[20]
Exponentially-Modified Gaussian Mixture Model: Applications in Spectroscopy	[21]
A Comprehensive Mathematical Model of Motor Unit Pool Organization, Surface Electromyography, and Force Generation	[22]
Mathematical Model for Skeletal Muscle to Simulate the Concentric and Eccentric Contraction	[23]
Quantitative evaluation of muscle synergy models: a single-trial task decoding approach	[24]
Estimation and Application of EMG Amplitude During Dynamic Contractions	[25]
Surface electromyogram signal modelling	[26]
A Comprehensive Mathematical Model of Surface Electromyography and Force Generation	[27]
Physiology and Mathematics of Myoelectric Signals	[28]
Motor unit recruitment strategies investigated by surface EMG variables	[29]
A Nonstationary Model for the Electromyogram	[30]
Surface EMG models: properties and applications	[31]
The Analysis of Surface EMG Signals with the Wavelet-Based Correlation Dimension Method	[32]
Generation of synthetic surface electromyography signals under fatigue conditions for varying force inputs using feedback control algorithm	[33]
An EMG Classifying Method Based on Bayes' Criterion	[34]
Description and Analysis of the EMG Signal	[1]
Mathematically modelling surface EMG signals	[35]
Experimental simulation of cat electromyogram: evidence for algebraic summation of motor-unit action-potential trains	[36]
EMG Modeling	[7]

incorporation and coding of velocity, generation of intracellular action potential and the resulting EMG measurements. In [38] the behavior of the EMG signal is validated by a proposed model which has five phases: signal acquisition, pre-processing, feature extraction, classification and post-processing.

In [30] propose a model in elementary representation of the weighted sum of the action potentials of the active driving groups. To this end, they make a representation of the EMG block diagram as an amplitude modulation process. They conclude that in experimental situations the functions can be generated as a whole, where the promediation method can be used for non-stationary signals. The studies are validated in [39, 40] that in summary seek to trace the responses to electrical stimulation to determine total muscle activity (EMG) by means of the sum of the activity of all the UMs and the time domain of the electromyography signals.

In the review of [1] it is determined as one of the most referenced book chapters in the systematic review of literature. Therefore, their research is validated in the articles: [41]; as the main study is to develop a method to compensate the effect of muscle fatigue, it uses four characterizations for the extraction of the EMG signal, time domain, zero crossing, sign slope change, Wilson amplitude and percentage rate of myopulse. [42]; it bases its study on mathematics and information about discrete redundant transformed wavelet (RDWT) and singular value decomposition and CS (compressive detection) theory are presented. [43]; his research suggests that controlling motor neuron groups of a pair of muscle agonists, antagonists as a singular group when performing a task.

In [35] propose a mathematical model using the Laplace equation and the handling of three-dimensional non-linear elements. In addition, simulation can be performed in a domain for several time steps, simulating electrical activity during a dynamic induced contraction. The study is validated in [44] as part of the study of the potential action of the motor unit, of how the waves known as trains for potential action propagate along each fiber.

In [26] proposes a mathematically simple weighting function, based on the algebraic sum of [1]. The signal is the sum of the contributions of many motor units acting together during the production of force. His study is validated in [45] through the small force oscillations that occur during isometric and isotonic contraction. Concluding that the estimation of muscle strength from EMG requires the normalization of EMG amplitudes to maximum values.

In [27] validate their results through a block diagram with the main components of the proposed model of muscle strength generation, an estimated total of 42 equations or functions to generate muscle strength and EMG signal based on time. These authors continue their studies in [22], by each electromyographic measurement, they add the corresponding strength signal in the skeletal muscles as an integral mathematical model.

In [23] propose two elements, one elastic in series and another parallel elastic element, which are used for electrical simulation to predict non-linear properties of the skeletal muscle and additional force-velocity ratio is calculated. The validation of the results of the numerical methods can be used to improve the numerical procedure for the calculation of stresses for a three-element muscle model. At present, no information is available on the validation or referencing of this article, but the topics investigated by the author show mathematical models in the force-length, force-velocity relationship of active and passive muscle contraction.

In [24] propose a mathematical model that is based on a computational framework of muscle synergy. In addition, they use a somewhat complex algorithm that defines the new metric as the decoding performance, i.e., the percentage of correct decoding of individual tests, as a function of a single test measurement of their activation coefficients, applied to the simulated EMG data. The ratification of your study is done in [46] by determining whether the number and choice of EMG muscles in an experimental protocol will affect the synergies identified using matrix factorization algorithms. Additionally, in another study in [47] the time limitations of EMG signals can generally only be measured for a subset of muscles involved in the task, concluding that if no electromyographic signals exist, muscle synergy cannot be measured.

The review of the scientific literature shows that there is a wide mathematical field of applicability of different models composed of equations, diagrams or algorithms. In this context, based on the mathematical study the following searches are proposed [20, 22, 23, 26, 28, 30, 31, 35, 36], widely present a publication on adaptive mathematical models in a novel computational framework in the research project on Hand Signal Recognition Using Electromyographical Signals (EMG) and Artificial Intelligence. In the descriptive field of the composition and behavior of EMG signals there are the following reviews [21, 24, 25, 27, 29, 32, 33, 46], the initial rationale gives reference for current research. Overall, most of the mathematical studies and behavior of EMG signals are based on articles in [1, 7], which in their descriptive research seek to obtain electromyographic signals as a separation of EMG signal behavior. Therefore, the research on EMG signals will be established in a compressive mathematical model through block diagram applicable different equations and resolution algorithms to solve the stages composed by acquisition of EMG signals in pre-processing, feature extraction, classification and post-processing.

3 Centroid Method and AHP Model

For the case study, the Centroid Method is presented, which weights arbitrary values for multiple-criteria decision-making [48]. The weighting values are divided according to their level of importance, essentially seeking to take the implicit relationship by weights in each criterion. AHP is used to identify the measurable criteria, so that the centroid method minimizes the weight limits.

For the study, the articles that have a context in functionality, modularity, description and validated results are grouped according to the literature review. The study determines a selection of a mathematical model applied to the determination of the behavior of the EMG signal by means of the centroid evaluation, which is subject to criteria values such as implementation, measured electric impulse strength, propagation speed, noise, etc. Subsequently, items that will be used in the centroid method

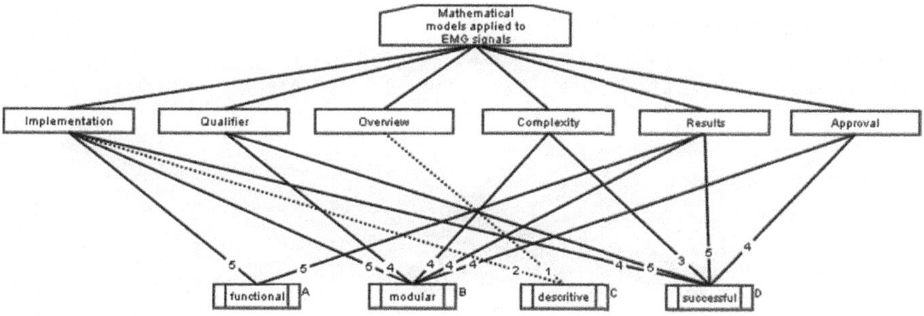

Fig. 1 AHP proposed for the section of mathematical models in EMG signals

Table 2 Centroid method values according to selected criteria

Implementation (I_g)	Qualifier (Q_g)	Overview (O_g)	Complexity (C_g)	Results (R_g)	Approval (A_g)
0,5	0,5	0	0	0	0
0,333	0,333	0,333	0	0	0
0,25	0,25	0,25	0,25	0	0
0,2	0,2	0,2	0,2	0,2	0
0,167	0,167	0,167	0,167	0,167	0,167
0,29	**0,29**	**0,19**	**0,1234**	**0,0734**	**0,0334**

should not be over-looked by model-based alternatives (see Fig. 1). This provides a wide range of information about the studies and the results of each model.

The selection of a mathematical model entails an in depth examination of the parameters for evaluating referees (case study), for which the hierarchical section of the AHP criteria is presented. The alternatives are compared in pairs according to their relative importance in the criteria considered. These data are expressed quantitatively analyzed by the researcher. These "matches" allow a value between 1 and 5 to be weighted. Therefore, for a value of 5 the articles are strongly valid, for 4 applied reference, for 3 adapted reference and for the values of 1 and 2 single reference.

Centroid method adjustable to mathematical models. The measurable criterion factor of the centroid method that uses complicated mathematical formulas [13]. These formulas take into account the criteria specified using a cost-benefit analysis to identify the best weighting values in Table 2. Criteria-based decision making is determined by an analysis of conflicting values. This means that the criteria and criteria vary according to the conflicting rationales within the research project. Each value is weighted according to a study determined in the [13] study, these values are incorporated into the alternatives and criteria that may affect the EMG signal filtering process.

The resulting values in Table 2 are then adjusted to the values with weights obtained in the AHP model for decision making through an assessment criterion in Table 3. The values are cross-referenced with the segments of each matches made

Table 3 Listing of items. Multiple criteria factors

Article	Ítem	I_g		Q_g		O_g		C_g		R_g		A_g	
[20]9, [21]10, [22]12, [25]14, [29]18, [31]20, [34]23, [35]24	A	5	0,29	1	0,29	1	0,19	1	0,1234	5	0,0734	1	0,0334
[7, 20, 26–28, 30, 32, 36]	B	5	0,29	4	0,29	1	0,19	4	0,1234	4	0,0734	4	0,0334
[23]12, [24]13, [26]15, [29]18, [33]22, [35]24	C	2	0,29	1	0,29	2	0,19	1	0,1234	1	0,0734	1	0,0334
[1, 13, 22, 25, 28, 30]	D	4	0,29	5	0,29	1	0,19	3	0,1234	5	0,0734	4	0,0334

in the literature review of the articles. Each value determines whether the selected article is used in the research of mathematical models for EMG signal filtering.

The articles reviewed that meet the most important characteristics to be considered for choosing the mathematical model are that it can be implemented and that it integrates all the processes of EMG signal research, the secondary characteristics would be overview, complexity, result and approval (see Fig. 2).

According to the evaluation parameters, item B obtains a higher average value, being this group of articles, the ones analyzed in the characteristics with their respective alternatives. In this context it is determined that the highest weights are given on the Implementation, Qualifier and Results. Therefore, articles [7, 20, 26–28, 30, 32, 36], are the most referenced ones for the research of the Hand Signal Recognition Using Electromyographical Signals (EMG) and Artificial Intelligence project. These articles are the most adjusted to the search parameters and the solution of EMG signal behavior problems.

Selective Mathematical Modeling Method: Literature Review Proposal. Clearly present the proposed mathematical model of selection, a step-by-step procedure is the first described:

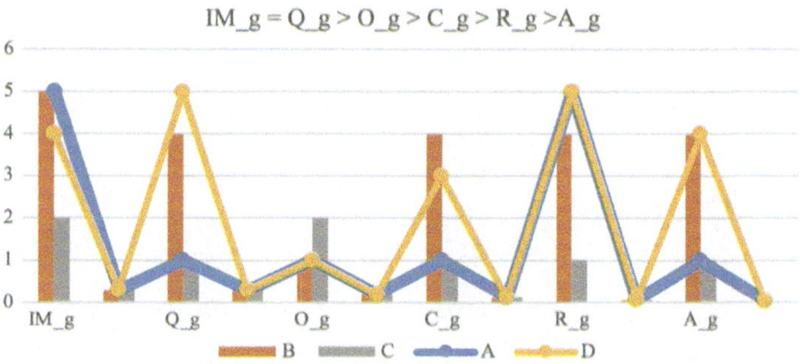

Fig. 2 Evaluation criteria for the selection of a mathematical model

Table 4 Multiple criteria factors-functional

Analysis/Implementation(AI_f)	Mathematics/Description (MD_f)	Applicability (A_f)
0,5	0,5	0
0,333	0,333	0,333
0,4165	0,4165	0,1665

Step 1. Form a project team and collect possible information about the mathematical model.
Step 2. Identify the characteristics of the mathematical models.
Step 3. Build an objectives structure to develop the hierarchy of fundamental objectives.
Step 4 and 5. Extract the attributes to evaluate the mathematical models with respect to the EMG structure of the objectives. Filter out unqualified articles by asking specific questions that are formulated according to the requirements of the research study.

4 Results

Correlation with various evaluation criteria;

The most important functional features are Analysis/Implementation and Mathematics/Description for the choice of a mathematical model in Table 4. This filtering study is based on the articles [7, 20, 26–28, 30, 32, 36].

The criteria AI_f, MD_f and A_f corresponds to the alternative of a functional mathematical model which can be subjected to a quantitative evaluation with respect to the previously selected articles. The condition criterion is based on determining whether the Analysis and Implementation equals the descriptive search and the applicability of the model with respect to EMG signals. This means that the articles

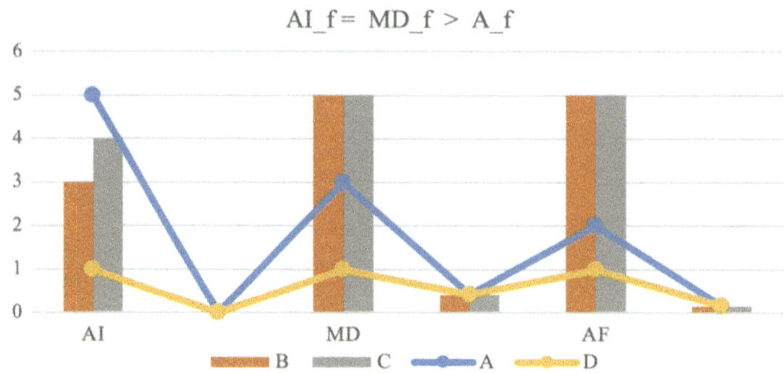

Fig. 3 Evaluation criteria for the selection of a mathematical model

Table 5 Multiple criteria factors-modular

Tests / Validation (TV_m)	Architecture (A_m)	Similarity (S_m)
1	0	0
0,5	0,25	0,25
0,75	0,125	0,125

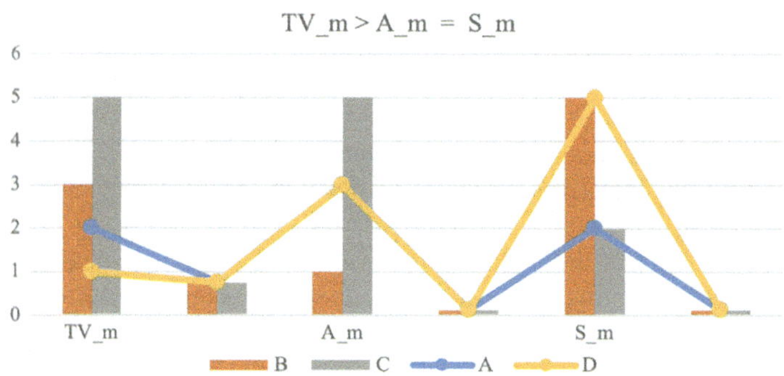

Fig. 4 Evaluation criteria of modularity with less determination for the selection of a mathematical model

reviewed in item C conform to a correct implementation of the proposed mathematical model (see Fig. 3).

The most important modular feature is Tests/Validation for the choice of a mathematical model in Table 5.

On the other hand, in the criteria TV_m, A_m and S_m applying a condition criterion it is determined what it infers to the proposed study on the mathematical model and if modularity Similarity to the other investigations (see Fig. 4). For this multi-criteria evaluation it is determined as the minimum value that tells us that the items of item A do not fit to the investigation of the behavior of EMG signals.

Table 6 Multiple criteria factors-successful

Acceptance (A_s)	Results (R_s)	Comments (C_s)
1	0	0
0,5	0,25	0
0,25	0,167	0,167
0.583	0,139	0,055

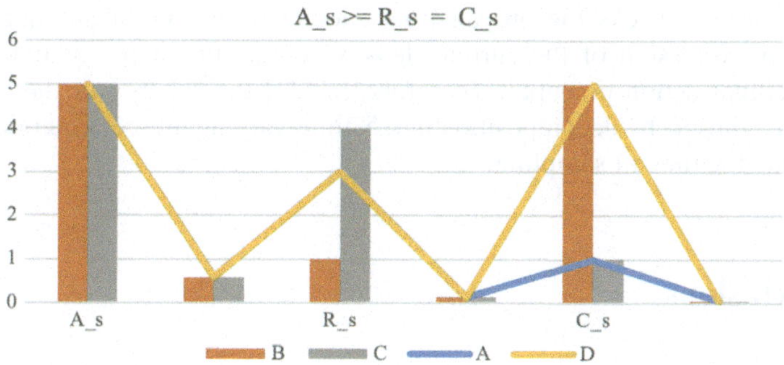

Fig. 5 Results evaluation criteria with greater or equal determination for the selection of a mathematical model

The most successful feature Important is the Acceptance for the selection of a mathematical model for EMG signals. These evaluation criteria allow the researcher to determine if the results are comparable to the research reviewed. The weights increase according to the importance that is determined in the alternative of successful (Table 6).

In most information searches, researchers focus primarily on whether or not the proposed mathematical model is accepted. To do this, they turn to the results section which, based on their research, can determine whether the results are clearly expressed with high acceptance rates. Later they self evaluate through comments if the proposed model is not the correct one to determine the behavior of the EMG signs. For the deductions it is obtained that items B and D are the most representative for the condition criterion proposed for decision making (see Fig. 5).

After discussion with the project team, we gathered the following results: Six articles of mathematical models that meet the requirements of the research study:

1 article on applied mathematics for each unit recruitment engine. 1 article on the Gaussian model applied to EMG signal acquisition and filtering. 1 article that determines the EMG signal as Myoelectric Signals, to apply an adaptive algebraic summation model. 1 article that through a stochastic process, obtains the spectral and probability characteristics of the EMG during a constant contraction. 2 articles analyzing models that describe how the morphological characteristics of the muscle are reflected in the multi-channel EMG measurements. 1 article that determines the extraction of time series characteristics, this article is determined as in later analysis.

In [7] it is determined by the linear sum of the MUAPTs for noise separation. Additionally, in [21] and [28] EMG filters or eliminates noise through two formulas of static and dynamic contraction.

In [30] it uses the non-stationary process and the Fourier transform. On the other hand, in [31] the use of the governing equations for the calculation of a virtual EMG signal and the Laplace generalized equation are determined. Therefore in [23], the total force generated by a muscle is calculated as the superposition of the individual force contractions.

In [20, 36], it checks the propagation velocity by means of an equation that provides an expression of the current density source, the step unit function and the intracellular action potential. Therefore, in [24] the synergy of the activation moment of a muscle by decomposition into N time-varying muscle synergies is used to determine the time of sampling.

5 Conclusions

In this work we analyzed the mathematical models that describe the behavior of EMG signals. Since the research shows a variety of models, either by their appreciation of the data in each study of the signal or by the estimation of EMG amplitude. The continuation of infestation will be based on a mathematical interpretation to solve based on the time domain, which better describes the non-stationary behavior of superficial EMG signals [26, 30], validated in [38, 40] and [45].

The present research allowed us to recapitulate the mathematical models found in the literature that helped us to determine the process of obtaining and the behavior of a resulting EMG signal. Adding a systematic review of the scientific literature where different authors solve the inconveniences at the time of determining the behavior of the resulting EMG signal. Obtaining as a result of the articles [7, 20, 26–28, 30, 32, 36] a review of the literature that presents different mathematical models between the main ones we can mention: the noise produced by the sensors, the measured force of the electrical impulse of the motor units by a long time, and the speed of propagation of the electrical impulse.

Used from the AHP model and the centroid method as a multi-criteria evaluation. It ensures that the structure of the objectives is consistent with the goals and information seeking strategies. The project team can understand the relationships between the different objectives and evaluate their influence by modernizing them according to the hierarchical and selection structures for a mathematical model applied to Hand Gesture Recognition Using Electromyographic Signals.

Acknowledgements The Corporación Ecuatoriana para el Desarrollo de la Investigación y la Academia (CEDIA) for the development of the research project CEPRA-2019-13-Reconocimiento de Gestos.

References

1. Basmajian J, De Luca CJ (1985) Description and analysis of the EMG signal. In: Muscles Alive: their Functions Revealed by Electromyography, pp 65–100. https://ci.nii.ac.jp/naid/10011397436/
2. Hewson DJ, Hogrel JY, Langeron Y, Duchêne J (2003) Evolution in impedance at the electrode-skin interface of two types of surface EMG electrodes during long-term recordings. J Electromyogr Kinesiol 13(3):273–279. https://doi.org/10.1016/S1050-6411(02)00097-4
3. Espinosa RV, Soto M, Garcia MV, Naranjo JE (2021) Challenges of implementing cleaner production strategies in the food and beverage industry: literature review. In: García MV, Fernández-Peña F, Gordón-Gallegos C (eds) Advances and Applications in Computer Science, Electronics and Industrial Engineering, vol 1307. AISC. Springer, Singapore, pp 121–133. https://doi.org/10.1007/978-981-33-4565-2_8
4. García MV, Irisarri E, Pérez F, Estévez E, Marcos M (2016) OPC-UA communications integration using a CPPS architecture. In: 2016 IEEE Ecuador Technical Chapters Meeting (ETCM), pp 1–6
5. Arief Z, Sulistijono IA, Ardiansyah RA (2016) Comparison of five time series EMG features extractions using Myo armband. In: Proceedings–2015 International Electronics Symposium: Emerging Technology in Electronic and Information, IES 2015. Institute of Electrical and Electronics Engineers Inc., pp 11–14
6. Parsaei H, Stashuk DW (2013) EMG signal decomposition using motor unit potential train validity. IEEE Trans Neural Syst Rehabil Eng 21(2):265–274. https://doi.org/10.1109/TNSRE.2012.2218287
7. Rodriguez-Falces J, Navallas J, Mal A (2012) EMG modeling. In: Computational Intelligence in Electromyography Analysis–A Perspective on Current Applications and Future Challenges. In Tech
8. García MV, Irisarri E, Pérez F, Estévez E, Orive D, Marcos M (2016) Plant floor communications integration using a low cost CPPS architecture. In: 2016 IEEE 21st International Conference on Emerging Technologies and Factory Automation (ETFA), pp 1–4
9. Irisarri E, García MV, Pérez F, Estévez E, Marcos M (2016) A model-based approach for process monitoring in oil production industry, pp 1–4. https://doi.org/10.1109/ETFA.2016.7733633
10. Chowdhury RH, Reaz MBI, Bin Mohd Ali MA, Bakar AAA, Chellappan K, Chang TG: Surface electromyography signal processing and classification techniques. MDPI AG. http://www.mdpi.com/1424-8220/13/9/12431
11. García MV, Pérez F, Calvo I, Moran G (2015) Developing CPPS within IEC-61499 based on low cost devices, pp 1–4. https://doi.org/10.1109/WFCS.2015.7160574
12. García MV, Pérez F, Calvo I, Morán G (2014) Building industrial CPS with the IEC 61499 standard on low-cost hardware platforms, pp 1–4. https://doi.org/10.1109/ETFA.2014.7005272
13. Solymosi T, Dombi J (1986) A method for determining the weights of criteria: the centralized weights. Eur J Oper Res 26(1):35–41. https://doi.org/10.1016/0377-2217(86)90157-8
14. Vaidya OS, Kumar S (2006) Analytic hierarchy process: an overview of applications. Eur J Oper Res **169**(1):1–29. https://doi.org/10.1016/j.ejor.2004.04.028
15. Vargas J, Mayorga F, Guevara D, Álvarez E (2019) Analytic hierarchy process of selection in version control systems: applied to software development. Adv Intell Syst Comput **884**:75–85
16. Juan P, Analuiza O, Carpio M, Castillo W (2021) Optimization of the mechanical properties responses of SBR 1502 rubber/amorphous silicon dioxide/others by DOE-MSR methodology. In: García MV, Fernández-Peña F, Gordón-Gallegos C (eds) Advances and Applications in Computer Science, Electronics and Industrial Engineering, vol 1307. AISC. Springer, Singapore, pp 215–231. https://doi.org/10.1007/978-981-33-4565-2_14
17. Angulo K, Gil D, Espitia H (2020) Method for edges detection in digital images through the use of cellular automata. In: Nummenmaa J, Pérez-González F, Domenech-Lega B, Vaunat J, Oscar Fernández-Peña F (eds) CSEI 2019, vol 1078. AISC. Springer, Cham, pp 3–21. https://doi.org/10.1007/978-3-030-33614-1_1

18. Benalcázar ME et al (2017) Real-time hand gesture recognition using the Myo armband and muscle activity detection. In: 2017 IEEE Second Ecuador Technical Chapters Meeting (ETCM), pp 1–6
19. Nogales R, Benalcazar F, Guilcapi J, Vargas J (2021) Selection of classifiers for hand gesture recognition using analytic hierarchy process: a systematic literature review. In: García MV, Fernández-Peña F, Gordón-Gallegos C (eds) Advances and Applications in Computer Science, Electronics and Industrial Engineering, vol 1307. AISC. Springer, Singapore, pp 265–278. https://doi.org/10.1007/978-981-33-4565-2_17
20. Petersen E (2016) A mathematical model of surface electromyographic measurements. https://www.ime.uni-luebeck.de/fileadmin/files/Publications/PDFs/Pe16.pdf
21. Ament S, Gregoire J, Gomes C (2019) Exponentially-Modified Gaussian Mixture Model: Applications in Spectroscopy. arXiv:1902.05601
22. Petersen E, Rostalski P (2019) A comprehensive mathematical model of motor unit pool organization, surface electromyography, and force generation. Front Physiol **10**(MAR):1–19. https://doi.org/10.3389/fphys.2019.00176
23. Kuthe C, Uddanwadiker RV, Padole PM, Ramteke AA (2015) Mathematical model for skeletal muscle to simulate the concentric and eccentric contraction. MCB Mol Cell Biomech 12(1):1–16. https://doi.org/10.3970/mcb.2015.012.001
24. Delis I, Berret B, Pozzo T, Panzeri S (2013) Quantitative evaluation of muscle synergy models: a single-trial task decoding approach. Front Comput Neurosci **7**(FEB):8. https://doi.org/10.3389/fncom.2013.00008
25. Clancy EA, Bouchard S, Rancourt D: Estimation and application of EMG amplitude during dynamic contractions. http://www.ncbi.nlm.nih.gov/pubmed/11838258
26. McGill KC: Surface electromyogram signal modelling. http://link.springer.com/, https://doi.org/10.1007/BF02350985
27. Petersen E, Rostalski P (2018) A comprehensive mathematical model of surface electromyography and force generation. bioRxiv, p 273458. https://doi.org/10.1101/273458
28. De Luca CJ (1979) Physiology and mathematics of myoelectric signals. IEEE Trans Biomed Eng **BME-26**(6):313–325. https://doi.org/10.1109/TBME.1979.326534
29. Farina D, Fosciand M, Merletti R (2002) Motor unit recruitment strategies investigated by surface EMG variables. J Appl Physiol **92**(1):235–247. https://doi.org/10.1152/jappl.2002.92.1.235
30. Shwedyk E, Balasubramanian R, Scott RN (1977) A nonstationary model for the electromyogram. IEEE Trans Biomed Eng **BME-24**(5):417–424. https://doi.org/10.1109/TBME.1977.326175
31. Stegeman DF, Blok JH, Hermens HJ, Roeleveld K.: Surface EMG models: properties and applications. http://www.ncbi.nlm.nih.gov/pubmed/11018441
32. Wang G, Zhang Y, Wang J (2014) The analysis of surface EMG signals with the wavelet-based correlation dimension method. Comput Math Methods Med 2014:1–9. https://doi.org/10.1155/2014/284308
33. Venugopal G, Deepak P, Ghosh DM, Ramakrishnan S (2017) Generation of synthetic surface electromyography signals under fatigue conditions for varying force inputs using feedback control algorithm. Proc Inst Mech Eng Part H J Eng Med **231**(11):1025–1033. https://doi.org/10.1177/0954411917727307
34. Lei M, Wang Z-Z, Cai L-Y, Zhang H-H, Cai H (2002) An EMG classifying method based on Bayes' criterion. Inst Electr Electron Eng (IEEE) 20:2625–2626
35. Mordhorst M, Heidlauf T, Röhrle O (2014) Mathematically modelling surface EMG signals. Pamm 14(1):123–124. https://doi.org/10.1002/pamm.201410049
36. Day SJ, Hulliger M (2001) Experimental simulation of cat electromyogram: evidence for algebraic summation of motor-unit action-potential trains. J Neurophysiol **86**(5):2144–2158. https://doi.org/10.1152/jn.2001.86.5.2144
37. Dimitrov GV, Dimitrova NA (1998) Precise and fast calculation of the motor unit potentials detected by a point and rectangular plate electrode. Med Eng Phys **20**(5):374–381. https://doi.org/10.1016/s1350-4533(09)00014-9

38. Jaramillo AG, Benalcazar ME (2018) Real-time hand gesture recognition with EMG using machine learning. In: 2017 IEEE 2nd Ecuador Technical Chapters Meeting, ETCM 2017. vol 2017-January. Institute of Electrical and Electronics Engineers Inc., pp 1–5
39. Sheng R, Zhang Z (2019) A hand gesture recognition using single-channel electrodes based on artificial neural network. In: Proceedings of 2019 IEEE 8th Joint International Information Technology and Artificial Intelligence Conference, ITAIC 2019. Institute of Electrical and Electronics Engineers (IEEE), pp 209–213
40. Hashemi SE, Ali Ahmadi-Pajouh M, Shamsi E (2018) Modeling and simulation of EMG signal and H-Reflex in healthy human subject. In: 2018 25th Iranian Conference on Biomedical Engineering and 2018 3rd International Iranian Conference on Biomedical Engineering, ICBME 2018. Institute of Electrical and Electronics Engineers Inc
41. Triwiyanto, Wahyunggoro O, Nugroho HA, Herianto (2018) Adaptive threshold to compensate the effect of muscle fatigue on elbow-joint angle estimation based on electromyography. J Mech Eng Sci **12**(3):3786–3796. https://doi.org/10.15282/jmes.12.3.2018.3.0334
42. Obiols EP, Mañanas MA, Ramos Barcelona J (2019) Analysis of electrode arrays for multichannel surface electromyography. https://upcommons.upc.edu/handle/2117/134030
43. Mason J et al (2019) Modulation of intracortical inhibition and excitation in agonist and antagonist muscles following acute strength training. Eur J Appl Physiol **119**(10):2185–2199. https://doi.org/10.1007/s00421-019-04203-9
44. Zea Guachamín JA (2018) Implementación de un sistema de clasificación de gestos del brazo humano utilizando Myo Armband para mando a distancia de un brazo robótico de 3GDL (Bachelor's thesis, Quito, 2018)
45. Staudenmann D, Roeleveld K, Stegeman DF, Van Dieën JH (2010) Methodological aspects of SEMG recordings for force estimation-a tutorial and review. J Electromyogr Kinesiol 20(3):375–387
46. Steele KM, Tresch MC, Perreault EJ (2013) The number and choice of muscles impact the results of muscle synergy analyses. Front Comput Neurosci (JUL). https://doi.org/10.3389/fncom.2013.00105
47. D'Andola M, Cesqui B, Portone A, Fernandez L, Lacquaniti F, D'Avella A (2013) Spatiotemporal characteristics of muscle patterns for ball catching. Front Comput Neurosci (JUL). https://doi.org/10.3389/fncom.2013.00107
48. Flores BE, Olson DL, Dorai VK (1992) Management of multicriteria inventory classification. Math Comput Model **16**(12):71–82. https://doi.org/10.1016/0895-7177(92)90021-C

Automation Framework for Analog Water Meters in the City of Cañar

Christian Vinicio Fernández Romero, Marco Avila Calle, and Juan-Carlos Cobos-Torres

Abstract The present investigation sought to reduce the errors caused by the current method of collecting water meter readings, using a computer system for the collection and processing of data collected from analog water meters in the city of Cañar. At present, in the world there are automatic or intelligent water consumption measurement systems, which are very expensive to implement. Also, there are still analog metering systems that are still installed in third world countries because of their cost, and they have not reached their useful life. The intention of reducing errors in analog systems is based on a meta-analytical study, which allows to relate the current reality of the data collection methodology with the effectiveness of automatic systems, in order to establish a hybrid solution that allows the use of analog meters but that reading errors are reduced by implementing technology. To achieve this, we proceeded to design a reading system using mobile devices and QR code tags that were placed on each analog meter in the city of Cañar. With the implementation of this pilot plan, the costs of migrating to a smart metering system have been reduced, which didn't affect the users or the institution in charge of water distribution. There has been a positive response from users since errors caused by human intervention in the collection of information have been reduced by 99%. At the same time, the institution has better control of the personnel who carry out the reading, which ensures that the data obtained are real and reliable.

Keywords Water meter · Residential water management · Smart metering · Water utility

C. V. F. Romero · M. A. Calle · J.-C. Cobos-Torres (✉)
Catholic University of Cuenca, Via a Patamarca y Cojimies (Estacia Cordero), Cuenca, Ecuador
e-mail: juan.cobos@ucacue.edu.ec

C. V. F. Romero
e-mail: chrisitan.fernandez1@est.ucacue.edu.ec

M. A. Calle
e-mail: mavila@ucacue.edu.ec

© The Author(s), under exclusive license to Springer Nature Switzerland AG 2022

M. V. Garcia et al. (eds.), *Advances and Applications in Computer Science, Electronics, and Industrial Engineering*, Lecture Notes in Networks and Systems 433,
https://doi.org/10.1007/978-3-030-97719-1_17

1 Introduction

Article 55 of the Organic Code of Territorial Organization, Autonomies and Decentralization (COOTAD) establishes as one of the exclusive competencies of the Intercultural Decentralized Autonomous Government of Canton Cañar has established a short to medium term plan to increase administrative operations and solve a series of problems. Among the problems are the excessive use of logistical resources and man-hours for taking readings; errors in the data recorded for readings; inconsistencies in the system of records that make it impossible to provide adequate attention to users in the event of complaints; delays in the post-processing of issuance (processing the data obtained in the field in the computer system for subsequent billing and collection).

The first drinking water system for the city of Cañar was built in 1961, then in 1982 and 1987 studies were conducted to improve and reinstate it. In 2014, the I [1] and II stages of the Drinking Water Master Plan were implemented, which enabled a substantial improvement in the quality of service, including increased coverage and population served with the installation of analog water meters throughout the canton.

With this improvement, the system of analog drinking water meters was implemented throughout the canton. There are automatic or smart metering systems [2], for water consumption [3], as well as analog systems [4]. The intention of reducing errors in analog systems is based on a meta-analytical study, which allows us to relate the current reality of the data collection methodology with the effectiveness of automatic systems, in order to establish a solution that allows us to use the meters currently installed in the city of Cañar and thus avoid the investment that would mean replacing all analog meters in the city with smart meters.

In large cities, smart water meters are being used [5], which offer all the advantages of analog meters [6] but with several technological advantages such as warning of insufficient water, emergency water, consumption limitation, automatic storage of reading data, data transmission, data security, consumption and event logging for up to 10 years; however, they also have disadvantages such as high investment for equipment replacement, need an electrical power source and, as in the case of the analog meter, need a reader, which is not a great advantage.

The use of computer tools such as mobile devices [7], currently has an internet connection which makes them useful tools to collect information. On the other hand, the use of analog meters [8] has been well proven throughout the region, so it is the most widely used. It is usually used in the water supply pipelines of residential and industrial facilities to make the pertinent charges to the users. With this we propose to combine the two technologies through the development of a framework to collect readings from analog water meters, using a computer system for the collection and processing of data collected from analog water meters in the city of Cañar.

2 Methodology

The research was developed as applied research. The Design Science Research (DSR) Methodology was used, which is schematized (see Fig. 1).

2.1 Identification of the Problem and Motivation

The following problems have been identified for the present investigation:

- Incorrect visual reading taking.
- Incorrect user code.
- No reading is taken in the corresponding month by the reader and random data is entered.
- The service user's complaints cannot be correctly addressed.
- There is no means of verification or control of the work performed by the readers.
- There is no record of incidents to deal with user complaints.
- Post-processing time of information is very long and susceptible to errors.

Fig. 1 Diagram of DSR methodology

2.2 Definition of the Goal of a Solution

To develop a set of computer solutions that combine the efficiency of smart water meters, but at the same time allow to maintain the analog water meters to reduce errors produced by human operation, with a minimum investment for the institution.

2.3 Design

The aim of the study is the application of technological tools in the collection and processing of data in analog water meters, in order to reduce the human factor that causes different types of errors when issuing the water bill [9] to the end-user.

The research will be descriptive in nature because the information analysis of the current methods of taking readings of analog meters will be carried out, as well as experimental because through a series of technological solutions it will be possible to verify the hybridization model (analog-digital) of taking readings of analog water meters.

Within the quantitative variables, we will analyze the percentage decrease of users who have filed complaints due to bad data reading by the readers, as an in-dependent variable we have to use the application and software used, through mo-bile devices and computers for data collection, and the dependent variable is the reduction of errors and the shorter processing time of monthly emissions.

The use of the system will be evaluated throughout the city of Cañar over a period of 60 days, working in all sectors and sequences established for the readers, collecting reading and location data in real-time. Additionally, the quality of the service provided to the user will be determined through the new analog water meter reading system, using data provided by the potable water and sanitation department of the city of Cañar.

With these parameters and with the results obtained in an automated way from the data collection of analog meters, the degree of effectiveness of the research will be evaluated, which will even be self-evaluated by the readers themselves as direct actors of the measurement process.

In order to implement the software as a solution to the errors that occur in the collection of information from analog water meters, the following considerations will be taken into account:

- Compatibility between hardware and software for instant access to the database of the municipal servers.
- Real-time access for the supervisor of the readers, allowing to verify the user, location, date, and time of taking the reading of the analog water meter.
- The supervisor must maintain the necessary permissions for any incident that may occur in the field, issuing the authorization to continue with the readings.
- Simplify the monthly consumption issuance process by automatically consolidating the information in the shortest possible time.

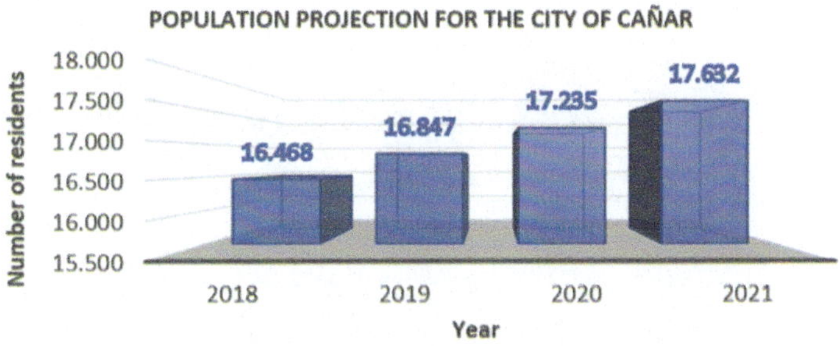

Fig. 2 Population projection for the city of Cañar

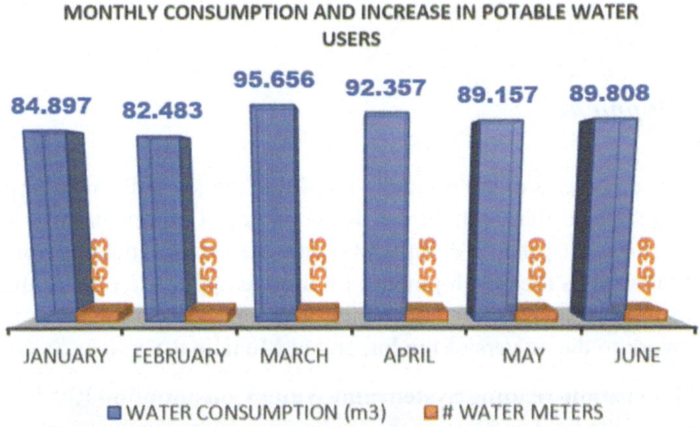

Fig. 3 Monthly consumption and increase in potable water users

Population and Sample. As part of the potable water supply system improvement project, in 2014 the municipality of Cañar implemented stages I [2] and II of the Drinking Water Master Plan, benefiting 17,632 residents. Currently, the number of users of the drinking water system has been increasing, (see Fig. 2).

For the study sample, we have considered the consumption in cubic meters of water in the months of January through June 2021 and the number of users that the city of Cañar has in that period, these data have been obtained from the digital platform of the city of Cañar, (see Fig. 3).

The potable water department of the municipality of the city of Cañar is in charge of the administration of the potable water supply system, and in order to control consumption by users has divided the city's potable water distribution system into 24 sectors, (see Fig. 4), with their respective sequences, which in turn are operated by 3 readers, covering the entire urban area of Cañar Canton.

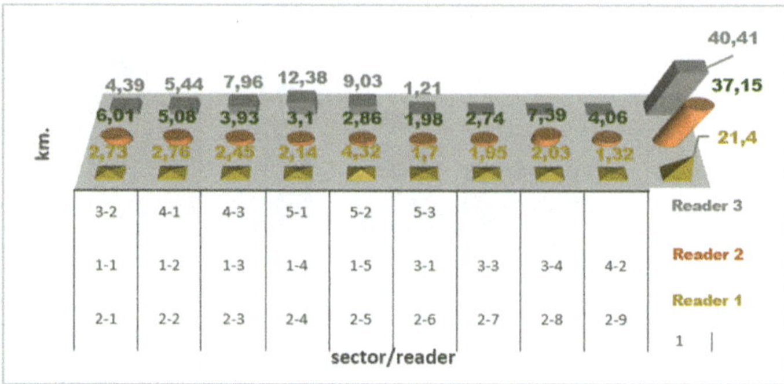

Fig. 4 Assignment of sectors for each reader

2.4 *Development*

To achieve the objectives of the research, a series of stages and computer process-es have been established that include the development of computer software on the Windows platform [10] that is compatible with the Integrated Municipal Services System [9], as well as the development of a mobile application under the Android platform for mobile devices [11, 12] which will send the data through the mobile internet network to the servers of the Integrated Municipal Services System.

QR Code Generation-reading System and Water Consumption Reading. In order to collect data in the field and so that the reader does not make mistakes when taking the reading, the QR code tag system was chosen for each analog water meter in the city. This follows the recommendations and experience applied in the municipality of Pimampiro, province of Imbabura [13, 14]. These will be generated using computer software that allows linking the QR codes with the database of drinking water users in the canton Cañar, (see Fig. 5).

Consolidation of Information, Sending and Management of Information. The reader, through a mobile device connected to the 4G data network, will scan the QR code of each meter assigned to its sector, and immediately will have the information in the server database, (see Fig. 6). Each reader will check the data in the field, upload the new readings and these will be sent back to the server and automatically consolidated, thus eliminating the manual process of information verification by the supervisor. At the end of each data collection period, the supervisor will have all the information to immediately issue monthly consumption data for billing and collection from the system users.

Reduction of Complains and Satisfaction Level. In order to follow up and control the readers and to ensure that they perform the measurements correctly, the following considerations are made, (see Fig. 7).

When the data are available on the server, they are immediately retrieved by the Integrated Municipal Services System [9] where the supervisor can review and verify

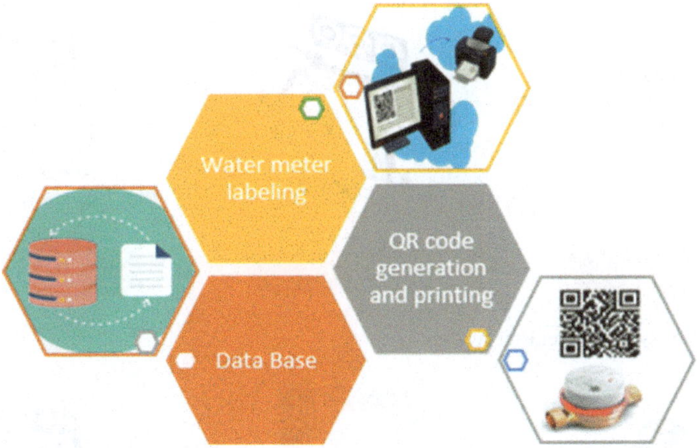

Fig. 5 Water meter labeling

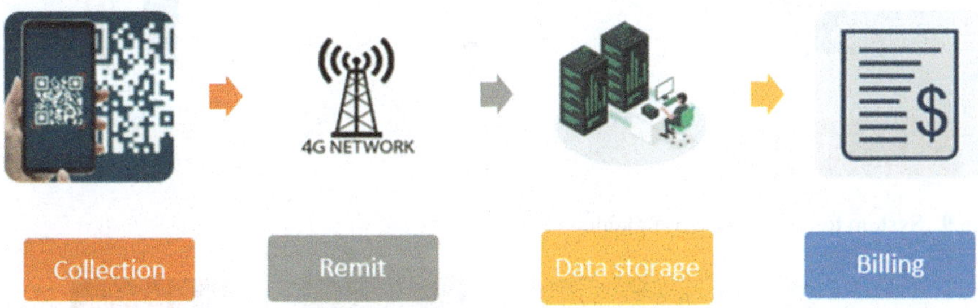

Fig. 6 System data flow

Fig. 7 Readers control scheme

whether the readers made the readings correctly (see Fig. 8), the software provides all the information of the field data collection, and as a security measure. The reader cannot modify the information or collect data without scanning the meter code.

Once the supervisor validates the information obtained, it is automatically consolidated in the system and reports are generated on the users' consumption with the respective date, time, and location. Finally, the register reader prepares the report for that period with the respective data support, notifications to the user, and photographic files of the reading history. Thus, if a user files a complaint, he can immediately be provided with a file containing all the information regarding his meter readings. (see Fig. 9).

Fig. 8 Control system to register readers

Fig. 9 System to reduce user complaints

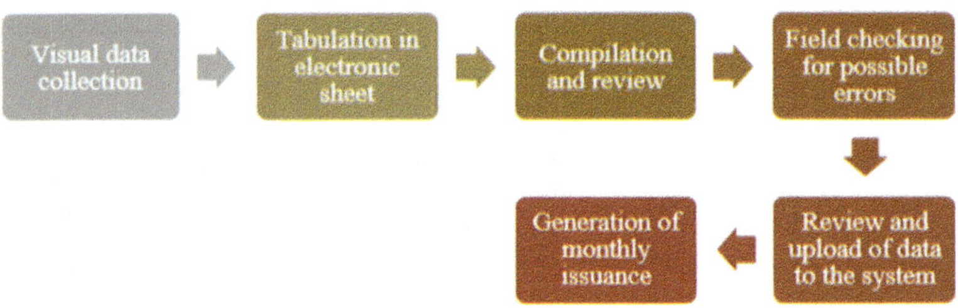

Fig. 10 Traditional method of taking readings

2.5 *Evaluation*

Traditional Method of Taking Readings. This process is detailed in (Fig. 10), each register reader visually collects the in-formation on a monthly basis, on an electronic sheet, the information is compiled by a supervisor, who after reviewing the information sends a report of consumption with zero and high readings. This is done so that the information can be verified again, prior to the monthly issuance of water consumption in the Integrated Municipal Services System [9] and the respective billing to the users of the city of Cañar.

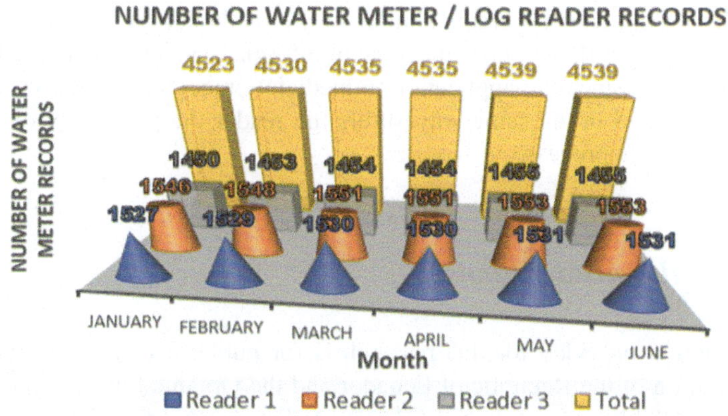

Fig. 11 Number of water meters records/month

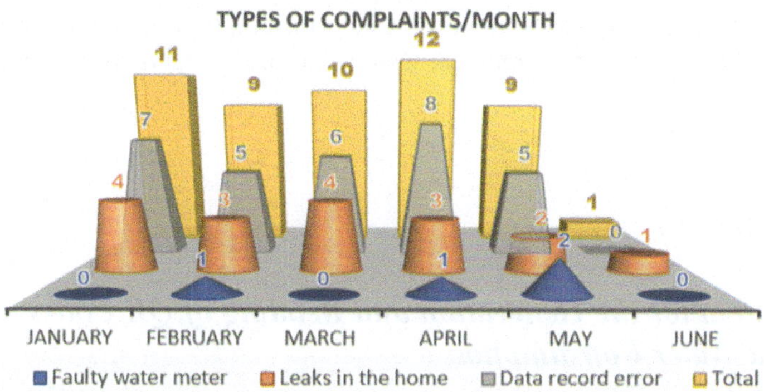

Fig. 12 Types of complaints/month

Fig. 13 Types of complaints/month

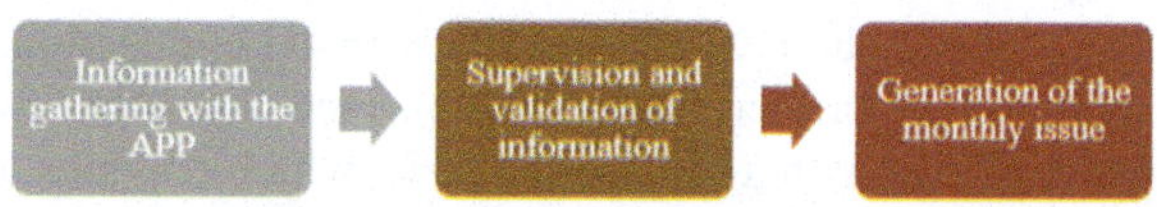

This process has caused several errors, mainly due to the manpower factor, since many readers, not complying with the measurements, make a manual projection of the readings. This can be observed by relating the information in (Fig. 11) and (Fig. 12) because regardless of the number of readings per sector, there are several complaints from users for different causes.

Automated Method for Analog Water Meters. Once established the development of the research to reduce the errors in the readings of analog water meters, it is expected that this process will help to reduce in a positive way, since as it is (see Fig. 13); in comparison to the traditional method, this method doesn't allow the readers to project the reads and the supervisor will validate the information that the readers upload to the system.

The use of computer programs similar to the methodology used by the smart meters proposed by [5], maintains the logic of information gathering but reduces the margins of errors, since the application immediately recognizes the user, location, date, and time; and in the same way, offers the reader the possibility of instantly verifying the whole process.

2.6 Ethical Considerations

The information provided for this research is for public use and is supported by the Constitution of the Republic of Ecuador and the Organic Law of Transparency and Access to Public Information (LOTAIP), and is found in the files of the potable water department, as well as in the financial department of the Municipality of the city of Cañar. The data provided by the Department of Potable Water and Sanitation of the City of Cañar provides monthly water consumption emissions for 100% of the registered users and beneficiaries, which in turn will serve as a sample for this research.

3 Results

3.1 System for the Generation and Reading of QR Codes and Water Consumption

The Intercultural Decentralized Autonomous Government of Cañar Canton, through the potable water service improvement plan, established an investment of 2,100.00 USD for the purchase of 3 tablets with 4G connection capacity with Internet access and a Zebra printer (see Fig. 14a), (see Fig. 14b); for labeling the 4,539 potable water meters with a unique QR code, which contains all the user information that is previously registered in the municipal servers when the user requests the service.

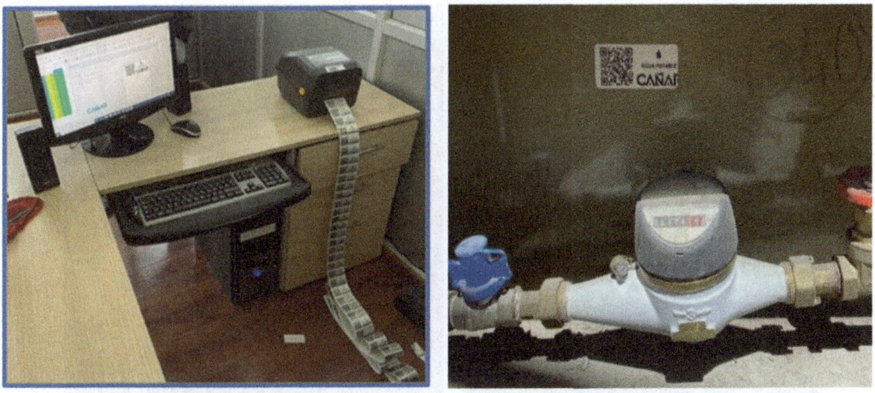

(a) (b)

Fig. 14 Hardware Architecture. **a** QR code printing system. **b** QR code in water metering device

3.2 Consolidation of Information, Sending and Management of Information

Collection and Sending of Information. With the tablet and the application developed on the Android platform Studio [15, 16] in agreement with the IT department of the Municipality of Cañar, each register reader scans the QR Code installed on the meters (see Fig. 15a) immediately obtaining the menu with the following information: user data, pending emissions, sector, meter number, the last reading of the previous period and once the reading is taken in the analog water meter, the current values can be entered; and in the case of excessive consumption, a warning is displayed in the menu.

Once all the information has been verified, the application issues the last question, (see Fig. 15b), asking the reader to validate and ratify the data entered. Immediately, the information is sent to the server and can be viewed by the supervisor, through the integrated system of municipal services [9]. If the Reader is unable to retrieve the information due to loss of the QR code label, it can communicate this to the supervisor who issues a temporary code, (see Fig. 15c), so that the data can be selected manually.

The application controls the reader according to the sector where it is located so that it completes the entire sequence; in this case, it is shown in green in the menu, and in the case of missing users in white, (see Fig. 15d). For high consumption, a photographic file is made, and the corresponding report is attached so that it can be presented to the user when he/she has a complaint (Fig. 17).

Information Management. To establish the link with the integrated system of municipal services, the IT department of the municipality of Cañar implemented a [2] system in which the supervisor has the capacity to visualize all the information obtained in the field by the readers, and can immediately establish the following control parameters, (see Fig. 16).

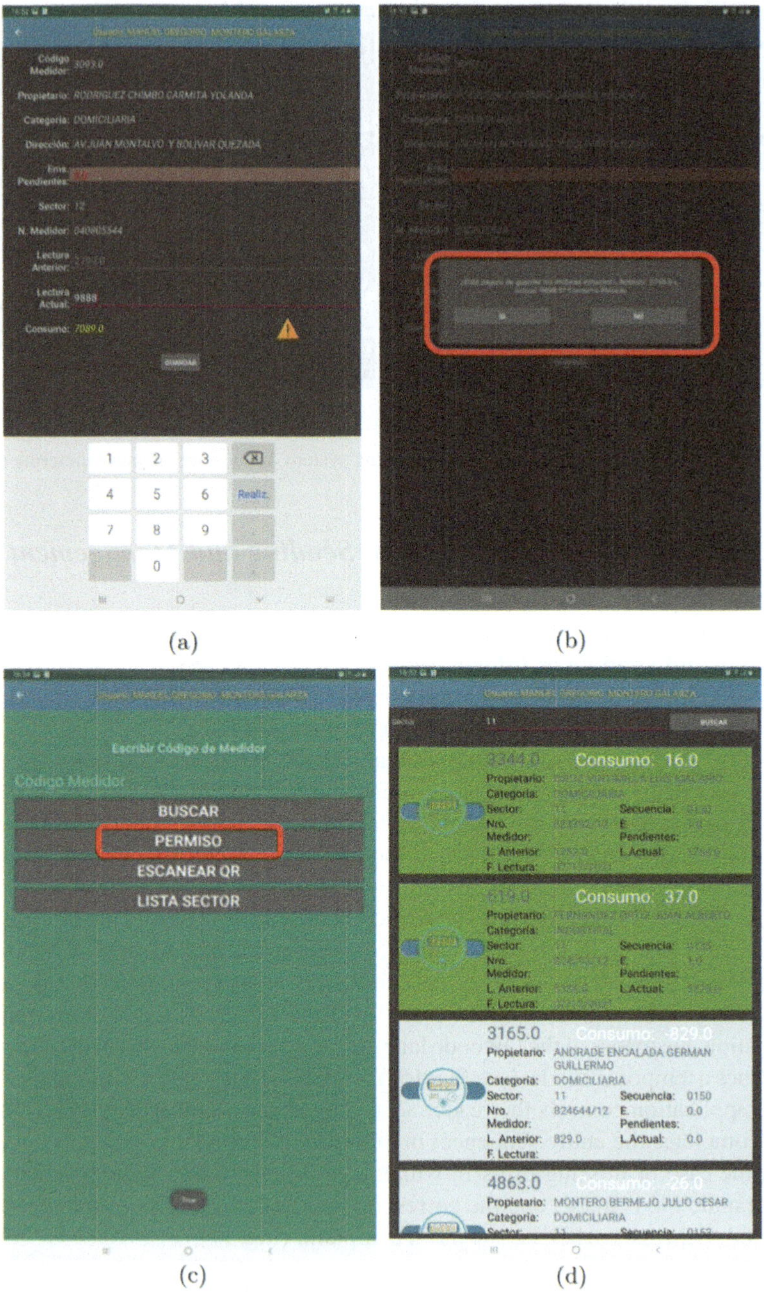

Fig. 15 Android App. **a** Mobile application menu. **b** Confirmation of data entered screen. **c** Temporary code menu screen. **d** QR code printing system screen

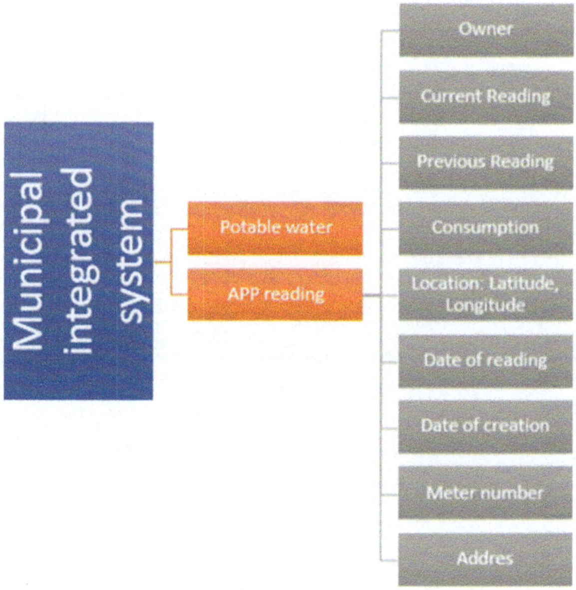

Fig. 16 Integrated municipal system menu

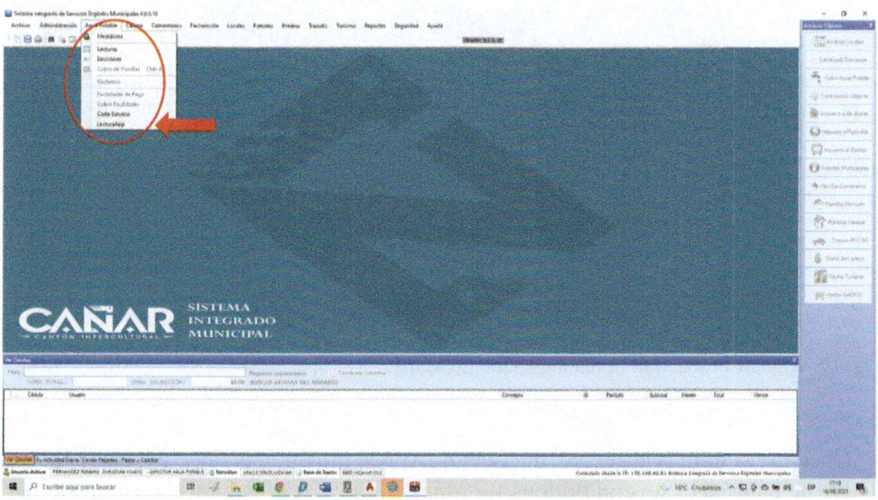

Fig. 17 Data filter

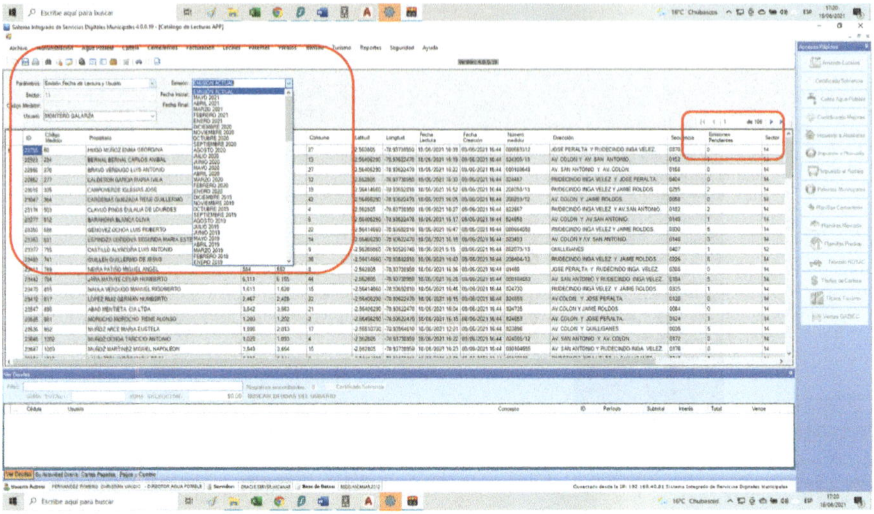

Fig. 18 Data filtering

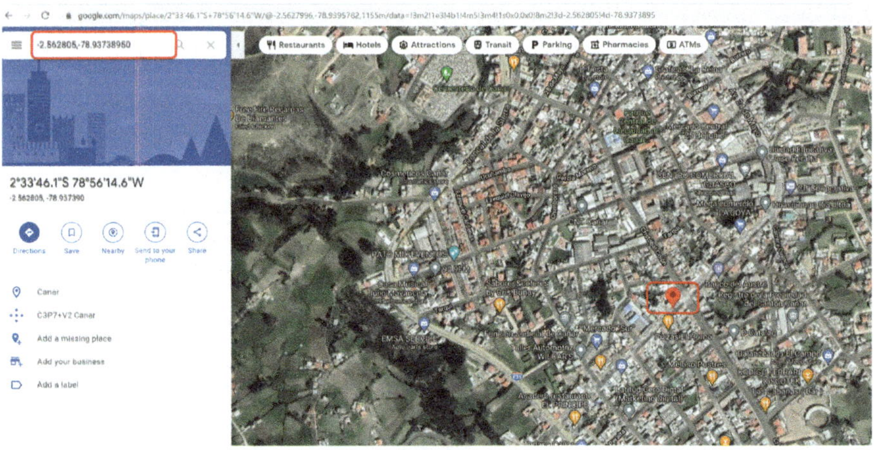

Fig. 19 Review of location of readings-Google Maps

Once the system is accessed, the LecturaApp application is selected, (see Fig. 18), where he can select several options for quick queries, retrieve information, and perform the following controls to each of the readers in their respective sectors and sequences. Additionally, control the number of readings, date, time, and location respectively.

In addition, with the help of the Google Maps web page, the supervisor will be able to enter the location data obtained from the Municipal Integrated System and verify if the information was actually collected in the property of each of the users. (see Fig. 19).

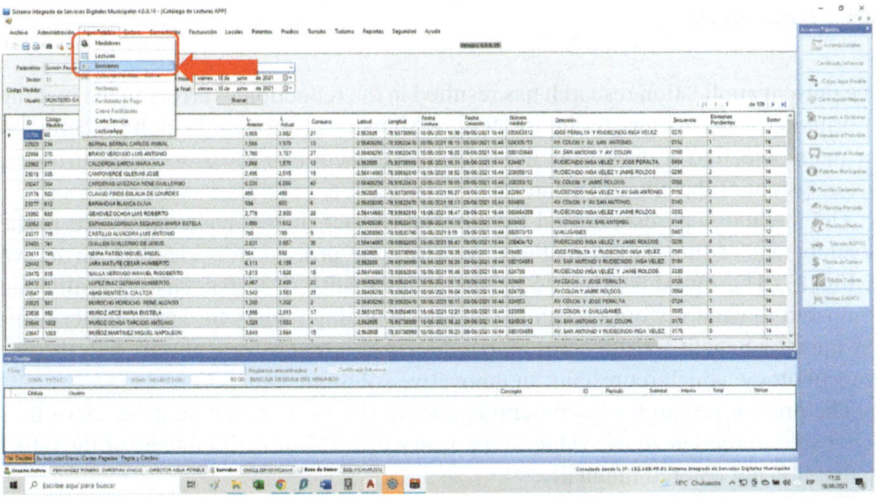

Fig. 20 Issuance menu

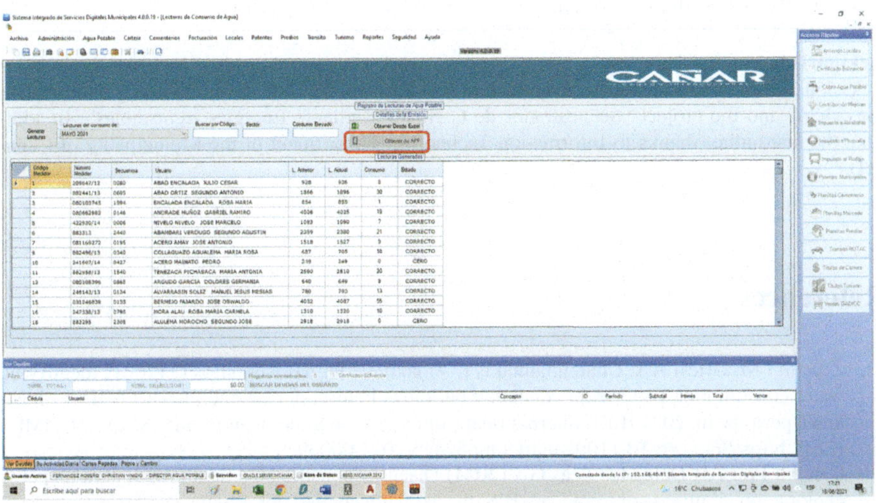

Fig. 21 Obtaining data from the App

Finally, to close the process, the Integrated Municipal System program facilitates the consolidation of information for the monthly issuance of drinking water consumption in Cañar Canton, (see Fig. 20).

The supervisor retrieves the data from the APP, verifies them and finally validates them, with this the issuance is done in a single step and immediately, closing the monthly period and leaving the information ready for the following period, (see Fig. 21).

3.3 Conclusions

The present application research has resulted in the reduction of errors in the reading of drinking water meters by up to 99%, as shown in (Fig. 12). Which has fulfilled the main objective of increasing the level of satisfaction of the users of the drinking water system of the city of Cañar.

The methodology established by the smart water meters has contributed to the development of this hybrid method, reducing the process considerably, as shown in Fig. 13. The investment made by the Municipality of Cañar has been minimal compared to what it would mean to change all analog meters for smart meters, this has meant a benefit for the entire population with low investment and with the use of technology, which represents an innovative solution.

The present research can contemplate more stages in its automation, as well as being able to recognize the value of the water meter reading by means of the tablet so that it is not entered manually.

The research is also opened to provide a solution to the water cuts made by the readers, so that this information can also be handled systematically and be able to have control of it, thus increasing the level of user satisfaction.

Acknowledgements This research is part of the work of research group "Sistemas Embebidos y Visión Artificial en Ciencias Arquitectónicas, Agropuecuarias, Ambientales y Automática (Seva4ca)" and the project "Secuestrador de Carbono Mediante Energías Renovables". In addition, the first author thanks to Information Technology Department of the Municipality of Cañar (TI).

References

1. Zecchini M, Griesi AA, Chatzigiannakis I, Amaxilatis D, Akrivopoulos O (2021) Identifying water consumption patterns in education buildings before, during and after COVID-19 lockdown periods. In: 2021 IEEE International Conference on Smart Computing (SMARTCOMP). IEEE. https://doi.org/10.1109/smartcomp52413.2021.00069
2. Cabrera Flor AP, Delgado Oleas GA (2014) Diseño y construcción de un robot para mapeo y exploración de minas subterráneas [B.S. thesis]. Universidad del Azuay
3. Mileiko S, Cetinkaya O, Yakovlev A, Balsamo D (2021) A non-intrusive ultrasonic sensor system for water flow rate measurement. In: 2021 IEEE Sensors Applications Symposium (SAS). IEEE. https://doi.org/10.1109/sas51076.2021.9530165
4. Canales P, Bryan K (2018) Desarrollo de un sistema de telemetría para el control y supervisión del consumo de agua en hogares residenciales. [B.S. thesis]. Universidad Católica de Santiago de Guayaquil
5. Zese R, Bellodi E, Luciani C, Alvisi S (2021) Neural network techniques for detecting intradomestic water leaks of different magnitude. IEEE Access 9:126135–126147. https://doi.org/10.1109/access.2021.3111113
6. Farinango Tuquerres JC (2018) Sistema de medidor inteligente de agua potable domiciliario para la junta administradora de agua "El Abra" [B.S. thesis]. Universidad Técnica de Norte

7. Rohmah RN, Supriyono H, Supardi A, Asyari H, Rahmadi R, Oktafianto Y (2021) IoT application on agricultural area surveillance and remote-controlled irrigation systems. In: 2021 9th International Conference on Information and Communication Technology (ICoICT). IEEE. https://doi.org/10.1109/icoict52021.2021.9527438
8. Gusqui Macas GR, Chaglla Moina AG (2016) Implementación de un sistema electrónico inteligente de asignación, medición y alerta del consumo de agua potable para hogares residenciales de la ciudad de Riobamba. [B.S. thesis]. Escuela Superior Politécnica de Chimborazo
9. Rodríguez-Flores J, Herrera VI, Morocho-Caiza A, Merino C (2020) Fractal control design with anti-windup effect for optimal operation of a power flyback source. In: Nummenmaa J, Pérez-González F, Domenech-Lega B, Vaunat J, Oscar Fernández-Peña F (eds) CSEI 2019, vol 1078. AISC. Springer, Cham, pp 308–328. https://doi.org/10.1007/978-3-030-33614-1_21
10. Cobos X, Oñate W, Caiza G (2021) Interactive application for learning electric motors using leap motion controller. In: García MV, Fernández-Peña F, Gordón-Gallegos C (eds) Advances and Applications in Computer Science, Electronics and Industrial Engineering, vol 1307. AISC. Springer, Singapore, pp 233–248. https://doi.org/10.1007/978-981-33-4565-2_15
11. Flores E, Cumbajin M, Sanchez P (2021) Design of a synchronous generator of permanent magnets of radial flux for a pico-hydropower station. In: García MV, Fernández-Peña F, Gordón-Gallegos C (eds) Advances and Applications in Computer Science, Electronics and Industrial Engineering, vol 1307. AISC. Springer, Singapore, pp 135–151. https://doi.org/10.1007/978-981-33-4565-2_9
12. Garcia CA, Naranjo JE, Ortiz A, Garcia MV (2019) An approach of virtual reality environment for technicians training in upstream sector, vol 52. Elsevier BV, pp 285–291. https://doi.org/10.1016/j.ifacol.2019.08.222
13. Garcia CA, Caiza G, Naranjo JE, Ortiz A, Garcia MV (2019) An approach of training virtual environment for teaching electro-pneumatic systems, vol 52. Elsevier BV, pp 278–284. https://doi.org/10.1016/j.ifacol.2019.08.221
14. García MV, Armentia A, Pérez F, Estévez E, Marcos M (2018) Vertical integration approach for the intelligent oil & gas field. at - Automatisierungstechnik 66(10):859–874. https://doi.org/10.1515/auto-2018-0033
15. Cartagena PD, Naranjo JE, Garcia CA, Beltran C, Castro M, Garcia MV (2018) Virtual reality-based system for hand rehabilitation using an exoskeletal orthosis, pp 105–117. https://doi.org/10.1007/978-3-319-95282-6_8
16. Naranjo JE, Ayala P, Altamirano S, Brito G, Garcia MV (2018) Intelligent oil field approach using virtual reality and mobile anthropomorphic robots, pp 467–478. https://doi.org/10.1007/978-3-319-95282-6_34

Air to Air Communication System for Collaborative Drone Work in Rural Areas

Fernando Caicedo, Hugo Moreno, and Janeth Moreta

Abstract This paper proposes the development and implementation of a communications system for multiple drones working in synchronization on the same task, as well as, the identification of optimal frequencies for optimal system performance. SDR technology was used by means of USRP NI-2932 devices and log-periodic antennas. The system was developed in LabVIEW. QAM modulation was used in the creation of a graphical interface for said system's transmitter and the receiver. The performance tests were carried out in two stages. The first stage was held in a laboratory whilst the second was held in the field in a rural area. It was possible to identify that there is the highest amount of interference produced by drones in operation between 2 and 3 GHz and that 1.8 GHz is the optimal performance frequency for the communication system.

Keywords Drone communication · Synchronized execution · SDR

1 Introduction

Wireless communication systems have grown exponentially in recent decades [1], making it possible to satisfy the demand for access to high-speed multimedia data [2, 3]. A basic communication system consists of a communication medium, a transmitter and a receiver, which are composed of subsystems such as filters, modulators, demodulators, encoders and decoders respectively [4, 5]. Given the large number of applications which require a communication system, work environments and the different operating parameters, it is difficult to have specific communication equip-

F. Caicedo (✉) · J. Moreta
Universidad Técnica de Ambato, 180103 Ambato, Ecuador
e-mail: fcaicedo5020@uta.edu.ec

J. Moreta
e-mail: jmoreta8974@uta.edu.ec

H. Moreno
Escuela Superior Politécnica de Chimborazo, 060155 Riobamba, Ecuador
e-mail: h_moreno@espoch.edu.ec

© The Author(s), under exclusive license to Springer Nature Switzerland AG 2022

M. V. Garcia et al. (eds.), *Advances and Applications in Computer Science, Electronics, and Industrial Engineering*, Lecture Notes in Networks and Systems 433,
https://doi.org/10.1007/978-3-030-97719-1_18

ment to perform performance tests [6, 7]. Software Defined Radio (SDR) is a very important modern technology in the area of communications, it allows users to control device communication through software and work with different frequencies and types of modulation [8]. It is for this reason that SDR devices are the most convenient option for developing systems in which the best operating parameters are identified under specific conditions [6].

As technology has advanced by leaps and bounds, robotics and the use of electronic devices to provide comfort and facilitate people's work has also undergone considerable advances. One of the most popular electronic systems used in both residential and industrial markets is the unmanned aerial vehicle (UAV), commonly known as a drone [9]. They are used for different purposes, such as military uses, surveillance systems, delivery services to domicile, crop fumigation, etc. [10–13]. There are applications that require very forced labor to be executed by a single drone, such as the transport of material or home delivery packages [14, 15]. In these kinds of situations, it is best to work collaboratively between two or more drones, as well as, attach robotic arms to said drones. This amalgam of an unmanned aerial vehicle and robotic extremities is known as an aerial mobile manipulator robot (AMR) [16]. The biggest challenge that the scientific community has faced in order to control this type of aerial robot is fluid communication between the drone and the robotic extremity to achieve coordination between them. This has led to many accidents and incidents costing thousands of dollars [11, 17].

Communications and control systems of robots and drones normally function on free frequency bands such as the 2.4 and 5 Ghz bands [18]. However, these frequency bands are not highly recommended because they present unstable communication and high latency [19] as there is already too much interference generated by residential WiFi systems in urban areas, and wireless internet service provider (WISP) systems in rural areas [20]. Several research projects have tried to establish different communication systems that allow correct communication be-tween drones, as is the case of [21] which proposes the use LoRaWAN technology as a secondary telemetry communication system to optimize consumption and improve the scope of communication. In [19] a multi-hop communication system is proposed in which routers are used as intermediaries for the control of robots or drones in end device mode using TDMA (time division multiple access) implemented in the data link layer (layer 2).

The present work is based on SDR and aims to design and implement a communication system for industrial drones that carry out collaborative work, as well as, identify a working frequency with the least amount of interference generated by remote controls, drone propellers and WiFi systems which in turn will allow more stable communication between the drones. We worked with two USRPs: one for transmission and the other for reception. Each one was configured through LabVIEW and equipped with two logarithmic antennas. Thanks to the functional and stable communication system, we were able to carry out tests to receive packets at different frequencies, first without the interference of the drones and later with the drones in operation.

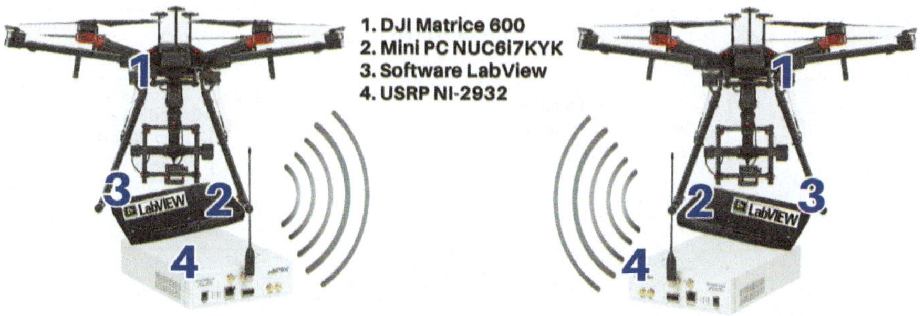

Fig. 1 Diagram of the equipment of the proposed communications system

2 Methodology

The proposed system consists of generating stable communication between co-working drones, for which one, controlled from the ground station, will send geopositioning data to the other drone to maintain a specific position or indicate movements to be carried out so both may move in synchronization. Communication is done through USRPs connected to Skull Canyon NUC mini computers integrated into each of the drones.

The matrix 600 has a weight of 9.6 kg including six TB48S batteries, the maximum take-off weight is 15.1 kg, having a payload capacity of 5.5 kg. The additional load that the drones are supporting is 3 kg between the mini computer, the USRP and the antenna that have a weight of 1.6, 1.2 and 0.2 kg respectively. Figure 1 shows the diagram of the equipment used in the collaborative work system between drones with the proposed communication system.

Figure 2 shows the flow chart of the proposed algorithm with each of the main stages considered for the design of the transmission and reception systems.

2.1 Transmitter

The communications system works with two USRP-2932 mounted on two Matrice 600 drones and was designed using LabView software. It consists of a transmitter made up of several blocks such as device configuration, data framing, data coding and modulation; and a receiver formed by a block of demodulation, decoding, de-framing and data presentation. Figure 3 shows the configuration block for the transmitter.

Synchronization is a very important part of a communications system [22]. The USRP clock signal has been taken as a basic parameter to generate synchronism in both the transmitter and the receiver. Figure 4 shows the framing and coding block of the data. A convolutional encoder VI (virtual instrument) with a code rate of $\frac{1}{2}$

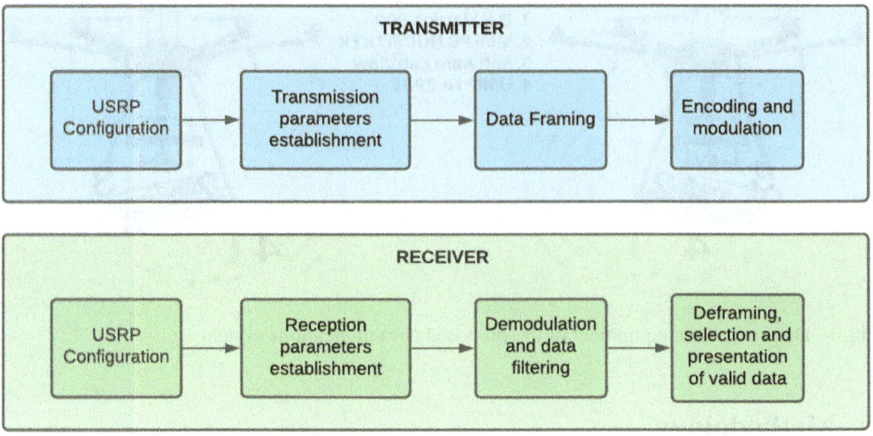

Fig. 2 Configuration block of the USRP in the transmitter

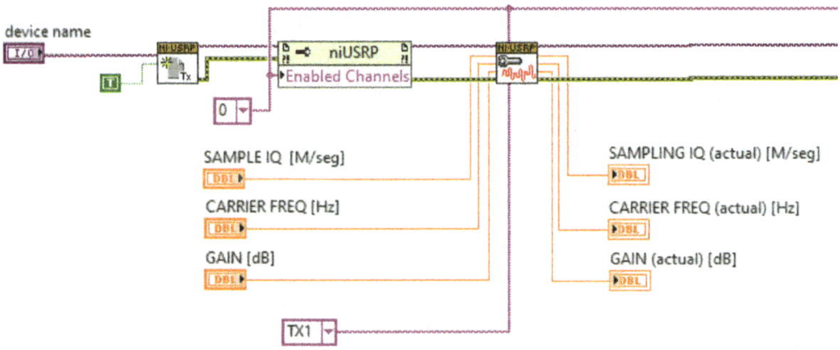

Fig. 3 Configuration block of the USRP in the transmitter

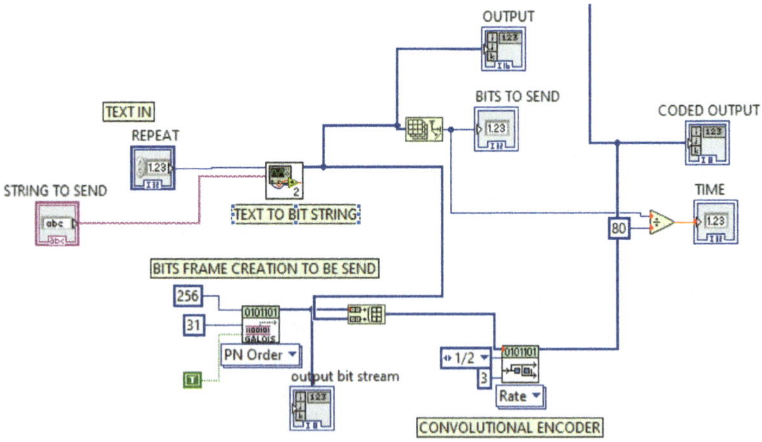

Fig. 4 Data framing and encoding block

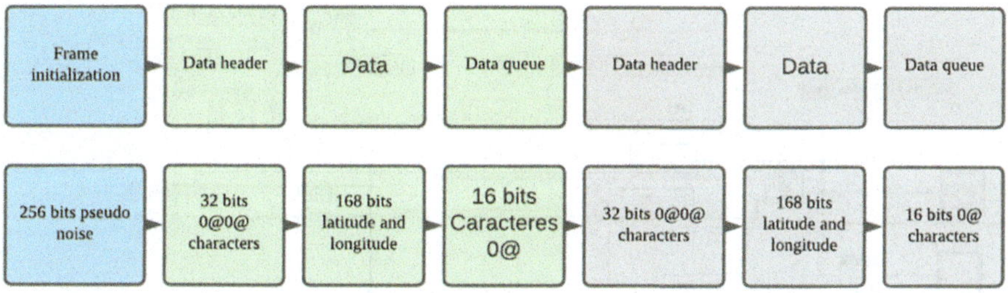

Fig. 5 Data framing structure

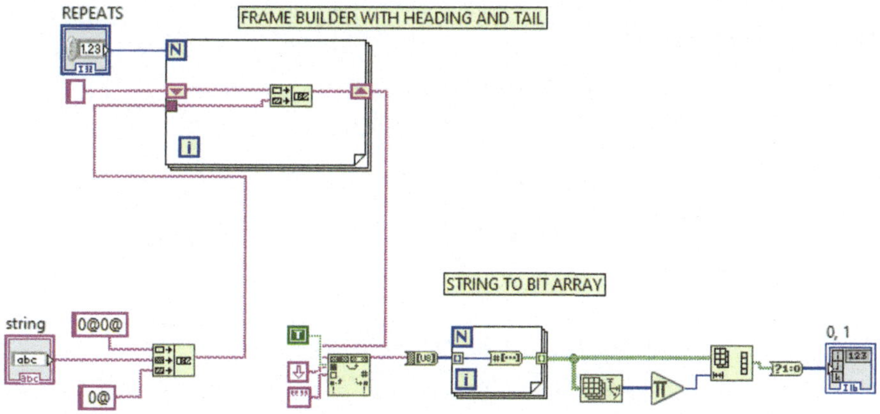

Fig. 6 Data block formed by the header, data and queue

was used for the encoding. This VI generates a stream of coded bits by means of a preconfigured matrix.

The data frame is made up of an initialization block, consisting of a sequence of 256 bits of Galois pseudo noise that allows better detection of the frame at the receiver, followed by the data block containing 32 header bits, 168 bits of corresponding data, 21 characters representing the geographic coordinates in latitude and longitude, and finally a block of 16 queue bits for the completion of the frame. The data block is sent twice to guarantee that it has been received correctly by the receiver. Figure 5 shows the structure of the frame that is transmitted in the system, while Fig. 6 shows the creation of the data block.

We have chosen to work with QAM for the modulation block since several investigations, such as [23, 24], have demonstrated that the quadrature amplitude modulation performs better compared to OFDM and that lower values of the constellation obtain lower BER values (erroneous bit rates) for having a higher bit rate. Figure 7 shows the modulation block which contains a stage for configuring modulation parameters, filter parameters, modulation and constellation graph.

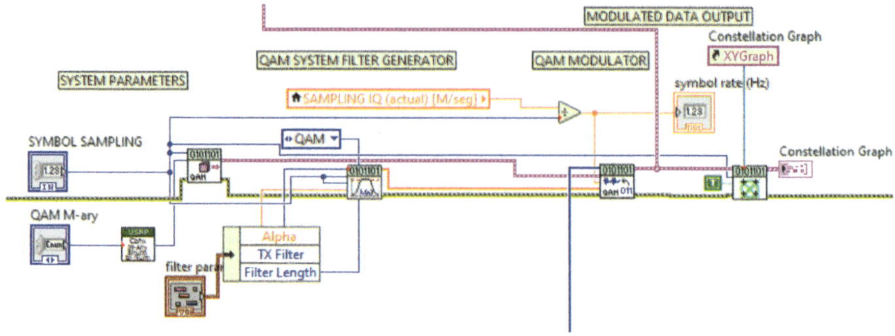

Fig. 7 QAM modulation block

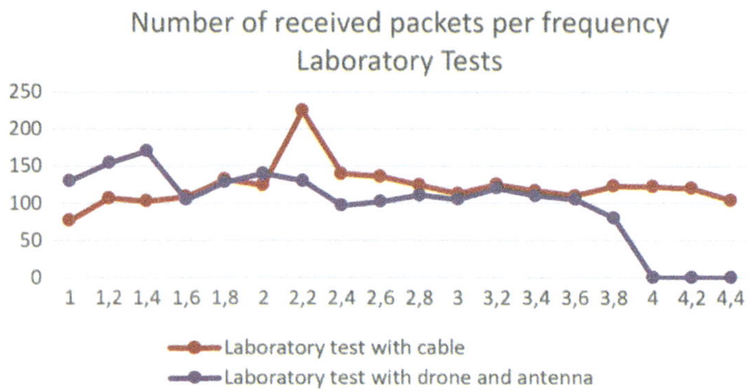

Fig. 8 Galois Pseudo Noise (PN) bit sequence and demodulation parameter block

2.2 *Receiver*

The receiver must be configured with the same modulation and coding parameters used in the transmitter, as well as the initialization bits, header and data queue. Figure 8 shows the configuration stage of the demodulation parameters and the generation of Galois pseudo noise bits for the detection of the initialization of the frame.

Once the initialization bits have been obtained, the frame is detected in the demodulator. The user can select the type of filter they want to work with from a drop-down list both in the modulator and in the demodulator. The available filters are Gaussian, root raised cosine, and raised cosine. In this case, we worked with the raised cosine filter since this type of filter reduces the side lobes of the signal spectrum, allowing the transmission power to be optimized by restricting it only to the bandwidth of interest [25]. Figure 8 and Fig. 9 show the configuration stages of the filter parameters, signal filtering and frame detection using the QAM demodulator.

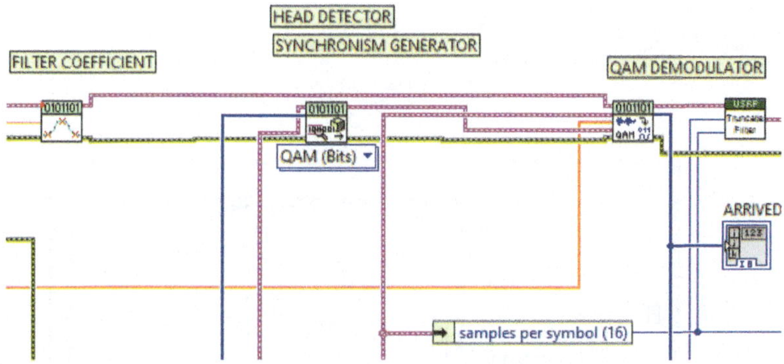

Fig. 9 Filter and synchronism coefficient block for the QAM demodulator

Fig. 10 Constellation diagram and eye diagram configuration block

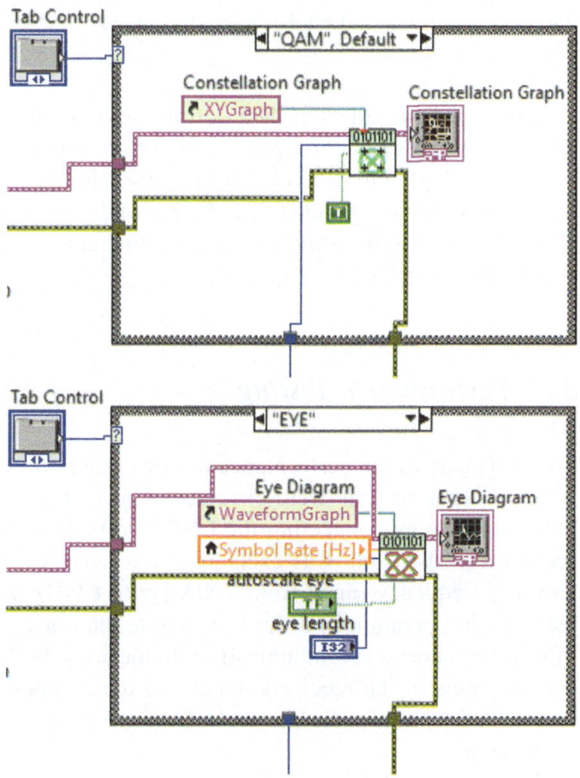

Figure 10 shows the blocks programmed for the display of the constellation and eye diagrams respectively. These diagrams are used to get an idea of the quality of the signal being received through the clarity they present [26]. These diagrams will be analyzed in the results stage to verify the best reception at certain frequencies.

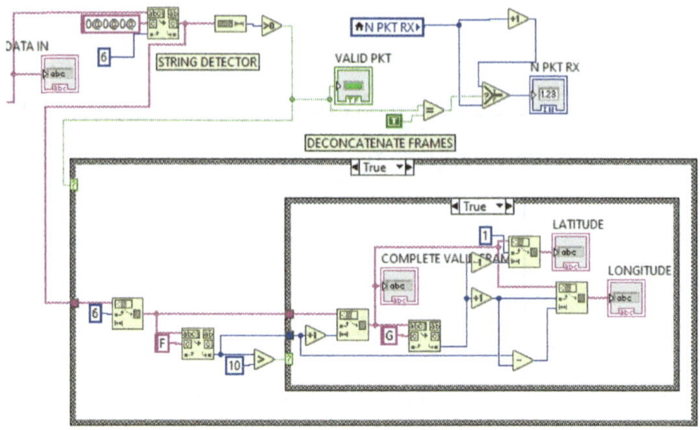

Fig. 11 Decatenation and valid data selection block

Lastly, we come to the data deframing and data presentation block. Figure 11 shows the frame detection processes and their decatenation using comparators to identify the beginning and end of the information corresponding to the latitude and longitude of the geographical coordinates of the transmitting drone. The demodulator also uses the raised cosine filtering parameters. The same parameters used in the transmitter modulator.

2.3 *Performance Testing*

The first stage of the performance tests was carried out in a laboratory environment where 3 types of tests were executed. The first test consisted of transmitting and receiving data by connecting the USRP ports directly using coaxial cable. For the second test, the cable was exchanged for a pair of periodic logo antennas with a working frequency range from 700 MHz to 6 GHz. For the third test, a drone was included in operation in order to generate the noise and interference characteristic of its propellers and communication frequencies. In the three performance tests, the eye diagrams of the received signal and the number of packets received after one minute at frequencies from 1 to 4 GHz were analyzed. Figures 12, 13, and 14 show these lab tests.

The second stage of the performance tests was carried out in the field at a separation distance of up to 300 m. Said tests were carried out in a rural area where there is no interference from radio frequency systems such as WiFi networks. Figure 15 shows the Matrice 600 drone controlled by the mini computer.

In order to carry out the operational tests with the drone in flight, it was necessary to mount both the USRP and the antenna on each of the drones. The minicomputer was used to run the LabVIEW program. The same one that executed the actions carried

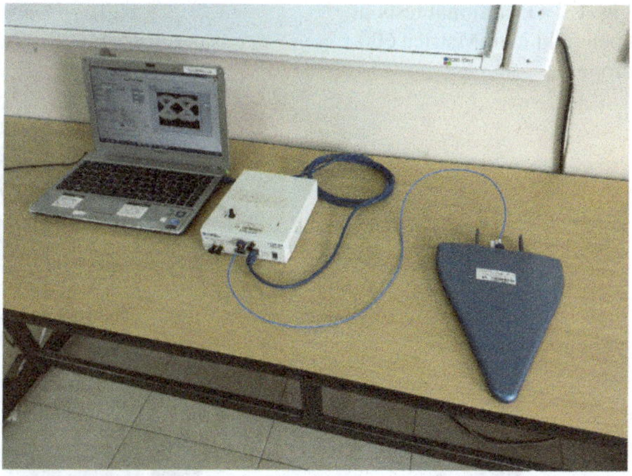

Fig. 12 Performance tests in laboratory with logoperiodic antenna

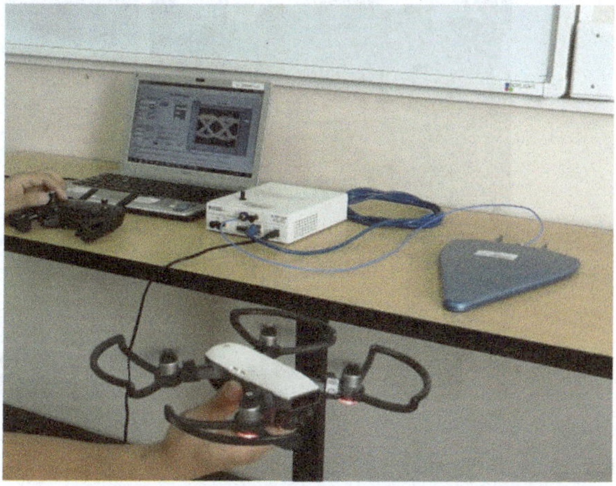

Fig. 13 Performance tests in laboratory with logoperiodic antenna and drone in operation

Fig. 14 Performance tests in laboratory with logoperiodic antenna and Matrice 600 drone in operation

Fig. 15 Operational tests in open field with Matrice 600 drone

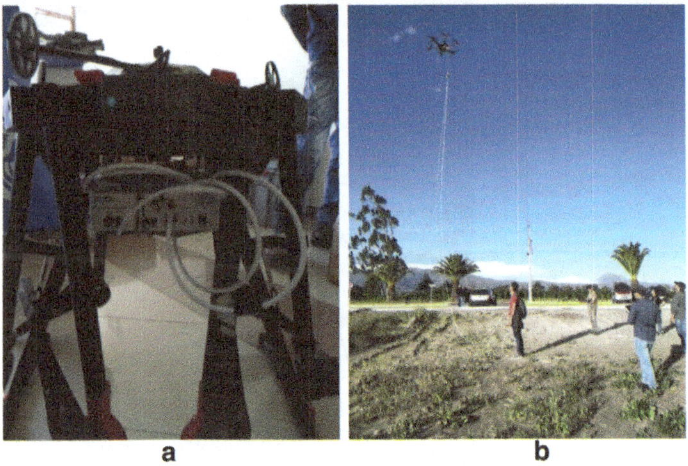

Fig. 16 **a** USRP and antenna mounted on Matrice 600 drone **b** Drone in flight for field performance tests in rural areas

out by the USRP in the transmitter, and, in the same way, allowed the analysis of the data captured in the receiver. A cable connected to a grounded outlet was used to power the mini computer and the USRP. Figure 16 a) shows the USRP being installed in the drone, while Fig. 16 b) shows the Matrice 600 drone in flight during operational tests.

3 Results

In both the transmitter and the receiver, a GUI was created for the entry of configuration parameters, data entry and visualization of received data. Figure 17 shows a

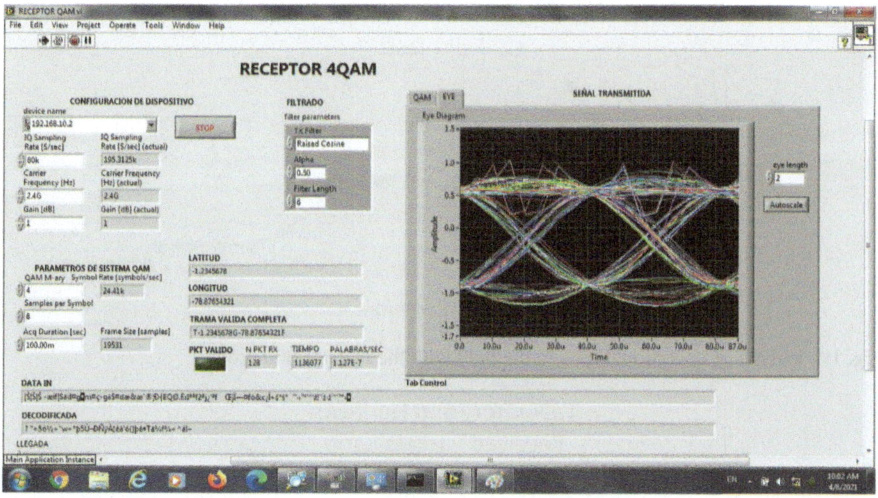

Fig. 17 Receiver GUI in use

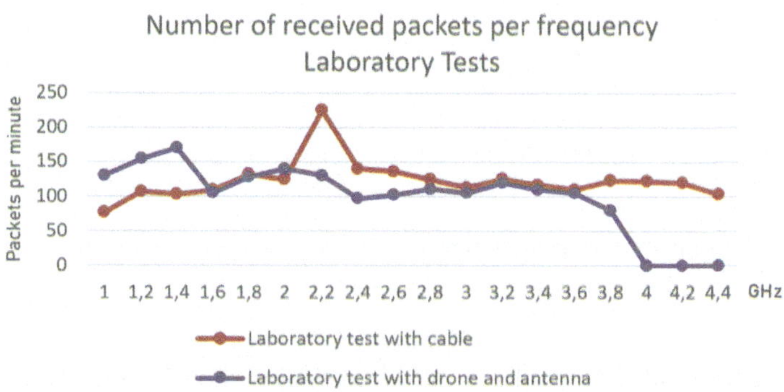

Fig. 18 Results of performance tests in a laboratory environment

screenshot of the receiver's graphical user interface in operation, in which the eye diagram of the received signal is evident.

The tests were carried out in the communications and microwave research laboratory of the Polytechnic School of Chimborazo (ESPOCH). Said operation tests of the system were carried out in two configurations: first, a system using wired communication via a coaxial cable coupled to 50 ohms of resistance, and a wireless system with logarithmic antennas in a far-field environment. Figure 18 shows the results obtained in the first phase of performance tests. The different carrier frequencies used are shown on the X axis, while the Y axis shows the number of packets received during a one-minute sample.

The lab tests were carried out in a frequency range from 1 to 4.4 GHz, which is the maximum operating frequency of the USRP card, with a variation of 200 MHz

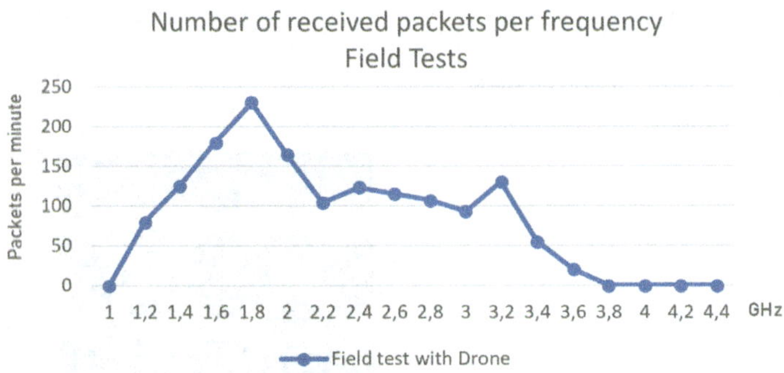

Fig. 19 Results of operational tests in the field with the matrice 600 drone

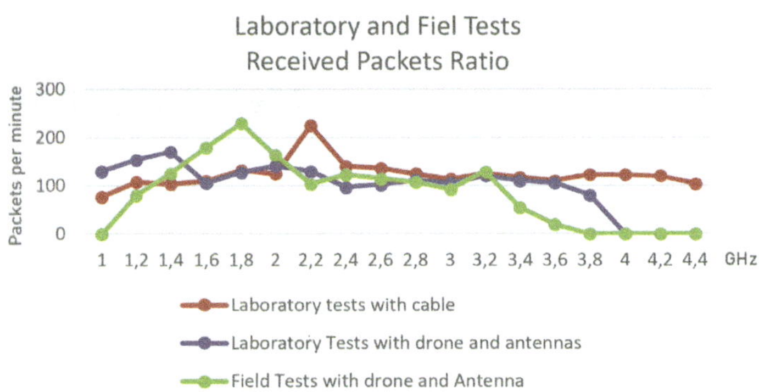

Fig. 20 List of results obtained in the different performance tests carried out both in the laboratory and in the field

in each measurement. The results of the wired system show better performance at a 2.2 GHz frequency while the wireless system is optimal at 1.4 GHz.

The second stage of testing was carried out in ab open field on the ESPOCH campus with the Matrice 600 drone at a height of 2,840 m above sea level. The results of these tests are shown in Fig. 19 where the system demonstrated a better performance while working at a frequency of 1.8 GHz.

The results obtained in the different scenarios mentioned above can be better ap-preciated in Fig. 20. It is clearly shown that more valid packets are received per minute when using a coaxial cable as a transmission medium than when using a wireless communication. It also shows that the interference generated by the drone in operation occurs between 2 and 3 GHz.

4 Discussion

The results of the proposed system show that the frequencies between 2 and 3 GHz suffer the greatest amount of interference generated by the control system and by the drone propellers. Thus, the use of the afore mentioned frequencies is not recommended for the communications system between co-wroking drones. Frequencies above 3 GHz are not recommended either, since the system becomes unstable causing considerable packet loss. The optimal frequency for the system to work would be 1.8 GHz, since at this frequency there is a fairly stable communication with a considerable number of valid packets received. The main drawback for the development of the system was the synchronization between the transmitter and the receiver in order to detect the start and end of valid data frames. It is recommended that the amount of bits sent in a frame be taken into consideration when selecting the type of modulation to be used.

5 Conclusions

A communication system was developed which permits communication between drones in flight so as to share geopositioning data. The optimal operating frequency of 1.8 GHz was established which made it possible for drones in flight to transmit and receive up to 230 valid packets per second.

The data frame used together with QAM modulation guarantees the correct reception of geographic coordination data in latitude and longitude, which in turn optimizes coordinated work between drones. Tests carried out in an open environment show that the optimal communication frequency is 1.8 GHz with 230 packets received correctly out of a total of 250. This represents a 925 level of system efficiency.

The drones used generated environmental noise affecting data communication starting from 2 GHz. This subsequently affected the transmission and reception of geographical coordinates thus resulting in an average packet reception level of 125 with an efficiency of 50%. These results makes it impossible to use this frequency range for drone communication.

Acknowledgements The authors of this research paper would like to thank the Ecuadorian Corporation for the Development of Research and Academia - CEDIA for their contribution to innovation through CEPRA projects, as well as the Polytechnic School of Chimborazo and the Technical University of Ambato for their support in the development of this research.

References

1. Chowdhury MZ, Shahjalal M, Ahmed S, Jang YM (2020) 6G wireless communication systems: applications, requirements, technologies, challenges, and research directions. IEEE Open J Commun Soc 1:957–975. https://doi.org/10.1109/ojcoms.2020.3010270
2. Yu X, Xu D, Schober R (2019) MISO wireless communication systems via intelligent reflecting surfaces: (Invited Paper). In: 2019 IEEE/CIC International Conference on Communications in China (ICCC). IEEE. https://doi.org/10.1109/iccchina.2019.8855810
3. Barona-Lopez LI, Valdivieso-Caraguay AL, Benalcazar ME, Aguas X, Zea JA (2021) Feature evaluation of EMG signals for hand gesture recognition based on mutual information, fuzzy entropy and RES index. In: García MV, Fernández-Peña F, Gordón-Gallegos C (eds) Advances and Applications in Computer Science, Electronics and Industrial Engineering, vol 1307. AISC. Springer, Singapore, pp 101–119. https://doi.org/10.1007/978-981-33-4565-2_7
4. Devi RP, Nishat H (2017) BER analysis of digital modulation schemes using LabVIEW. CVR J Sci Technol 13:41–45
5. Cartagena PD, Naranjo JE, Saltos LF, Garcia CA, Garcia MV (2019) Multifunctional exoskeletal orthosis for hand rehabilitation based on virtual reality. In: Botto-Tobar M, Barba-Maggi L, González-Huerta J, Villacrés-Cevallos P, S. Gómez O, Uvidia-Fassler MI (eds) TICEC 2018, vol 884. AISC. Springer, Cham, pp 209–221. https://doi.org/10.1007/978-3-030-02828-2_16
6. Muslimin J, Asnawi AL, Ismail AF, Jusoh AZ (2016) SDR-based transceiver of digital communication system using USRP and GNU radio. In: 2016 International Conference on Computer and Communication Engineering (ICCCE). IEEE. https://doi.org/10.1109/iccce.2016.100
7. Garcia CA, Salinas G, Perez VM, Salazar L. F, Garcia MV (2019) Robotic arm manipulation under IEC 61499 and ROS-based compatible control scheme. In: Botto-Tobar M, Barba-Maggi L, González-Huerta J, Villacrés-Cevallos P, S. Gómez O, Uvidia-Fassler MI (eds) TICEC 2018, vol 884. AISC. Springer, Cham, pp 358–371. https://doi.org/10.1007/978-3-030-02828-2_26
8. Abid NB, Souani C (2020) SDR-based transmitter of digital communication system using USRP and GNU radio. In: Bouhlel MS, Rovetta S (eds) SETIT 2018, vol 147. SIST. Springer, Cham, pp 373–381. https://doi.org/10.1007/978-3-030-21009-0_36
9. Suzuki S (2018) Recent researches on innovative drone technologies in robotics field. Adv Robot 32(19):1008–1022. https://doi.org/10.1080/01691864.2018.1515660
10. Ortiz JS, Zapata CF, Vega AD, Andaluz VH (2018) Path planning based on visual feedback between terrestrial and aerial robots cooperation. In: Zeghloul S, Romdhane L, Laribi MA (eds) Computational Kinematics, vol 50. MMS. Springer, Cham, pp 96–105. https://doi.org/10.1007/978-3-319-60867-9_12
11. Okutake T, Uchida N, Yamamoto N (2016) A collaborative safety flight control system for multiple drones. In: 2016 10th International Conference on Innovative Mobile and Internet Services in Ubiquitous Computing (IMIS). IEEE. https://doi.org/10.1109/imis.2016.128
12. Greenwood WW, Lynch JP, Zekkos D (2019) Applications of UAVs in civil infrastructure. J Infrastruct Syst **25**(2):04019002. https://doi.org/10.1061/(asce)is.1943-555x.0000464
13. Naranjo AE, Alarcon-O A, Amancha-P G, Ortiz-Espinosa J, Naranjo JE (2021) Low-cost assistive system for deaf people based on artificial vision. In: García MV, Fernández-Peña F, Gordón-Gallegos C (eds) Advances and Applications in Computer Science, Electronics and Industrial Engineering, vol 1307. AISC. Springer, Singapore, pp 249–264. https://doi.org/10.1007/978-981-33-4565-2_16
14. Leica P, Rivera K, Muela S, Chavez D, Andaluz G, Andaluz VH (2019) Consensus algorithms for bidirectional teleoperation of aerial manipulator robots in an environment with obstacles. In: 2019 IEEE Fourth Ecuador Technical Chapters Meeting (ETCM). IEEE. https://doi.org/10.1109/etcm48019.2019.9014872
15. Quintero Lorza DP, Duque Méndez ND, Gómez Soto JA (2020) GLORIA: a genetic algorithms approach to tetris. In: Nummenmaa J, Pérez-González F, Domenech-Lega B, Vaunat J, Oscar Fernández-Peña F (eds) CSEI 2019, vol 1078. AISC. Springer, Cham, pp 111–126. https://doi.org/10.1007/978-3-030-33614-1_8

16. Aguirre OA, Nacato JC, Andaluz VH (2020) Virtual simulator for collaborative tasks of aerial manipulator robots. In: 2020 15th Iberian Conference on Information Systems and Technologies (CISTI). IEEE. https://doi.org/10.23919/cisti49556.2020.9141092
17. Uchida N, Okutake T, Yamamoto N (2017) Image recognitions of collaborative drones' security controls for FPV systems. Int J Space-Based Situated Comput 7(3):129. https://doi.org/10.1504/ijssc.2017.088998
18. Hell P, Mezei M, Varga PJ (2017) Drone communications analysis. In: 2017 IEEE 15th International Symposium on Applied Machine Intelligence and Informatics (SAMI). IEEE. https://doi.org/10.1109/sami.2017.7880305
19. Kagawa T et al (2017) A study on latency-guaranteed multi-hop wireless communication system for control of robots and drones. In: 2017 20th International Symposium on Wireless Personal Multimedia Communications (WPMC). IEEE. https://doi.org/10.1109/wpmc.2017.8301849
20. Singh PJ, De Silva R (2018) Design and implementation of an experimental UAV network. In: 2018 International Conference on Information and Communications Technology (ICOIACT). IEEE. https://doi.org/10.1109/icoiact.2018.8350739
21. Rahmadhani A, Richard, Isswandhana R, Giovani A, Syah RA (2018) LoRaWAN as secondary telemetry communication system for drone delivery. In: 2018 IEEE International Conference on Internet of Things and Intelligence System (IOTAIS). IEEE. https://doi.org/10.1109/iotais.2018.8600892
22. Kocamaz UE, Çiçek S, Uyaroğlu Y (2017) Secure communication with chaos and electronic circuit design using passivity-based synchronization. J Circuits Syst Comput 27(04):1850057. https://doi.org/10.1142/s0218126618500573
23. Alhasani MM, Nguyen QN, Ohta GI, Sato T (2019) A novel four single-sideband M-QAM modulation scheme using a shadow equalizer for MIMO system toward 5G communications. Sensors 19(8):1944. https://doi.org/10.3390/s19081944
24. Farzamnia A, Hlaing NW, Mariappan M, Haldar MK (2018) BER comparison of OFDM with M-QAM modulation scheme of AWGN and Rayleigh fading channels. In: 2018 9th IEEE Control and System Graduate Research Colloquium (ICSGRC). IEEE. https://doi.org/10.1109/icsgrc.2018.8657503
25. Hmida MA, Braham A (2019) Raised cosine filter embedded in stationary wavelet packet transform for induction motor fault detection. In: 2019 19th International Conference on Sciences and Techniques of Automatic Control and Computer Engineering (STA). IEEE. https://doi.org/10.1109/sta.2019.8717222
26. Li CY et al (2018) A 5 m/25 gbps underwater wireless optical communication system. IEEE Photonics J 10(3):1–9. https://doi.org/10.1109/jphot.2018.2842762

Real-Time Video Transmission and Communication System via Drones over Long Distances

Janneth Moreta, Hugo Moreno, and Fernando Caicedo

Abstract One of the major technological trends worldwide is the development and use of Unmanned Aerial Vehicles (UAV) civil and military fields. Nowadays, video transmission has become a trend, and as technology advances, the human being demands more sophisticated technological tools. In this case we are looking at better quality real-time video transmission which increases even more with the help of a drone allowing the user to have an overview of areas difficult to access. This article shows the design and implementation of a wireless communications system which allows transmission of video in real time (Streaming) from the DJI Matrice 600 Pro industrial drone in flight controlled from a ground base station. Software with SDR (Software Defined Radio) technology was used for this project, which has helped complete several stages of the system. With the exception of elements such as antennas or amplifiers, all of the blocks have been implemented normally. On the other hand, the use of hardware was minimized to the use of a computer with LabVIEW, USRPs, a drone, a camera and the antennas. For the verification of the results, the respective tests were carried out in a lab.

Keywords Video · Transmission · Interface · Image · Drone

1 Introduction

The development of information, communication and electronics technology has allowed the world of telecommunications to advance by leaps and bounds, generating advances and improvements in the transmission of data through wireless channels, in

J. Moreta (✉) · F. Caicedo
Universidad Tecnica de Ambato, UTA, 180103 Ambato, Ecuador
e-mail: jmoreta8974@uta.edu.ec

F. Caicedo
e-mail: fcaicedo5020@uta.edu.ec

H. Moreno
Escuela Superior Politécnica de Chimborazo, 060155 Riobamba, Ecuador
e-mail: h_moreno@espoch.ec

© The Author(s), under exclusive license to Springer Nature Switzerland AG 2022
M. V. Garcia et al. (eds.), *Advances and Applications in Computer Science, Electronics, and Industrial Engineering*, Lecture Notes in Networks and Systems 433,
https://doi.org/10.1007/978-3-030-97719-1_19

the field of robotics, and giving way to the emergence of new electronic devices [1, 2], which today make life easier for humans.

Communications systems work with state-of-the-art equipment, allowing information, such as data, voice or video [3, 4], to be sent and received from anywhere in the world. This is due to the fact that the signal overcomes problems presented in different environments. Thus, they also satisfy the demand for access to high speed multimedia data transmission [5].

Nowadays, video transmission has become a trend, and as technology advances, human beings demand more sophisticated technological tools. In this case better quality in real-time video transmission which improves even more if said data transmission is boosted with an unmanned aerial device (UAV) [6]. This gives the user a panorama of all the areas of difficult access in their area. But continuous transmission of video from a drone requires a strong connection with its wireless network. Live streaming is less efficient in terms of bandwidth, as confirmed by [7] where the measured bandwidth from a drone is 10 Mbps for 25 FPS.

Drones have achieved great success in research and development. In many industrial fields, they have generated great impact today due to the advantages they offers [8]. They have gained great popularity in recent years thanks to their lower cost, smaller size, lower weight, higher capacity and advances in batteries and motors [2]. The technical term is Unmanned Aerial Vehicles (UAVs), however, it is more common to hear the word drone to refer to an automated tool that has expanded its uses and applications in different sectors [9].

Unmanned aerial vehicles (UAVs) are a technology that has military [10], public and civil [11] applications. The military has been using UAVs for over 25 years and consists mainly of border surveillance, reconnaissance and attack. Public use is mainly in public services such as the police, public security and public transport management [12]. For 15 years the use of UAVs in the civil sector has been in for various activities, such as rescue and evacuation operations, cartography, geology, agriculture, photography, filming, surveillance, private security, etc. Therefore, the reliability of UAV connections and the accuracy of the information exchanged between them are critical parameters [11]. The pervasiveness of drones among the civilian population can lead to technical, security and public safety issues that need to be addressed, regulated and prevented [13].

Multi-UAV systems can collaboratively complete missions more efficiently and economically compared to individual UAV systems. However, there are many problems that must be resolved before UAVs can be effectively used to provide stable and reliable context-specific networks [14]. On the other hand, a wide set of applications coincide in providing a UAV with a video camera and the appropriate means to capture and/or transmit images to a ground station [15]. The ability to acquire images make to tackling tasks such as inspection, surveillance or search much easier.

This article shows the design and implementation of a wireless communications system that allows the transmission of video in real-time (Streaming) from an unmanned aerial vehicle in flight (DJI Matrice 600 Pro industrial drone) to a fixed land-based station. This system is beneficial by allowing air control due to the reduced size of the drone, visualization of areas difficult to access, and avoids putting the safety and physical integrity of a person at risk.

2 Methodology

This section details the methodology followed to design the communications system. Each functional block that makes up a video transmission system is detailed, as well as the analysis process for selecting the appropriate operating frequency, channel capacity and transmission power necessary for said communications system.

The blocks that make up the communications system for the transmission of video in real-time are shown in Fig. 1. There the stages of the transmitter can be observed, as well as the block diagram for the receiver. The process of each block is captioned below.

For the transmission system stages, the information source block represents the acquisition of the consecutive images acquired through the video camera, and makes data available to the system in the form of images. These are subsequently digitized into bits to be transmitted. The video camera uses compression techniques, providing fewer bits without loss of image quality. The convolutional encoder is the block responsible for providing robustness and information recovery capacity to the information signal against interfering signals. This is accomplished through redundancy bits. Two actions are carried out in the modulation block: source coding and digital modulation. In this section, the digitized information produced by the convolutional encoder is taken and transformed into a continuous signal of baseband pulses. It then proceeds to capture this information in a band-pass frequency using digital modulation techniques. The high-power amplifier (HPA) and transmission antenna represent the analog block, where the modulated signal travels through the air via radio waves (radiocommunications).

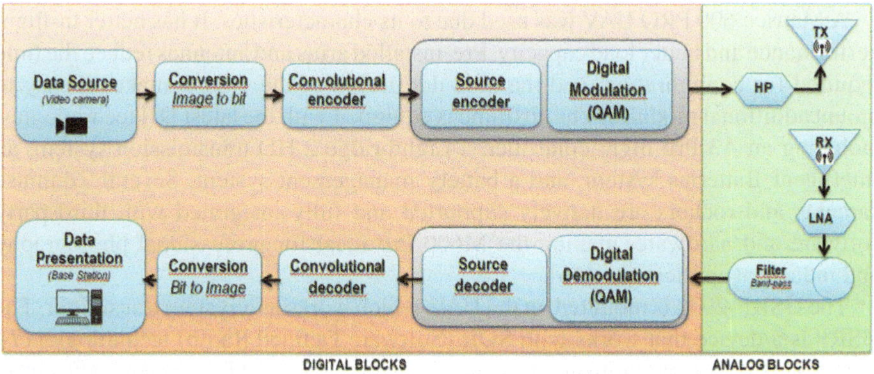

Fig. 1 Block diagram for the video transmission and communications system

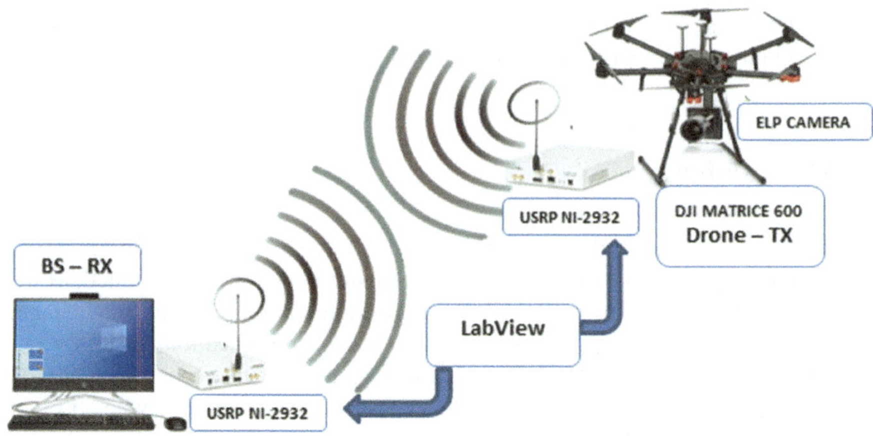

Fig. 2 Equipment diagram of the video transmission and communications system

For the reception system stages, a reception antenna, a low noise amplifier and a band-pass filter are used. These devices extract the information signal present in the air and recover the original signal as much as possible after it has been contaminated with noise due to the environment. The demodulation stage is responsible for trans-forming the received modulated signal into a continuous pulse signal and then trans-forming it back into a stream of bits. In the convolutional decoder block, errors are corrected and the original information is recovered. This is achieved by separating the redundancy bits from the obtained bit stream, and only the bits containing the information are extracted. Finally, we have the bit-to-image conversion and the presentation of the information on the computer. In this block, the bit-level digital information is converted back to images or frames for later presentation.

The devices and software used in the design and implementation of the wireless communications system for the transmission of video in real-time (Streaming), from an unmanned aerial vehicle to a base station, are shown in Fig. 2.

A Matrice 600 PRO UAV was used due to its characteristics. It has better in-flight performance and better load capacity. Pre-installed arms and antennas reduce the time required for configuration, and the modular design of the system makes it easy to mount additional modules. The airframe is equipped with the latest DJI technologies, including an A3 Pro flight controller, a Lightbridge 2 HD transmission system, an Intelligent Batteries system, and a battery management system. Several Zenmuse cameras and rockers are natively supported and fully integrated with third-party software and hardware, making the M600 Pro ideal for professional photography and industrial applications [16].

The USRP-2932 is mounted on the UAV, which works as a system transmitter. The USRP is a device that works with SDR (Software Defined Radio) technology [17]. It uses a 20 MHz bandwidth and has two channels for Tx and Rx at frequencies from 400 to 4.4 GHz. This adjustable RF transceiver has a high speed analog-to-digital converter and a digital-to-analog converter for transmitting I and Q baseband signals to a host PC over 1 Gigabit of ethernet. In addition, an ELP video camera developed by the company Ailipu Technology was used, which works with resolutions (pixels) of 320×240, 640×360, 640×480, 1280×720 and 1920×1080. In this case a low definition setting was used to capture 320p of video. It also has a frame rate capture of up to 30 fps and a built-in H.264 compression technique [13].

The base station was docked on the ground in a fixed position, which permitted the researchers to view the video transmitted by the UAV. A PC operating Microsoft Windows was used together with a second URSP-2932 as system receivers. Omnidirectional antennas were used for signal transmission and reception, such as the HyperLOG 3080 and HyperLOG 7060 antenna, for lab tests. Said atennas function at frequencies of 380 MHz–18 GHz and 700 MHz–6 Ghz respectively. The use of this type of antennas provides the system with a continuous connection, thus it is connected at all times [18, 19].

LabView was used for the communications system since this software program is suitable for the design and implementation of communications algorithms. In addition to signal processing functions and its modulation kit, LabView also uses the NI USRP library to manage and configure NI USRP devices, facilitating the development of radiocommunications system prototypes. The NI FPGA library also uses the USRP FPGA and NI VISUAL ACQUISITION for the video capture section. The entire process is described in blocks below.

2.1 Transmission System

The transmission interface consists of three sections:

- USRP Tx and QAM Modulation configuration.
- Video configuration.
- Visualization of transmitted data.

For the USRP to work as a transmitter it must be given initial parameters, such as: the name of the device or the IP address (Device ID), the transmission channel, the transmission sampling frequency, which can vary between 100 kbps and 2 Mbps, the transmission frequency of the USRP (carrier frequency, which in this case will depend on its technical specifications), the gain applied to the RF signal and the port destined for the antenna to work.

Data must be entered manually before running the program, and then displayed while the program is running. In Fig. 3 the programming in LabView is shown, where the USRP configuration parameters are first entered and at the same time the camera information is obtained, which are then entered into the system to be processed. Once completed, the parameters are correctly configured. Additionally, you can see the initialization of the queue (Init Queue) that is used for the queuing process, and that is explained later.

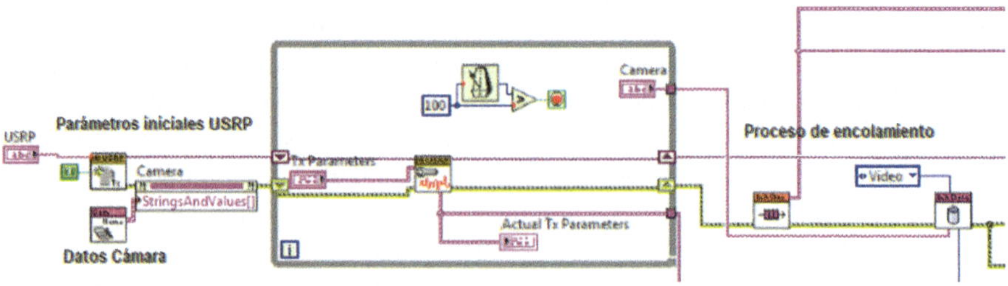

Fig. 3 USRP transmitter configuration

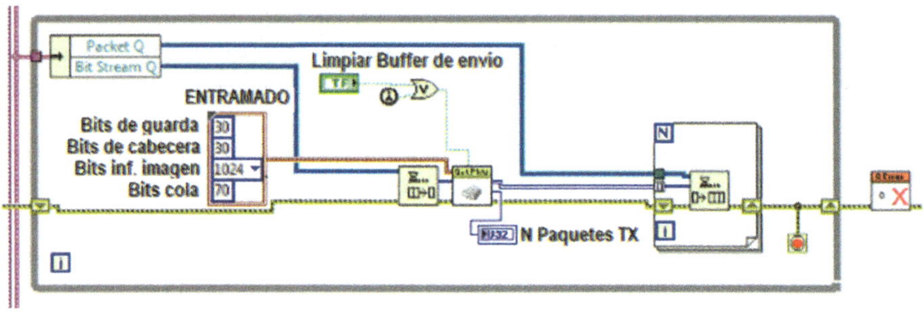

Fig. 4 Bit-framing block

The information obtained from the video camera is subjected to the framing process to later be glued, modulated and transmitted. When working with video, the in-formation must be divided into small fragments, in other words, split into small frames of up to 1024 bits with at least 30 guard bits, 30 header bits and 70 tail bits. This way, loss of information is avoided. The framing block Fig. 4 shows the number of packets to be transmitted and to reset the system if necessary.

Once the data has passed through the framing process, it enters the queuing block. Here the frames are transformed into bits in accordance with the required image quality. Then, the generated bits are directed to the element queue where they are transformed into an arrangement of bits concatenated into a single one-dimensional arrangement. Once all the bits have been queued, the process is closed. Later the bits are sent to the modulator, see Fig. 5.

The modulation block works in four phases: The first is the modulator parameter configuration where the samples per symbol and the number of modulator states are established. This data is entered manually. The second stage consists of a filter coefficient generator that calculates the filter coefficients for the shaping of pulses. The third stage unqueues the data bits to be sent to the modulator taking into account the configured parameters. The fourth shows the constellation of the QAM modulation for reference and comparison with the constellation of the receiver. See Fig. 6.

Once the modulated signal is obtained, it is received by the writing module of the USRP transmitter. This module is in charge of writing the modulated signal on the channel specified for transmission. As it is about video transmission, it works in a

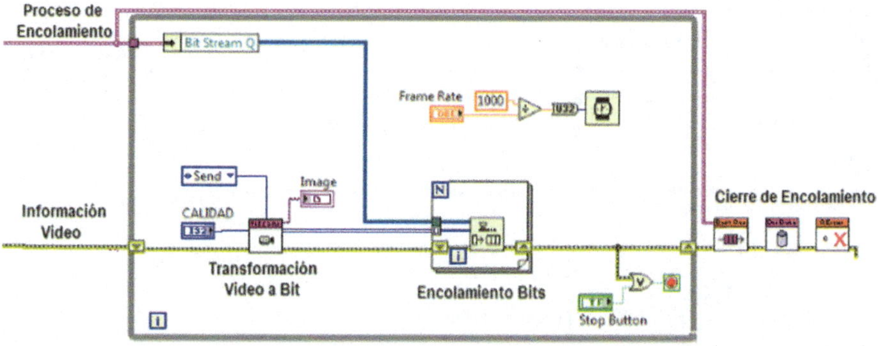

Fig. 5 Bit queue

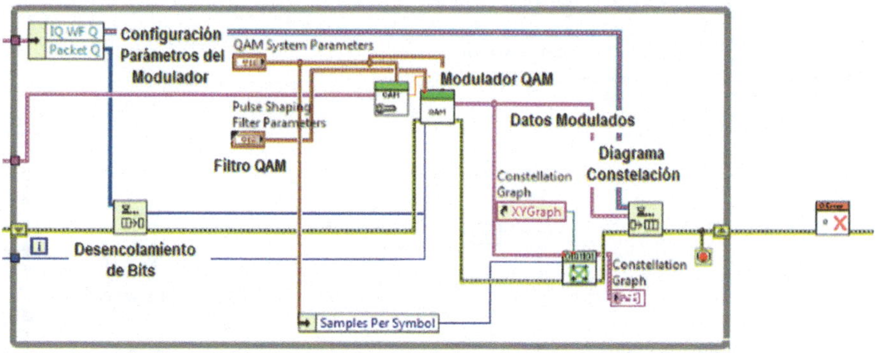

Fig. 6 QAM modulation block

loop. If an error occurs the system stops, otherwise it will continue processing. In Fig. 7 the transmission module is shown together with the number of sent packets, error packets and transmitted bits.

2.2 *Reception System*

The transmission interface consists of three sections:

- USRP Rx and QAM modulation configuration.
- Configuration of the reception frame.
- Visualization of received data.

In the receiving USRP the user must configure the initial parameters, such as: the name of the device or the IP address (Device ID) and the reception channel. There are also parameters that must be directly related to the parameters established in the transmitter, such as: the sample rate, carrier frequency, and RF signal gain. This data

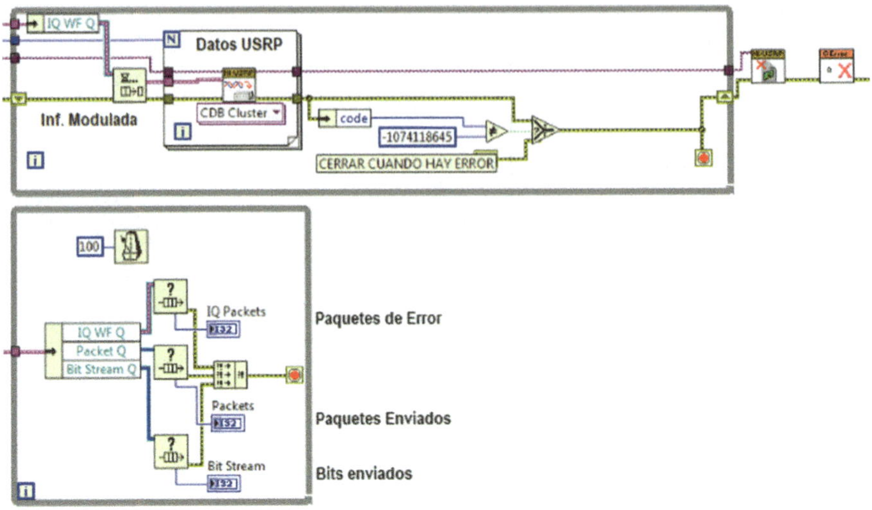

Fig. 7 Transmission block

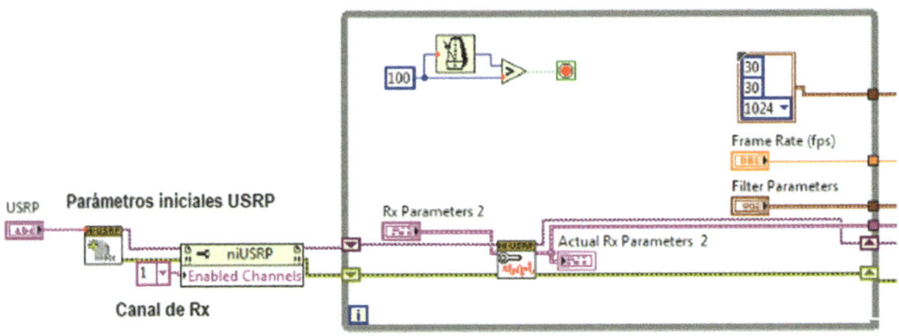

Fig. 8 Receiver USRP configuration block

is entered manually. It is inputted to an evaluation loop and then allowed to continue to the next block. See Fig. 8.

In the reading block, the USRP initialization module is activated as shown in Fig. 9. This allows the passage of data to the reading module, then orders the system to start reading the data received from the RF module. This also generates the creation of queues where the arriving bits will be transformed into an arrangement of bits concatenated into a single one-dimensional arrangement. When there is noise or interference that causes information peaks, the detector activates and is responsible for eliminating it.

The QAM demodulation block is made up of three phases as shown in Fig. 10. The first is the modulator parameter configuration, where the samples per symbol and the number of modulator states are established. This data is entered manually. In the second stage, the data that arrives from the reading block and the bits are

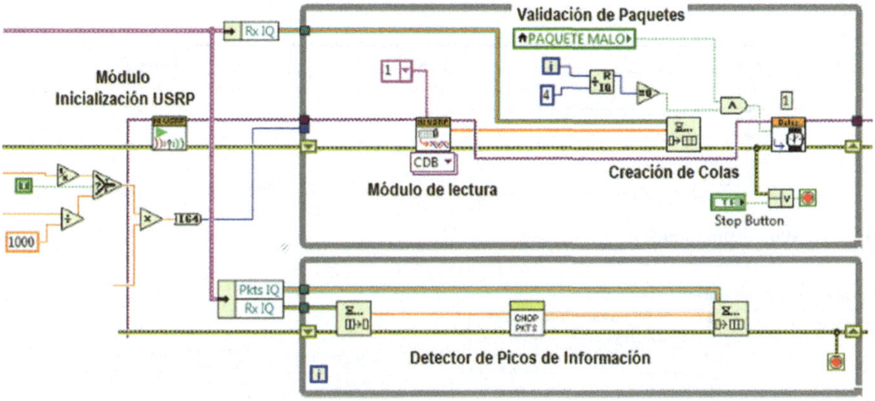

Fig. 9 USRP data read block

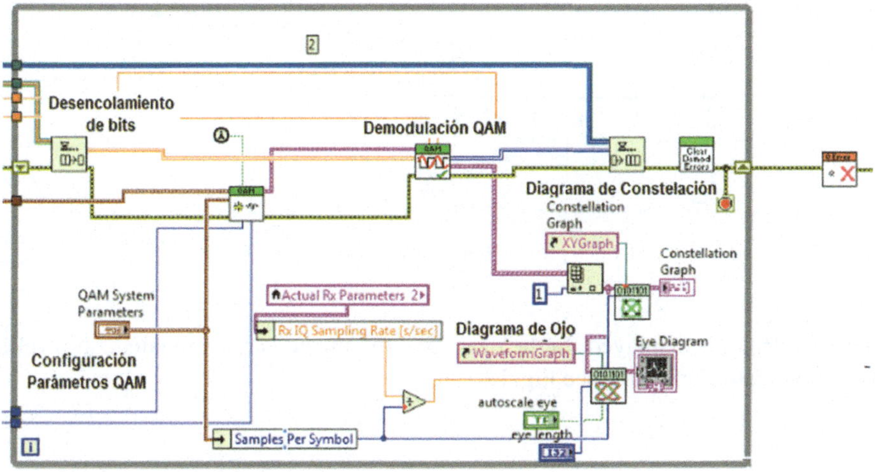

Fig. 10 QAM demodulation block

unqueued to be sent to the demodulator. The third stage shows the constellation of the QAM modulation and the eye diagram of the received signal. In these diagrams the quality of the received data and the signal-to-noise ratio can be analyzed.

The process continues with the sync block, where the sync bits are detected, as shown in Fig. 11. Internally it works as a pseudo-random noise generator, which concatenates a chain of 0 and 1 at the beginning and end of the frame being transmitted.

The final part of the reception system is presented, where the queued information arrives and is reconstructed into its original state, taking into account that the objective is a 32-bit RGB image to later be sent to the video decoder. Once the image has been validated, it is displayed on a viewer and the effective rate of packets received

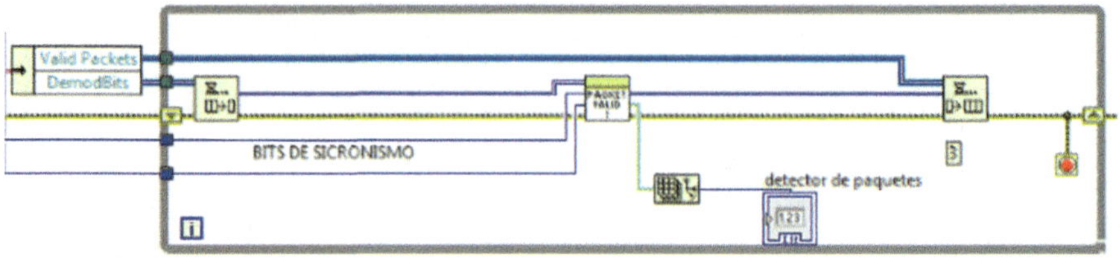

Fig. 11 Synchronism block

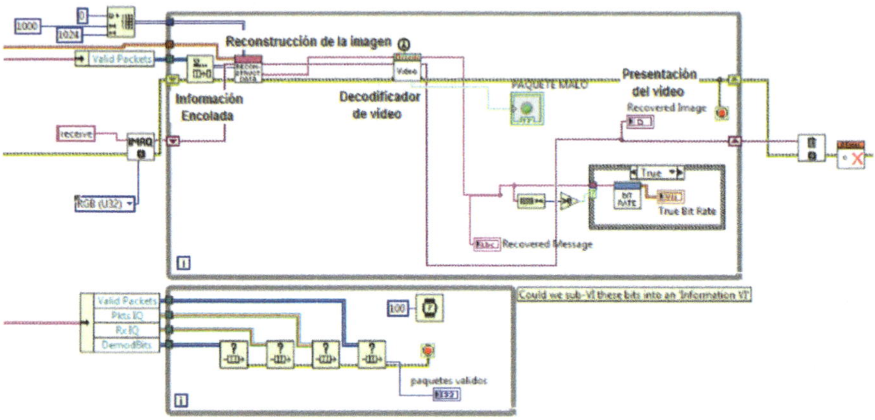

Fig. 12 Reconstruction and video presentation block

(True Bit Rate) is displayed. In the event that the image is not validated, an invalid packet indicator lights up. See Fig. 12.

2.3 Performance Testing

Three types of tests were carried out in a lab. In the first the video was transmitted and received by connecting the USRP directly through a coaxial cable, as shown in Fig. 13.

For the second test, the coaxial cable was replaced with log-periodic antennas Fig. 14 and 15. A frequency range from 950 MHz to 3.5 GHz was also used.

The last test was carried out using the drone in operation, so its propellers could generate noise and interference, and communication frequencies. See Fig. 16.

During the execution of the tests, the eye diagrams of the received signal and the images received within the established frequency ranges were analyzed.

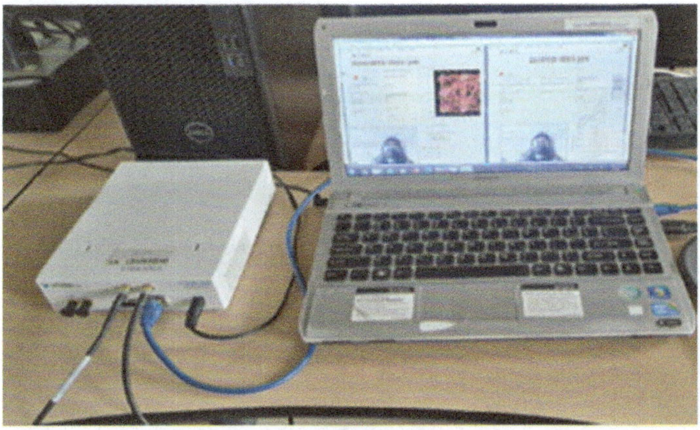

Fig. 13 Coaxial cable performance tests

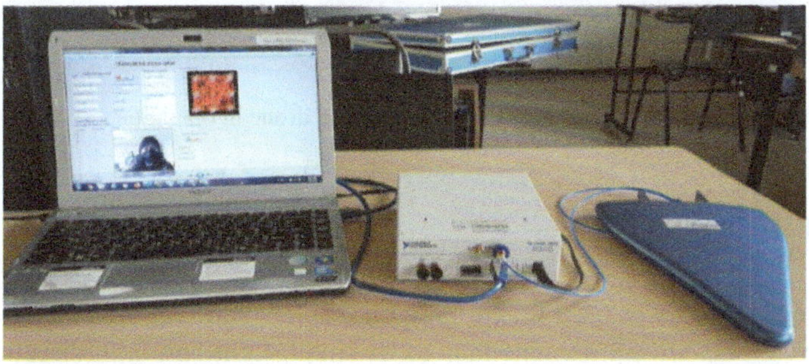

Fig. 14 Coaxial cable performance tests

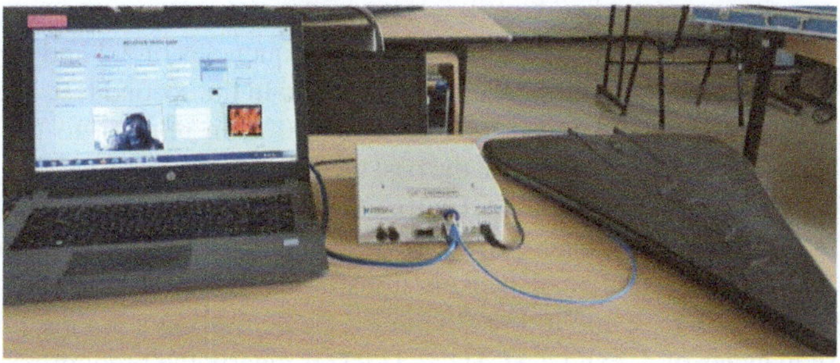

Fig. 15 Performance tests with a log-periodic receiver antenna

Fig. 16 Performance tests with log-periodic antenna and drone in operation

3 Results

Wireless technology was used functioning on a 5 GHz band for the video transmitter and receiver. This allows the antennas to be small, light and omnidirectional. The operating frequency was selected (fc) taking into account certain parameters in the three tests carried out. The first test compared the estimated reception time in seconds to the range of frequencies used, such as 950 to 3.5 GHz. Figure 17, indicates that the best video-reception time is less than 100 s at frequencies of 2.2 and 2.3 GHz. At frequencies higher than this, the time is much longer. In the second test, the received image quality is considered in a range of 1 to 100%, in relation to the range of frequencies used. Figure 18 shows that the best received image quality is found in 80% at a frequency of approximately 2.3 GHz. In the third test, the transmitted PKTs are analyzed in relation to the range of frequencies considered. Figure 19 demonstrates that there is a greater number of packets transmitted in the frequency ranges that go from 2.1 to 2.4 GHz, while in Fig. 20 the PKTs received in a greater number are between frequencies of 2.2 and 2.3 GHz. 2.3 GHz is clearly the predominate of these two frequencies. Table 1 presents a comparative analysis of the possible frequencies considered in regard to the interferences presented in the tests, resulting in an optimal carrier frequency of 2.3 GHz.

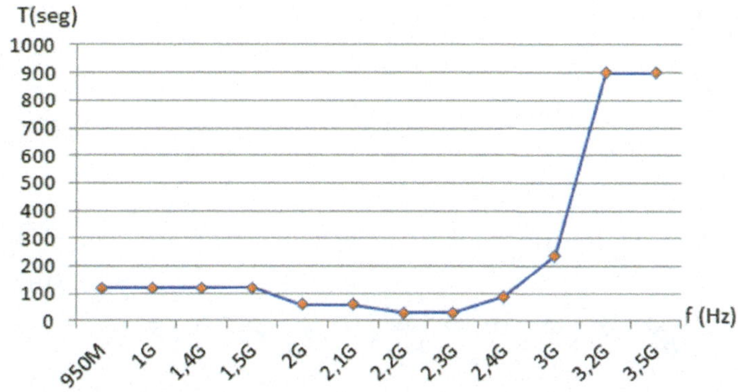

Fig. 17 Estimated RX time in seconds

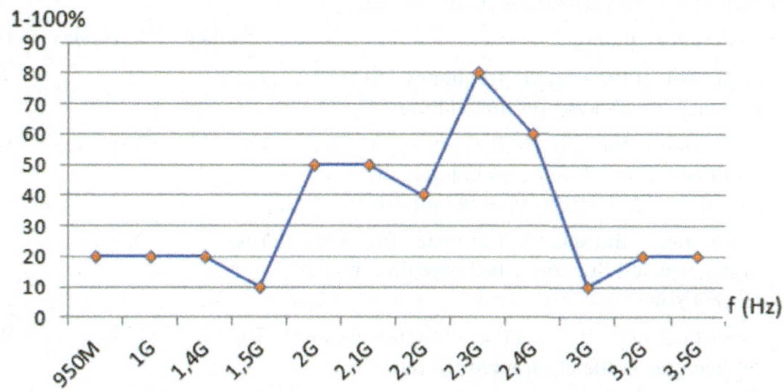

Fig. 18 Image quality received from 1 to 100% at different frequencies

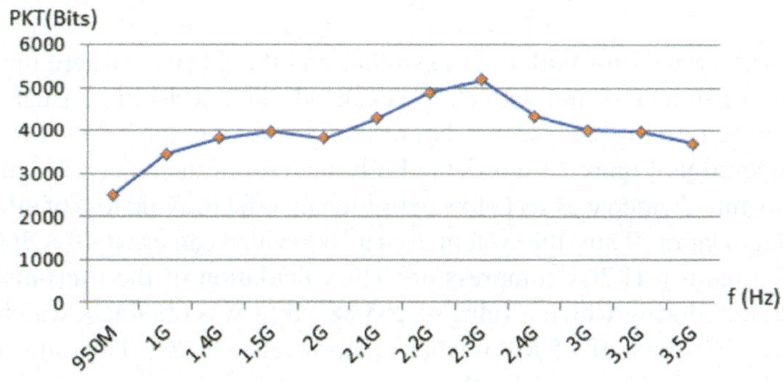

Fig. 19 Packets transmitted

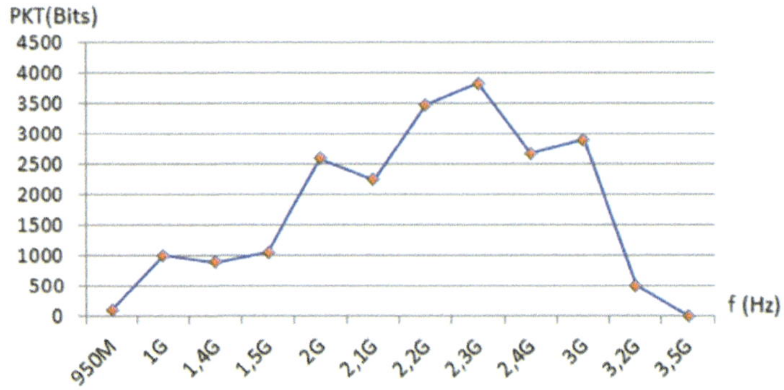

Fig. 20 Packets received, results obtained from performance tests in a lab

Table 1 Comparative analysis of frequencies vs. interference

N°	Criteria/Frequency	1.5 GHz	2.3 GHz	3 GHz
1	Is the link at the specific frequency affected by mechanical noise produced by motors?	Yes	No	No
2	Is the link at the specific frequency affected by devices or technologies that use the same frequency? Is there co-channel interference?	No	No	Yes
3	Is the link at the specific frequency prone to being affected due to obstacles within the first fresnel zone?	Yes	No	No
4	Does the link at the specific frequency require de-energizing the equipment for the frequency change?	No	No	Yes

A GUI was created for both the transmitter and the receiver, where the entry of configuration parameters and data entry is carried out, and for the visualization of the received data. In Fig. 21 a screenshot of the receiver's graphical user interface is shown in operation. Figure 22 shows the EYE diagram of the received signal.

The transmitted video was set to low definition in order to avoid loss of information or image degradation. Thus, the system used a 360p video camera (640 × 360p), with 30 fps and a built-in H.264 compression. The calculation of the transmission rate was carried out, during which a value of 165.88 Mbps was obtained, which gave an approximate equivalent of 1.5 Mbps when compressed at H.264. This value indicates the channel capacity necessary for the video transition.

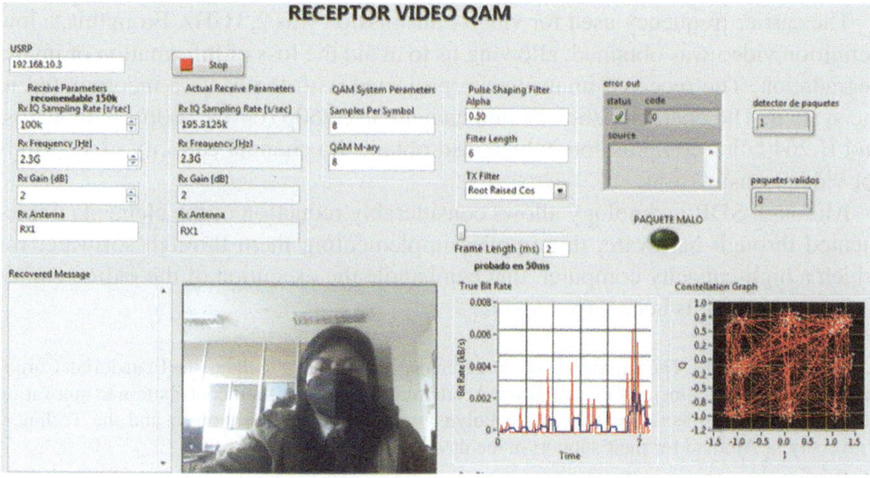

Fig. 21 Functioning receiver GUI

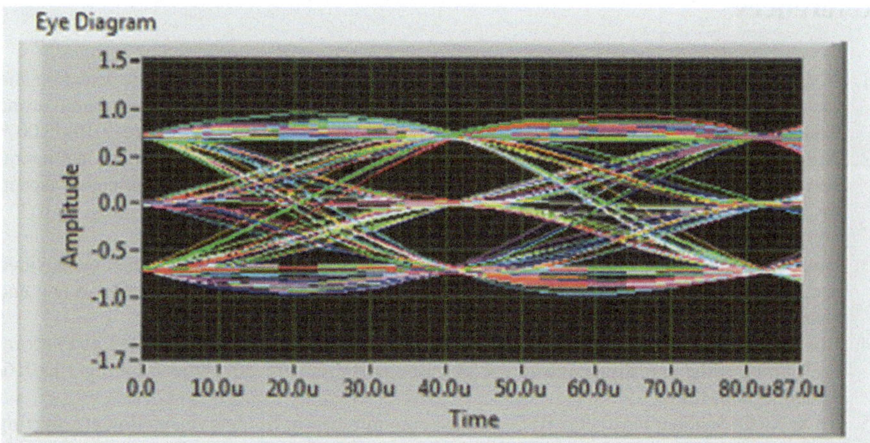

Fig. 22 EYE diagram of the received signal

4 Conclusions

A communications system for video transmission from an in-flight UAV to a ground base station, has been developed using software with SDR (Software Defined Radio) technology, which has allowed the completion of several stages of the system, Therefore, with the exception of elements such as antennas or amplifiers, the rest of the blocks have been implemented normally. On the other hand, the use of hardware was also reduced to the use of a computer with LabVIEW, USRPs equipment, a drone, a camera and antennas.

The carrier frequency used for video transmission was 2.3 GHz. From this, a low definition video was obtained, allowing us to avoid the loss of information or image degradation. The received images were projected with the camera incorporated in the system. The characteristics of said camera are: 360p (640 × 360p), with 30 fps, and H.264 color compression, which also obtained a channel capacity of 1,5 Mbps for video transmission.

Although SDR technology allows considerably reduction of the elements implemented through hardware, this implies implementing them through software, for which a high-capacity computer, that can handle the execution of the radiocommunications system, is necessary.

Acknowledgements The authors of this research paper would like to thank the Ecuadorian Corporation for the Development of Research and Academia - CEDIA for their contribution to innovation through CEPRA projects, as well as the Polytechnic School of Chimborazo and the Technical University of Ambato for their support in the development of this research.

References

1. Al-Sa'd MF, Al-Ali A, Mohamed A, Khattab T, Erbad A (2019) RF-based drone detection and identification using deep learning approaches: an initiative towards a large open source drone database. Futur Gener Comput Syst 100:86–97. https://doi.org/10.1016/j.future.2019.05.007
2. Arias E, Encalada P, Tigre F, Granizo C, Gordon C, Garcia MV (2019) Convolutional neural network applied to the gesticulation control of an interactive social robot with humanoid aspect. In: Advances in intelligent systems and computing. Springer International Publishing, pp 1039–1053. https://doi.org/10.1007/978-3-030-29513-4_76
3. Baseca CC, Diaz JR, Lloret J (2013) Communication ad hoc protocol for intelligent video sensing using AR drones. In: 2013 IEEE 9th international conference on mobile ad-hoc and sensor networks. IEEE. https://doi.org/10.1109/msn.2013.115
4. Bologna JK, Garcia CA, Ortiz A, Ayala PX, Garcia MV (2020) An Augmented Reality Platform for training in the industrial context, vol 53. Elsevier BV, pp 197–202. https://doi.org/10.1016/j.ifacol.2020.11.032
5. Yu X, Xu D, Schober R (2019) MISO wireless communication systems via intelligent reflecting surfaces: (invited paper). In: 2019 IEEE/CIC international conference on communications in China (ICCC). IEEE. https://doi.org/10.1109/iccchina.2019.8855810
6. Barmpounakis EN, Vlahogianni EI, Golias JC (2016) Unmanned aerial aircraft systems for transportation engineering: current practice and future challenges. Int J Transp Sci Technol 5(3):111–122. https://doi.org/10.1016/j.ijtst.2017.02.001
7. Wang J, Feng Z, Chen Z, George S, Bala M, Pillai P, et al (2018) Bandwidth-efficient live video analytics for drones via edge computing. In: 2018 IEEE/ACM symposium on edge computing (SEC). IEEE. https://doi.org/10.1109/sec.2018.00019
8. Suzuki S (2018) Recent researches on innovative drone technologies in robotics field. Adv Robot 32(19):1008–1022. https://doi.org/10.1080/01691864.2018.1515660
9. Díaz Saravia MW, Cáceres Chiquillo JJ, et al (2018) Aplicaciones de los drones para el diagnóstico de infraestructura eléctrica y de telecomunicaciones. Revista Tecnológica, no 11
10. Marra WC, McNeil SK (2013) Understanding the loop: regulating the next generation of war machines. Harv JL Pub Pol'y 36:1139
11. Abdallah A, Ali M, Mišić J, Mišić V (2019) Efficient security scheme for disaster surveillance UAV communication networks. Information 10(2):43. https://doi.org/10.3390/info10020043

12. Gupta L, Jain R, Vaszkun G (2016) Survey of important issues in UAV communication networks. IEEE Commun Surv Tutor 18(2):1123–1152. https://doi.org/10.1109/comst.2015.2495297
13. Ferranti L, Cuomo F, Colonnese S, Melodia T (2018) Drone cellular networks: enhancing the quality of experience of video streaming applications. Ad Hoc Netw 78:1–12. https://doi.org/10.1016/j.adhoc.2018.05.003
14. Andre T, Hummel K, Schoellig A, Yanmaz E, Asadpour M, Bettstetter C et al (2014) Application-driven design of aerial communication networks. IEEE Commun Mag 52(5):129–137. https://doi.org/10.1109/mcom.2014.6815903
15. Wien M, Cazoulat R, Graffunder A, Hutter A, Amon P (2007) Real-time system for adaptive video streaming based on SVC. IEEE Trans Circuits Syst Video Technol 17(9):1227–1237. https://doi.org/10.1109/tcsvt.2007.905519
16. Da-Jiang Innovationss: DJI Official. https://www.dji.com/. Accessed 07 Jan 2020
17. Samijayani ON, Gitomojati P, Astharini D, Rahmatia S, Pratama NIH (2017) Implementation of SDR for video transmission using GNU radio and USRP B200. In: 2017 5th international conference on cyber and IT service management (CITSM). IEEE. https://doi.org/10.1109/citsm.2017.8089288
18. Aaronia AG (2020) Logper/Directional Antennas - 380MHz-35GHz. https://aaronia.com/antennas/hyperlog-series-logper-directional/. Accessed 07 Jan 2020
19. Salazar FW, Núñez F, Buele J, Jordán EP, Barberán J (2019) Design of an ergonomic prototype for physical rehabilitation of people with paraplegia. Springer International Publishing

Preliminary Study on the Detection of Autonomic Dysreflexia Using Machine Learning Techniques

Nagore Sagastibeltza, Asier Salazar-Ramirez, Ainhoa Yera, Raquel Martinez, Javier Muguerza, Nora Civicos Sanchez, and Maria Angeles Acera Gil

Abstract This work focuses on identifying the occurrence of episodes of Autonomic Dysreflexia (AD) in patients suffering from spinal cord injury (SCI) using features extracted from both clinical and physiological information. For this purpose, in collaboration with the Cruces University Hospital, first, data from the clinical history of a set of patients. Second, blood was extracted from the patients to analyse their hormone levels. And finally, in the last stage of the test the patients' bladder was filled with a saline solution to simulate an episode of AD in a controlled environment while capturing different physiological signals, such as blood pressure, heart

This work was partially funded by the Department of Education, Universities and Research of the Basque Government (ADIAN, IT-980-16); and by the Spanish Ministry of Science, Innovation and Universities—National Research Agency and the European Regional Development Fund—ERDF (PhysComp, TIN2017-85409-P), and from the State Research Agency (AEI, Spain) under grant agreement No RED2018-102312-T (IA-Biomed).

N. Sagastibeltza · A. Yera · J. Muguerza
Department of Computer Architecture and Technology, University of the Basque Country (UPV/EHU), Donostia, Spain
e-mail: nagore.sagastibeltza@ehu.eus

A. Yera
e-mail: ainhoa.yera@ehu.eus

J. Muguerza
e-mail: j.muguerza@ehu.eus

A. Salazar-Ramirez (✉) · R. Martinez
Department of System Engineering and Automation, University of the Basque Country (UPV/EHU), Bilbao, Spain
e-mail: asier.salazar@ehu.eus

R. Martinez
e-mail: raquel.martinez@ehu.eus

N. C. Sanchez
Spinal Cord Injury Unit, Cruces University Hospital, Barakaldo, Spain
e-mail: nora.civicossanchez@osakidetza.eus

M. A. Acera Gil
Neurodegenerative Diseases Group, BioCruces Health Research Institute, Barakaldo, Spain
e-mail: m.angeles.aceragil@osakidetza.eus

© The Author(s), under exclusive license to Springer Nature Switzerland AG 2022
M. V. Garcia et al. (eds.), *Advances and Applications in Computer Science, Electronics, and Industrial Engineering*, Lecture Notes in Networks and Systems 433,
https://doi.org/10.1007/978-3-030-97719-1_20

rate and peripheral resistance. After the test, the researchers have extracted different features from those experimental stages to generate a classification system capable of identifying which subjects suffer from AD. Successfully classifying 80% of the cases, this preliminary study has proven that the application of these techniques to clinical and physiological data is a promising method for analysing patients with SCI and diagnosing which may suffer from episodes of AD.

Keywords Physiological computing · Supervised-learning techniques · Autonomic Dysreflexia detection · Feature extraction · eHealth

1 Introduction

Autonomic dysreflexia (AD) is a serious condition that is common in patients suffering from a SCI. Also known as autonomic hyperreflexia, AD is an abnormal response of the body that takes place when the body experiences pain or discomfort below the level of a SCI and it develops in more than half of patients with cervical or upper thoracic SCI (above T5–T6) [1].

In the case of AD, the nerve signals of pain and/or discomfort produced by any event below the level of the lesion do not reach the brain. Therefore, the brain knows that something is happening but does not know exactly what. As a result, the body reacts in an improper way forcing blood pressure begins to rise, which may lead to serious conditions such as stroke, brain injury and even death [2].

The consequences of not diagnosing DA and, consequently not treating it, can be fatal as explained in the previous paragraph. Several articles focus on the medical explanation of what AD is, what it consists in and the problems that can result from its occurrence [3–5]. However, apart from what presented in Suresh's and Duestock's work [6], in which they propose a system to detect AD onsets in patients already diagnosed with AD, the authors' have found that the automatic diagnosis of AD in SCI patients by applying machine learning methods to clinical and/or physiological features extracted from patients themselves is a task that has not been explored yet in depth in the literature.

In order to diagnose a possible case of AD, it is necessary to know what are the most common causes of this anomaly and to study them to obtain results that will allow the development of a model that will help in the diagnosis of the onset of AD. People with a SCI above T6 are at high risk of developing AD [7]. Different causes can lead to the condition, but the most common is bladder-related problems. When the SCI is above the bladder, signals that the bladder is filling do not reach the brain because of the SCI. Thus, the brain does not know when the bladder is filling. Therefore, it is not able to send the order to empty the bladder and causes the bladder to continue to fill. As a result, discomfort and pain are constantly increasing and, consequently, sweating, blurred vision, headache and sudden increases in blood pressure may take place. The latter is the most dangerous symptom, which can lead to vascular problems such as myocardial ischemia and cerebral haemorrhage if bladder filling is not controlled [8].

The serious health problems derived from the presence of dysreflexia underline the importance of its detection and diagnosis to treat and alleviate the effects that are triggered by the onset of AD. For this reason, this study aims to carry out a preliminary analysis that gathers different biosignals and clinical data from different patients intending to be able to predict the development of this condition. For this purpose, an experiment has been carried out at the Cruces University Hospital where a bladder filling has been performed on several patients and different physiological signals have been collected, such as the electrocardiogram (ECG), Total Peripheral Resistance (TPR) and arterial pressures. In addition, other data from the patients' clinical histories and different analytical tests carried out during the experimentation were also collected and used for this study.

As mentioned above, the main target of this work is to enable the classification of dysreflexic patients using the different medical features collected from the experimentation. To date, the use of Machine Learning (ML) techniques has proven to be of great help in the detection of different types of diseases [9–11]. For this reason, aiming to expand the literature around the automatic detection of this disease and complement the know-how presented in [6], this paper studies the usability of different ML techniques for the detection of patients with AD, giving a comparison of the performance of different algorithms.

The rest of this paper is organised as follows. First, Sect. 2 presents the experimental setup which has been used to collect the data and physiological signals. Then, this section explains the pre-processing and preparation of the data collected in the experiment. This article continues with Sect. 3, where it presents the proposed classification using ML algorithms and gives their respective results. Finally, the last section, Sect. 4, discusses the results obtained in the previous section and provides conclusions and future lines for this work.

2 Materials and Methods

First, this section will define and explain the experimental process designed to collect the medical data and the biosignals. After that, it will present the methodology used for the analysis of the collected data to build the database that will be used for classification in Sect. 3.

2.1 Experimental Setup

As described in the introduction, bladder filling is the most common cause of AD in patients suffering from SCI. Besides, the filling of the bladder is a process that can be easily controlled and reversed by the medical team in hospital facilities and

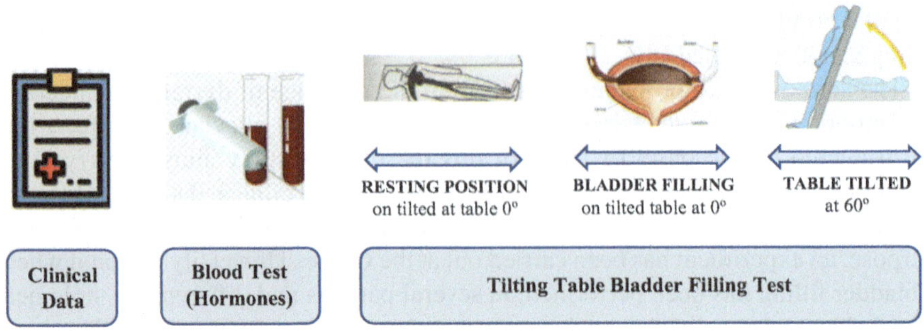

Fig. 1 Stages of the experiment

does not involve major risks for the patients. Therefore, the research team decided to simulate this situation in the laboratory for the study presented in this work.

Due to the high level of invasiveness of the experiment and the sensitivity of this type of patient, a sample of 5 patients with different symptomatologies is considered to be sufficient for this preliminary work. In the current sample of 5 patients, 3 of them experience onsets of dysreflexia and 2 do not develop it. Later, if the results of the study show to be promising, the researchers will propose an increase in the number of subjects to the medical team so that more conclusive results can be obtained.

The experiment took place in the Unit of Dysautonomia of the Cruces University Hospital under the supervision of a medical team composed of a doctor and a nurse. It is important to note that the experimental procedure explained in this subsection has been analysed by the Clinical Research Committee of the Basque Country and has received approval under the validation code PI2013132. First, the patients had to sign a consent form to formalise their participation in the study. Then, the medical team ensured the emptying of their bladder 24 h before the start of the experimentation. After those preparatory phases, the patients were ready for the three main stages of the experiment, which can be seen in Fig. 1 and are described below. The experiments were organised on 5 different days so that all attention could be focused on a single patient each day.

The first stage of the experiment consisted in the collection of certain data coming from the clinical history of the patients. The information of the clinical history considered for this experiment is shown in Table 1.

Then, the second phase of the experiment consisted of extracting two blood samples from the patients in the Day Hospital. First, patients were placed in a supine position for 40 min and, once that time had passed, a blood sample was taken from their arm. Three minutes after taking the first sample, the patients were raised to a 60° head-up position and another blood sample was taken again. The idea behind this test is to be able to assess the levels of catecholamines in the two positions in which the blood was drawn. Catecholamines, also called amino hormones, are hormones that are released into the bloodstream. They are composed of a large number of

Table 1 Data extracted from the clinical history

Clinical history data
Date of the injury
Gender
Type of injury
Level of the injury
Classification according to five grades determined by the absence or preservation of motor and sensory function (ASIA)
Daily sphincter functioning
Pre-test symptomatology

different substances. Among all of them, only the analysis of the levels of adrenaline, noradrenaline, dopamine and vasopressin were considered relevant for this study.

Subsequently, for the third phase of the experiment, patients were fitted with a urinary catheter and transferred to a tilting table with straps for the chest, abdomen and lower limbs. Once positioned on the table, different electrodes were placed on the patients' bodies to capture the desired physiological signals. First, patients were placed in resting supine position and the signals for systolic and diastolic blood pressure $[mmHg]$, heart rate (HR) $[bpm]$ and total peripheral resistance (TPR) $[dynes \cdot s/cm^5]$ were collected using Task Force Monitor device (from CNSystems in Graz, Austria). After 10 min, DA was induced on the patients by filling their bladder with 0.9% saline solution at a temperature of 37 °C. During this process, the values of the aforementioned signals continued to be recorded. Then, once the bladder was full and all the data had been recorded, the bladder was gradually emptied and the urinary catheter removed.

Then, after the removal of the catheter, the patients remained at rest lying on the tilting table in a supine position (0°) for 10 min. Then, after those minutes, the table was tilted 60° for another 10 min to continue collecting patient data. Finally, to conclude with the experiment, the patients were returned to a supine position, the electrodes were removed and the patients left the tilting table under the supervision of the doctor and nurse. Concerning this last part of the experiment, the authors discarded the signal excerpts corresponding to the resting 10 min between having the table at 0° and 60° because they lacked interest in this study.

2.2 *Data Pre-processing*

After data collection, the next phase of the study corresponds to the processing of the physiological signals collected during the experiment: systolic and diastolic blood pressures, HR and TPR.

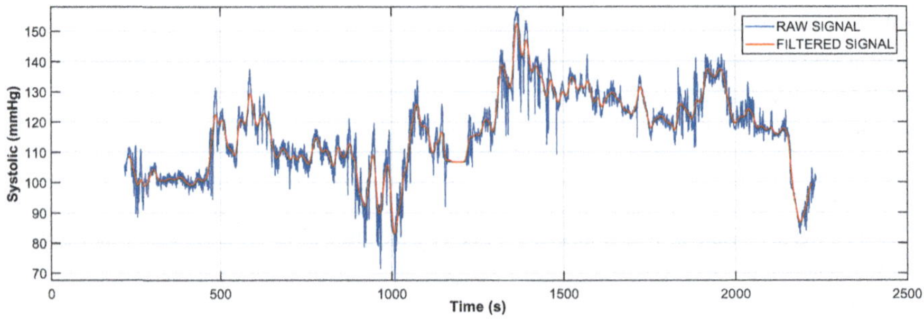

Fig. 2 Raw and filtered systolic pressure signals

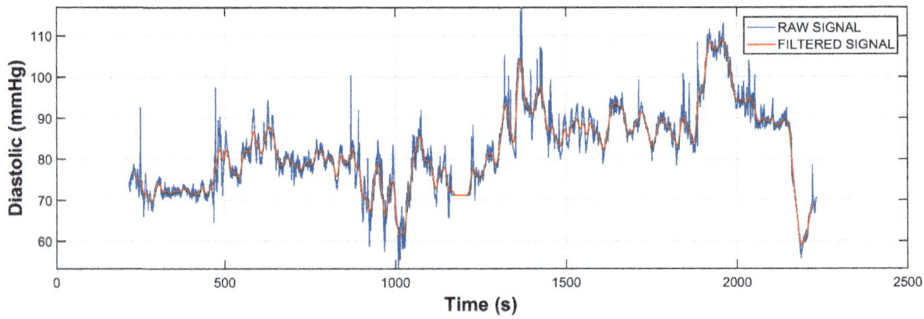

Fig. 3 Raw and filtered diastolic pressure signals

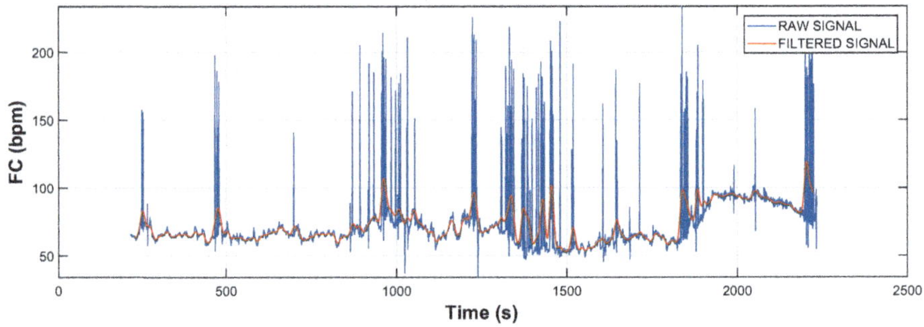

Fig. 4 Raw and filtered heart rate signals

Due to the existence of noise and the lack of signal segments due to gaps and artefacts in the acquisition, the physiological signals underwent a filtering process for getting a better reconstruction. For this purpose, the team used a low-pass to soften the signals. Both the filtering and the graphical representation of the signals were carried out with the mathematical tool Matlab. All the four raw signals (in blue) and their filtered versions (in red) are depicted in the following Figs. 2, 3, 4 and 5.

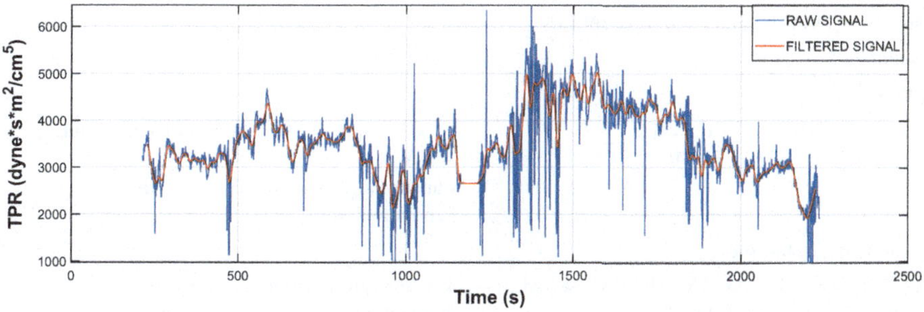

Fig. 5 Raw and filtered total peripheral resistance signals

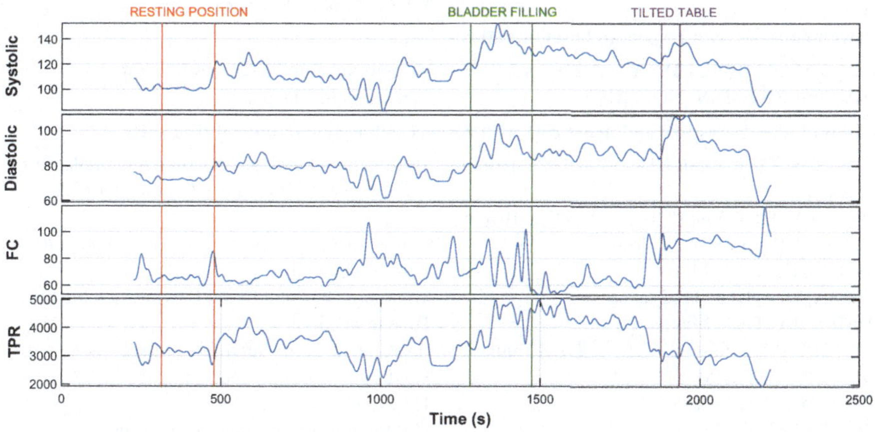

Fig. 6 Filtered signals

Figure 6 shows the four filtered signals of patient number 5, marking the data segments of the three stages of the tilting table/bladder filling part of the experiment.

Once the physiological signals had been processed, the team created a database in which all the data were grouped for subsequent analysis. This data is formed with features associated with different types of information: clinical history data, values of the blood extraction analyses done during the experiment, statistical values derived from the recorded physiological, etc. The names of the analysed features can be seen in Table 2 along with the description of each of them.

Finally, the team normalised all the features between 0.25 and 1 applying the linear normalisation represented by Eq. 1. The reason for this normalisation is none other than to give treatment to the features for which it has not been possible to obtain any value because they could not be recorded during the experiment. Therefore, the value 0 has been reserved to represent those missing feature values and thus, differentiate them from the real values of the features. It could be considered that this is not a wise decision. However, the team decided to follow this criterion because the fact of

Table 2 Feature names and descriptions

Naming	Definition
Clinical:	
TYPE_INJ, INJ_LEVEL, ASIA, SPHINCTER, PREV_SYMPT	Type of injury, Level of the spinal cord injury, ASIA, Functioning of the sphincter, Previous symptomatology
Basal Catecholamines:	
VASOP_BASAL, AD_BASAL, NA_BASAL, DOP_BASAL, SYS_BASAL, DIA_BASAL, HR_BASAL, TPR_BASAL	Vasopressin (decubitus), Adrenaline, Noradrenaline, Dopamine, Systolic Blood Pressure (resting), Diastolic Blood Pressure, Heart Rate, Total Peripheral Resistance
Catecholamines with respect to baseline ratios:	
DIV_VASOP_BIPE, DIV_AD_BIPE, DIV_NA_BIPE, DIV_DOP_BIPE	Vasopressin (supine 60°), Adrenaline, Noradrenaline, Dopamine
Ratios of blood pressure and heart rate with respect to baseline in table tilting stage:	
DIV_SYS_TILT_3 min/DIV_SYS_TILT_10 min [1.25ex] DIV_DIA_TILT_3 min/DIV_DIA_TILT_10 min [1.25ex] DIV_SYS_TILT_Min/ TIME_SYS_TILT_Min [1.25ex] DIV_DIA_TILT_Min/ TIME_DIA_TILT_Min DIV_HR_TILT_ min/DIV_HR_TILT_10 min DIV_HR_TILT_Max/ TIME_HR_TILT_Max DIV_HR_TILT_Min/ TIME_HR_TILT_Min	Systolic blood pressure 3/10 min after table tilting Diastolic Blood Pressure 3/10 min after table tilting Minimum Systolic Blood Pressure at table tilt/ Time to obtain it Minimum Diastolic Blood Pressure at table tilt/ Time to obtain it Heart rate 3/10 min after table tilting Maximum heart rate after tilting the table/ Time to obtain it Minimum heart rate after table tilting/ Time to obtain it
Ratios of blood pressure and heart rate with respect to baseline in bladder filling stage:	
DIV_SYS_FILL_Max/ TIME_SYS_FILL_Max DIV_DIA_FILL_Max/ TIME_DIA_ FILL_Max DIV_HR_FILL_Max/ TIME_HR_FILL_Max DIV_HR_FILL_Min/ TIME_HR_FILL_Min DIV_TPR_FILL_Max/ TIME_TPR_FILL_Max	Peak Systolic Blood Pressure at Bladder Filling/ Time to obtain it Peak Diastolic Blood Pressure at bladder filling/ Time to obtain it Maximum heart rate at bladder filling/ Time to obtain it Minimum heart rate at bladder filling/ Time to obtain it Maximum Total Peripheral Resistance at bladder filling/ Time to obtain it

Table 3 Estimation of the best classifier

	1R	DT	1-NN	3-NN	NB	RBF	SVM	LR	RF	MLP
Accuracy (%)	0	40	60	40	20	0	80	80	60	80
Errors class 0 (FP)	2	2	1	2	2	2	0	0	1	0
Errors class 1 (FN	3	1	1	1	2	3	1	1	1	1

not having values for some features also provides information that can be useful to perform the classification.

$$X_{norm} = \frac{X_i - X_{min}}{X_{max} - X_{min}} \tag{1}$$

3 Classification and Results

Once all the features related to clinical and physiological variables had been pre-processed, the team went on with the experimentation and took those features to classification. The goal of this phase was to determine whether it is possible to detect episodes of dysreflexia using the most discriminating among the previously extracted features.

To do so, the team used different supervised learning algorithms. All the experimental procedure related to classification was carried out using the implementation provided by the Weka data-mining platform (Waikato Environment for Knowledge Analysis) [12]. None of the algorithms used in this experimentation was fine-tuned; the team used the default structure and parameter values provided by Weka. All the experimentation was done according to the leave one out validation methodology.

The problem proposed in this study is a two-class classification (-0- for subjects without dysreflexia and -1- for patients suffering from dysreflexia). To tackle this problem, the research team opted for implementing ten algorithms, encompassing different types of classification paradigms: 1R rule, Decision Trees (DT), k-Nearest Neighbours (1-NN and 3-NN), Naive Bayes (NB), Radial-Basis Network (RBF), Support Vector Machine (SVM), Multilayer Perceptron (MLP), Logistic Regression (LR) and Random Forest (RF). The accuracy of the classifiers for each of the 5 patients is presented in Table 3. The best overall classifiers were the Logistic Regression, Support Vector Machine and Multilayer Perceptron, with an accuracy of 80%.

Table 4 Classification performance for the best classifiers

Subject	Class	SVM	LR	MLP
1	0	0 (1)	0 (1)	0 (0.789)
2	0	0 (1)	0 (1)	0 (0.942)
3	1	1 (1)	1 (1)	1 (0.986)
4	1	1 (1)	1 (1)	1 (0.88)
5	1	0 (1)	0 (0.638)	0 (0.906)
Accuracy	**80%**			
FP		0	0	0
FN		1	1	1

Besides, Table 4 also presents the result obtained for the three best classifiers. This table shows the class predicted by the classifier, as well as the probability of belonging to that class (between brackets). As it can be seen, the three classifiers guessed correctly the two cases of normality and the first two cases of dysreflexia.

Nevertheless, the last case of dysreflexia (subject 5) was incorrectly classified as an absence of dysreflexia, i.e., all classifiers produced a false negative. This classification error should have been avoided, as according to medical criteria, not diagnosing the disease could lead to severe consequences for the affected patient. Different working lines will be proposed in Sect. 4 to tackle this problem.

4 Conclusions and Future Lines

The results obtained in this work have confirmed that applying ML techniques to clinical and physiological information is a promising option for analysing patients with SCI and diagnosing which of them may suffer from episodes of Autonomic Dysreflexia (AD).

Therefore, considering that the results obtained are overall good, the natural next step in this research is to try to reduce to the maximum the number of false negatives (FN) produced during the classification, looking to even get a zero ratio. To achieve this, more complex classification structures will be explored, such as hierarchical structures or multiple classifiers that permit handling the classification problem in different steps. Furthermore, Feature Subset Selection (FSS) techniques will also be applied to determine which features are the most discriminant. This would facilitate classification, leading to better classification performance and lower FN rates.

Nevertheless, it is difficult to advance in this objective with a small experimental sample as the presented in this study. Accordingly, the research team has already started working together with the Cruces University Hospital towards this objective and have already planned a new set of experiments to expand the sample of the study. This way, it will become possible not only to improve and tune the used classification methods, but it will also make it possible to make a more thorough validation of the results obtained in this preliminary work.

Finally, another line to be explored is the classification of the features according to their invasiveness. It should be borne in mind that this type of patient lacks most of their mobility and that they are a very sensitive group of people. Therefore, it is considered essential to minimise the number of invasive tests that need to be performed to permit the automatic detection of the disease. To achieve this, more complex classification schemes will be explored, such as hierarchical structures, which will allow the classification of a part of the sample without having to use any information coming from invasive tests.

References

1. Sánchez NC, Acera M, Murueta-Goyena A, Sagastibeltza N, Martínez R, Cuadrado M, et al (2021) Quantitative analysis of dysautonomia in patients with autonomic dysreflexia. J Neurol 268(8):2985–2994. https://doi.org/10.1007/s00415-021-10478-w
2. Allen KJ, Leslie SW (2020) Autonomic dysreflexia. StatPearls Publishing, Treasure Island
3. Chiodo A, Crane DA, Reyes MR, Song S, Klebine P: Autonomic dysreflexia. Model Systems Knowledge Translation Center. https://msktc.org/sci/factsheets/autonomic_dysreflexia
4. Sharif H, Hou S (2017) Autonomic dysreflexia: a cardiovascular disorder following spinal cord injury. Neural Regen Res 12(9):1390. https://doi.org/10.4103/1673-5374.215241
5. Walter M, Knüpfer SC, Cragg JJ, Leitner L, Schneider MP, Mehnert U, et al (2018) Prediction of autonomic dysreflexia during urodynamics: a prospective cohort study. BMC Med 16(1). https://doi.org/10.1186/s12916-018-1040-8
6. Suresh S, Duerstock BS (2020) Automated detection of symptomatic autonomic dysreflexia through multimodal sensing. IEEE J Transl Eng Health Med 8:1–8. https://doi.org/10.1109/jtehm.2019.2955947
7. West CR, Squair JW, McCracken L, Currie KD, Somvanshi R, Yuen V, et al (2016) Cardiac consequences of autonomic dysreflexia in spinal cord injury. Hypertension 68(5):1281–1289. https://doi.org/10.1161/hypertensionaha.116.07919
8. Liu N, Zhou M, Biering-Sørensen F, Krassioukov AV (2015) Iatrogenic urological triggers of autonomic dysreflexia: a systematic review. Spinal Cord 53(7):500–509. https://doi.org/10.1038/sc.2015.39
9. Bahl M, Barzilay R, Yedidia AB, Locascio NJ, Yu L, Lehman CD (2018) High-risk breast lesions: a machine learning model to predict pathologic upgrade and reduce unnecessary surgical excision. Radiology 286(3):810–818. https://doi.org/10.1148/radiol.2017170549
10. McCarthy JF, Marx KA, Hoffmann PE, Gee AG, O'Neill P, Ujwal ML, et al (2004) Applications of machine learning and high-dimensional visualization in cancer detection, diagnosis, and management. Ann New York Acad Sci 1020(1):239–262. https://doi.org/10.1196/annals.1310.020
11. Zhan A, Mohan S, Tarolli C, Schneider RB, Adams JL, Sharma S, et al (2018) Using smartphones and machine learning to quantify Parkinson disease severity. JAMA Neurol 75(7):876. https://doi.org/10.1001/jamaneurol.2018.0809
12. Witten IH, Frank E, Hall MA (2011) Data mining: practical machine learning tools and techniques, 3rd edn. Morgan Kaufmann, Burlington

The manufacturer's authorised representative in the EU is Springer Nature Customer Service Centre GmbH, Europaplatz 3, 69115 Heidelberg, Germany. If you have any concerns regarding our products, please contact ProductSafety@springernature.com

Printed and bound by CPI Group (UK) Ltd, Croydon, CR0 4YY
30/09/2026
02225320-0014